Smart Aquaculture

Advances in Fish Farming

Smart Aquaculture

Advances in Fish Farming

Samuel Turner
Ellie Francis

Kruger Brentt
Publishers
2026

Kruger Brentt Publishers UK. LTD.
Company Number 9728962

Regd. Office: 68 St Margarets Road, Edgware, Middlesex HA8 9UU

ISBN: 978-1-78715-628-9

Disclaimer:

Every possible effort has been made to ensure that the information contained in this book is accurate at the time of going to press, and the publisher and author cannot accept responsibility for any errors or omissions, however caused. No responsibility for loss or damage occasioned to any person acting, or refraining from action, as a result of the material in this publication can be accepted by the editor, the publisher or the author. The Publisher is not associated with any product or vendor mentioned in the book. The contents of this work are intended to further general scientific research, understanding and discussion only. Readers should consult with a specialist where appropriate.

Every effort has been made to trace the owners of copyright material used in this book, if any. The author and the publisher will be grateful for any omission brought to their notice for acknowledgement in the future editions of the book.

For information on all our publications visit our website at http://krugerbrentt.com/

PREFACE

Smart Aquaculture: Advances in Fish Farming explores the integration of modern technologies into aquaculture to enhance productivity, sustainability, and resource efficiency. As the global demand for fish and seafood continues to rise, traditional aquaculture practices face challenges including disease outbreaks, environmental impacts, and resource limitations. This book provides students, researchers, and aquaculture professionals with a comprehensive understanding of smart aquaculture techniques, innovations, and their practical applications in modern fish farming.

The book begins with an overview of aquaculture systems, species selection, and the principles of sustainable fish farming. It then delves into smart technologies such as IoT-based monitoring, automated feeding systems, water quality sensors, remote sensing, AI-driven decision-making, and precision aquaculture practices that improve efficiency and minimize environmental impact.

A major focus of the text is on disease management, biosecurity, and environmental sustainability, highlighting how technology can assist in early detection, monitoring, and mitigation of risks. The book also addresses data management, predictive modeling, and the use of analytics for optimizing growth, feed conversion, and overall farm performance.

Case studies and real-world examples demonstrate successful implementation of smart aquaculture systems across diverse species and geographic regions. The book also discusses future trends, challenges, and policy considerations, providing a holistic view of the evolving field of digital and technology-enabled fish farming.

We hope that Smart Aquaculture: Advances in Fish Farming will serve as an essential resource for learners, practitioners, and policymakers, combining theoretical foundations with practical insights to promote efficient, sustainable, and technologically advanced aquaculture practices.

Samuel Turner

Ellie Francis

CONTENTS

1 Chapter

AQUACULTURE SYSTEM VIABILITY: A METHODOLOGICAL AND COMPARATIVE FRAMEWORK

1. INTRODUCTION: AN OVERVIEW

In the past 30 years, aquaculture has undergone remarkable advancement in worldwide animal production, achieving an average annual growth rate over 10% from 1980 to 2000 (FAO, 2009). During this period, capture fisheries had a progressive stagnation, with growth ceasing in 1995, resulting in total catches varying between 90 and 95 Mt every year, depending on the year. The expansion of aquaculture, while advantageous and essential for satisfying the projected demand for marine products estimated at 270 million tons by 2050 (Chevassus au Louis et Lazard, 2009; Wijkström, 2003), presents some challenges pertinent to its sustainable development. These include concerns regarding feed for cultivated creatures, biological variety, economic sustainability of farms, the effects of aquaculture development on social fairness, and the governing framework of the sector.

The use of feed is presently contentious, as illustrated by the seminal article by Naylor et al. (2000), which highlights the effects of extensive fish meal and fish oil utilization in fish and prawn aquaculture. The authors advocate for a reversion to less intensive aquaculture systems, drawing inspiration from traditional Asian practices that employ more extensive methods based on polyculture and fertilization, wherein artificial feed is regarded merely as a supplementary option. This diagnosis, while intriguing and sparking considerable debate, was ultimately incomplete and wrong; by concentrating on a singular criterion and a sole facet (environmental) of sustainability, the authors proposed solutions unlikely to be embraced by stakeholders.. De facto, farming systems have continued to intensify and this has led to a sustained increase in the use of fish meal and fish oils. Indeed, the aquaculture sector's consumption of fish meal and oils increased respectively from 2.9Mt to 3.7Mt and from 0.6Mt to 0.8Mt between 2000 and 2008 (Tacon and Metian, 2008). Over and above the issue relating to the use of feed with a high biological value for aquaculture production, Naylor *et al.* (2000) contrast two aquaculture models: the first one, an input-intensive system, in particular as regards fish meal and oils and *a priori* non sustainable, and the second one, classically described as extensive or semi-extensive, considered to be sustainable. This implicit or explicit assimilation of intensive and extensive/semi-extensive systems with models of respectively weak or strong sustainability can be found in many publications from the 1980s and 1990s (Billard, 1980; Edwards et al., 1988; Kautsky et al., 1997; Lazard, 1993; Veverica et Molnar, 1997), and even quite recently (Belton et al., 2009; Delgado et al., 2003; World Bank, 2007). Two other examples of approaches that take into account only one pillar of sustainable development, in the social domain this time, give contradictory results. The first example comes from work on the role of aquaculture as an activity with a **direct** impact on poverty alleviation efforts (Edwards, 1999, 2000) inspired by the analytical framework of sustainable livelihoods in the rural milieu (Carney, 1998). The conclusions and recommendations of this global reflection process which aim to take into account the assets (physical, natural, human, social and financial) of "poor and vulnerable populations" remain for the most part tentative as Edwards (2000) concludes that if aquaculture remains in theory an attractive way to improve the livelihoods of poor populations, "*there is a need to raise awareness of the large potential contribution of aquaculture*". This goal is far from having been attained. The second example comes from a study of coastal aquaculture in brackish lagoons in the Philippines where Irz et al. (2005) highlighted the significant role played in this country by a mostly "capitalist" aquaculture in income redistribution that benefits the poorest directly (salaries) or indirectly (services). These authors therefore recommend that when public policies are implemented, particular attention should be paid to maintaining such social redistribution outcomes, especially when promoting new technologies. These examples show that the real question is to find whether there are specific aquaculture systems that can contribute to poverty reduction in parallel with profit-orientated systems.

A final example is provided by a large-scale project initiated by ICLARM (now Worldfish) in the Philippines at the end of the 1980s ("GIFT Project", Genetically Improved Farmed Tilapia) which aimed to genetically improve the tilapia most frequently used in farming, *Oreochromis niloticus*. According to its promoters, *the overall development objective of the GIFT Project was to increase the quantity and quality of protein consumed in low income rural and urban populations in tropical developing countries in all regions of the world and increase the income of low-income producers. As with future agricultural and aquacultural developments, the objective was to aim for sustainable systems, in harmony with the natural environment, to benefit producers and consumers.* If, for the most part, this project is considered to be successful from biotechnical and micro-economic viewpoints (Dey et al., 2000; Gupta et Acosta, 2004), various analyses show that things are quite different at social (low usage by "poor fish farmers") and environmental levels (risk induced by the introduction of the GIFT strain in the original area of the Nile tilapia) (Lazard, 2009).

It is clear therefore that numerous discussions of aquaculture sustainability are based on a single component of sustainable development. Very little work has been undertaken on a global and comparative basis. The analysis of the main reference frameworks such as codes of conduct, guides of good practice, standards, labels etc. (Boyd et al., 2005, 2008; FAO, 1995; WWF, 2005, 2006, 2007, 2010 among others) and of initiatives for the construction of sustainable development indicators (Caffey et al., 2000; Consensus, 2005; GFCM, 2010; IUCN, 2005) in aquaculture, show that most of them are based on very unbalanced approaches concerning the dimensions of sustainable development that are taken into account. They are also often highly centralised with little reliance on participatory processes (Mathé et al., 2006; Rey-Valette et al., 2007a). This is why the approach suggested in this article is instead designed to cover all the dimensions of sustainability including the traditional pillars (economic, social and environmental) as well as the institutional one (governance). This aspect, in particular, imparts originality and innovation to this method. The approach is both multidisciplinary and participative, comparing various countries and aquaculture systems, culminating in a diagnostic and comprehensive recommendations.

This chapter aims to provide a comprehensive summary of the employed methodology and the outcomes of the diagnosis conducted. The results presented pertain to the EVAD project (Evaluation of Aquaculture System Sustainability), conducted from 2005 to 2009, aimed at assessing the sustainability of aquaculture systems. The primary objective of the EVAD project was to develop a universal methodology for analyzing sustainable development determinants and indicators in aquaculture, incorporating its territorial aspects and stakeholders' opinions. To ensure its generic nature, the technique was formulated utilizing extremely diverse locales concerning socio-geography, production systems, agricultural settings, and regulatory frameworks. In addition to the evaluation procedure, the project aimed to offer two categories of sustainable development indicators for aquaculture: basic indicators (both qualitative and quantitative) and, for environmental considerations, composite indicators derived from life cycle assessments of aquaculture systems.

2. THE RATIONALE UNDERPINNING THE APPROACH AND THE DIFFERENT WORK PHASES

The process used for the EVAD project is characterised by its transdisciplinary approach (Bürgenmeier, 2004), meaning that, for each phase of the project, it associates very closely not only human and biological sciences but also the stakeholders who are part of the procedural and participatory approach. The approach relies on the **co-construction** of indicators for the sustainable development of aquaculture which then become a tool to drive and legitimate sustainable development (Boulanger, 2007). The co-construction of indicators with broad-based groups of stakeholders makes it possible to initiate a participatory approach and a collective learning process and facilitates the appropriation of sustainable development (Mickwitz et al., 2006; Fraser et al., 2006; Hatchuel, 2000; Hilden et Rosenström, 2008; Rey-Valette et al., 2007b). This co-production also promotes the institutionalisation of the monitoring and the implementation of the indicator system[1], especially as it draws on local actors' knowledge. The approach is based on a systemic approach to sustainability which encompasses the four dimensions of sustainable development including the institutional dimension which is taken into account through the governance processes. Furthermore, the method favours a territorial approach to sustainability in the spirit of local agenda 21s recommended by the Agenda 21 during the Rio Earth Summit (Chapter 28) by combining two complementary scales of approach: the sustainability of farms and of the aquaculture sector (sectoral approach) and the contribution of fish farms to the sustainability of the areas where they are located (territorial approach). Taking the territorial level into account is a first step towards integrating the ecosystem services provided by aquaculture.

3. THE AREAS

Several carefully chosen areas were used to test the genericness of the method (table 1). These areas were as follows.

Table 1. Position of aquaculture systems under study according to three criteria: environment, regulation and intensification

	Rural area		Coastal area	
	Low density	High density	Low density	High density
Weak regulation	Ponds Indonesia (Tangkit) Ponds Cameroon		Coastal ponds Philippines	
Strong regulation		Cages Indonesia (Cirata) Raceways Brittany		Cages Mediterranean (France and Cyprus)

3.1 Rainbow trout farming in Brittany (France)

Rainbow trout farming is an intensive farming system based on a high input level and on an increased stocking rate. At present, in Brittany, the number of trout farms is decreasing, farms are being concentrated and the overall production is being reduced due to numerous constraints: environmental constraints, social constraints (farming activity acceptance, product image, etc.), regulatory and economic constraints (input cost variation, competition with salmon, etc.).

†The indicator system comprises an information system which gathers information in all its forms (oral, written, private, public...) together with the arrangements for the management of this information (for instance, an observatory).

3.2 Mediterranean sea bass and sea bream farming

In order to satisfy a strong demand (tourists and indigenous population), the production of aquaculture fish (mainly sea bass and sea bream) started in 1980 and increased by 25 % each year between 1990 and 2000 (the current production is estimated at 200,000 tonnes per year). Current production systems (consisting of sea-based cages or land-based raceways) are in conflict with tourism and other models will have to be developed (Rey-Valette et al., 2007c). Due to recent crises, aquaculture activity has become concentrated as fish farms have been bought by major groups.

3.3 Fish and shrimp farming in coastal ponds in the Philippines

Fish farming plays a major role in the economy of the Philippines and coastal ponds, consisting essentially of extensive milkfish based polyculture, represent around 60 % of the overall aquaculture production. Observing the development dynamics of Philippine aquaculture systems underlines the significant flexibility of extensive systems compared to the economic fragility of intensive fish farms when markets are saturated.

3.4 Small scale fish farming in Indonesia

In Indonesia, although freshwater fish farming is generally a small-scale activity, it nevertheless represents one of the highest yearly production rates in the world. Fish farming production systems with high input rates have rapidly developed locally over the last ten years: catfish in ponds in the Centre of Sumatra (Tangkit, Jambi province) and carps and tilapia in floating cages in the Cirata reservoir (West Java).

3.5 Commercial fish farming in family agricultural enterprises in Western Cameroon

Despite an increasing demand for fish, the history of fish farming in Cameroon (and more largely in Sub-Saharan Africa) remains characterized by a marginal production which is most likely due to the fact that their farming systems are not sufficiently efficient from technical and socioeconomic points of view. The high plateaux in the Western region, which are characterized by a very dynamic diversification of agricultural production systems, represent one of the areas in Cameroon where the greatest number of fish ponds have been constructed with numerous fish farming innovations involving an input intensification.

4. METHODOLOGY USED TO ESTABLISH THE CHECK-LIST OF SUSTAINABILITY INDICATORS

The process of establishing the check-list of sustainability indicators in aquaculture relies on a hierarchical nesting approach which makes it possible to link **indicators** with general sustainability **criteria and principles** (Prabhu et al., 2000). This type of nesting puts into context the definition of indicators and they can then be linked to territorial and sectoral issues. This approach differs from initiatives to construct indicators as inventories linked to pillars of sustainable development and uses instead a principle-guided method, which, through a cross sectional conception of social choice and by putting the general principles of sustainable development into context, helps in its appropriation by actors (Droz et Lavigne, 2006). The relationship between the implementation phases of the Principle Criterion Indicator approach is shown in figure 1. The preparatory phase should establish a diagnosis of the sector and study actors' representations in order to address them and the issues at stake. These representations are defined by Jodelet (1989) as *"forms of knowledge, socially developed and shared, of a practical nature and contributing to the construction of a reality common to a social group"*. In order to achieve this, two surveys were carried out during the first year of the project in the six areas, the first one concerned fish farms with 128 interviews, and the second one all the stakeholders involved in the aquaculture value chain with 168 interviews (table 2). These were face to face interviews using detailed questionnaires combining both closed questions, in particular in the survey on representations, and open questions where textual analysis was used (WordMapper 8.0 from Grimmer soft (p.5)). The first survey aimed to collect data that would enable aquaculture systems to be characterised in the technical, economic and relational senses, to identify the types of farm, the strengths and the constraints, the types of regulation as well as the challenges. Typologies were established and sustainability principles and criteria put into context (Chia et al., 2009; Lazard et al., 2009, 2010). The objective of the second survey was to collate the representations that actors had of sustainable development and the consequences they could foresee for the sector's dynamics (Lazard et al. 2009, 2010). More precisely, this survey made it possible to characterise collective representations, to identify local issues related to aquaculture, to analyse the coherence of the sustainable development model with the current situation of the actors, to analyse the relations and interactions (for example the power systems) within and between the groups and to identify traditional beliefs concerning aquaculture. By integrating international and national norms found in the existing sustainability reference frameworks for aquaculture, this work on representations established a final list of thirteen sustainability principles for aquaculture (fig. 2 and table 3). In addition to the analysis of representations, a systemic analytical framework for aquaculture systems was developed to account for sustainable development. This framework combines all the factors relating to productive systems, to regulatory systems and to the territory and hence enables sustainability to be addressed from two supplementary and interactive viewpoints, which relate respectively to the sustainability factors for aquaculture enterprises and the contribution of these systems to the sustainability of the territory where they are located (fig. 3).

From these principles, a check-list of criteria and indicators was established collectively by the multidisciplinary team of researchers in order to identify the key variables for aquaculture systems. This check-list of principles, criteria and indicators was then validated within the framework of collective working groups comprising the relevant stakeholders in each of the six areas. The purpose was to note, in decreasing order, the principles considered to be the most important and for each selected principle to rank the associated criteria and indicators according to the following categories: "priority", "important", "to be integrated later on", "secondary" or "don't know". By weighting each category with a coefficient (8 for "priority", 4 for "important", 2 for "to be integrated later on", 1 for "secondary"), it was possible to establish scores by country and by types of actor (figure 4). In order to develop indicators, it is necessary to have a good knowledge of the information systems considered to be important in each area so that the new monitoring system for sustainable aquaculture can be positioned relative existing systems. This condition both reduces information collection costs and facilitates understanding and usage of the information system, a part of which will comprise indicators already familiar to actors.

Table 2. Survey distribution according to the sites and types of surveys.

Brittany Cameroon		Indonesia			Mediter-ranean region				The Philip-pines	Total
		Tangkit	Cirata	Total	Turkey	Cyprus	France	Total		
Survey No 1: fish farmers										
8	13	29	27	56	9	4	8	21	30	128
Survey No 2: industry actors										
8	2	7	9	16	0	5	4	9	14	49
institutional actors										
18	8	11	7	18	0	7	17	24	15	83
fish farmers (subsample from Surveys No 1)										
4	5	5	9	14	0	3	4	7	6	36
Total of actors surveyed in relation to Survey No 2										
30	15	23	25	48	0	15	25	40	35	168
Total of Surveys No 1 + 2										
38	28	52	52	104	9	19	33	61	65	296

These results were then discussed collectively within the framework of working groups and the results, once adjusted by the group of researchers, were validated for each area. The discussions concerning the prioritisation/ selection of the PCIs constituted a collective reflexion process on the issues at stake and the practicalities of sustainable aquaculture. It highlighted, in particular, the principles and criteria that made "most sense" for the actors and the motives behind the rejection of those that were not selected. This procedure corresponds to a negotiated vision of what the actors consider to be sustainable development, of the way in which each one can and should contribute and of the rules used to "judge". It is an essential stage in the construction of a common language and project that is necessary for the implementation of sustainable development. Following the phase of developing sustainability evaluation tools in each country, sustainability diagnoses were undertaken and used to verify that the measurement of the selected indicators was feasible.

Table 3. The 13 aquaculture principles and their grouping according to the dimensions of sustainable development.

Technico- economic dimension	P6- Increase the capacity to cope with uncertainties and crises P7- Strengthen the long term future of exploitations P2- Develop approaches which promote quality
Environmental dimension	P3- Ensure that natural resources and the environmental carrying capacity are respected. P4- Improve the ecological yield of the activity P5- Protect biodiversity and respect animal well-being
Social dimension	P1- Contribute to meet nutritional needs P8- Strengthen sectoral organisation and identity P9- Strengthen companies' social investment
Institutional dimension	P10- Strengthen the role of aquaculture in local development P11- Promote participation and governance P12- Strengthen research and sector-related information P13- Strengthen the role of the State and of public actors in putting sustainable development into place

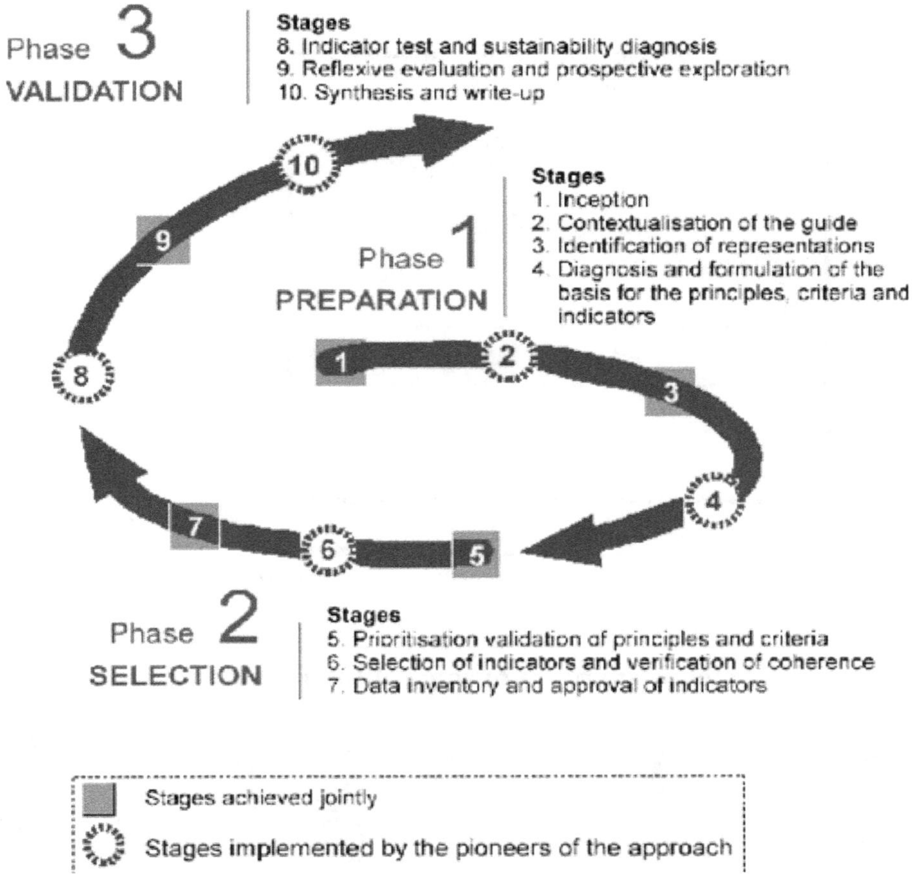

Fig. 1. Implementation process for the co-construction approach.

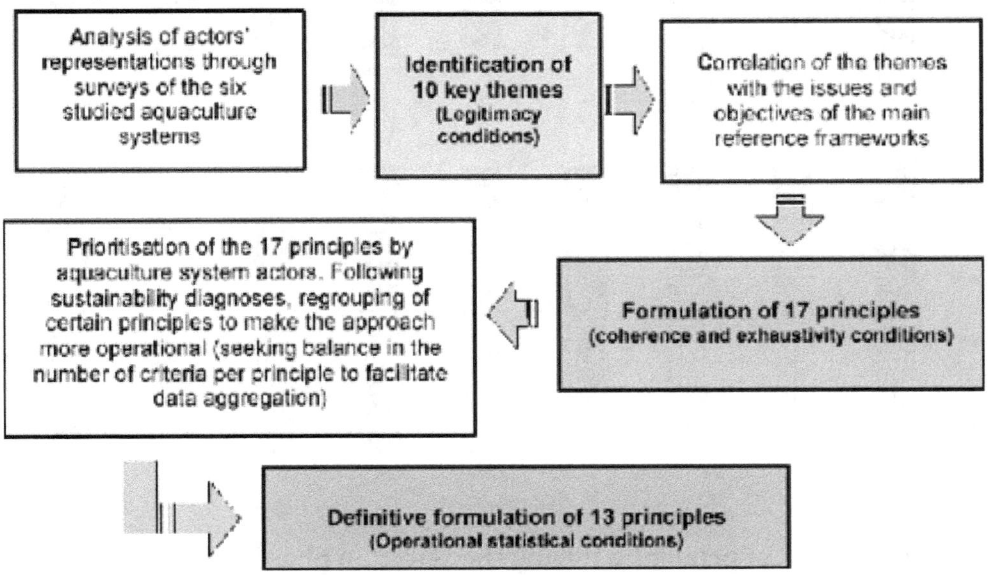

Fig. 2. Traceability of the 13 suggested principles.

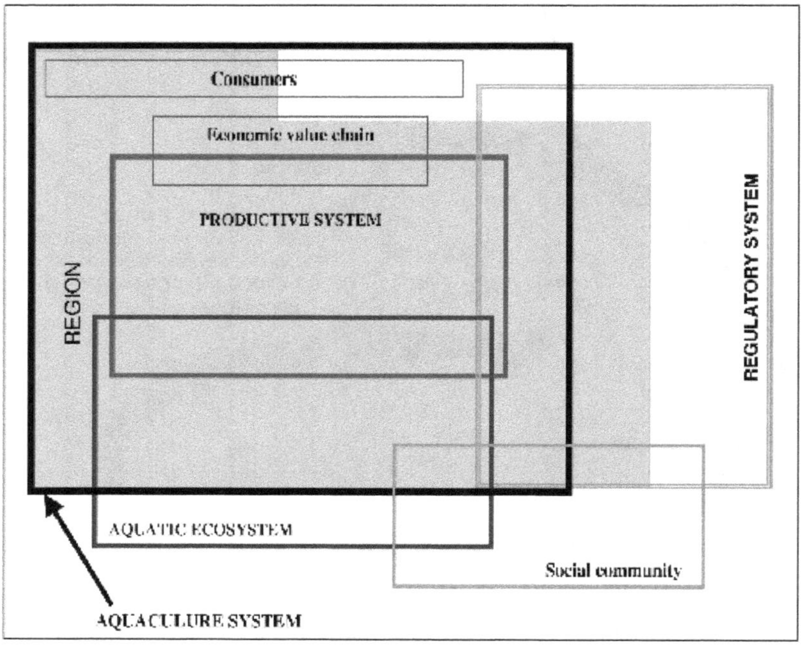

Fig. 3. A systemic approach to aquaculture production systems.

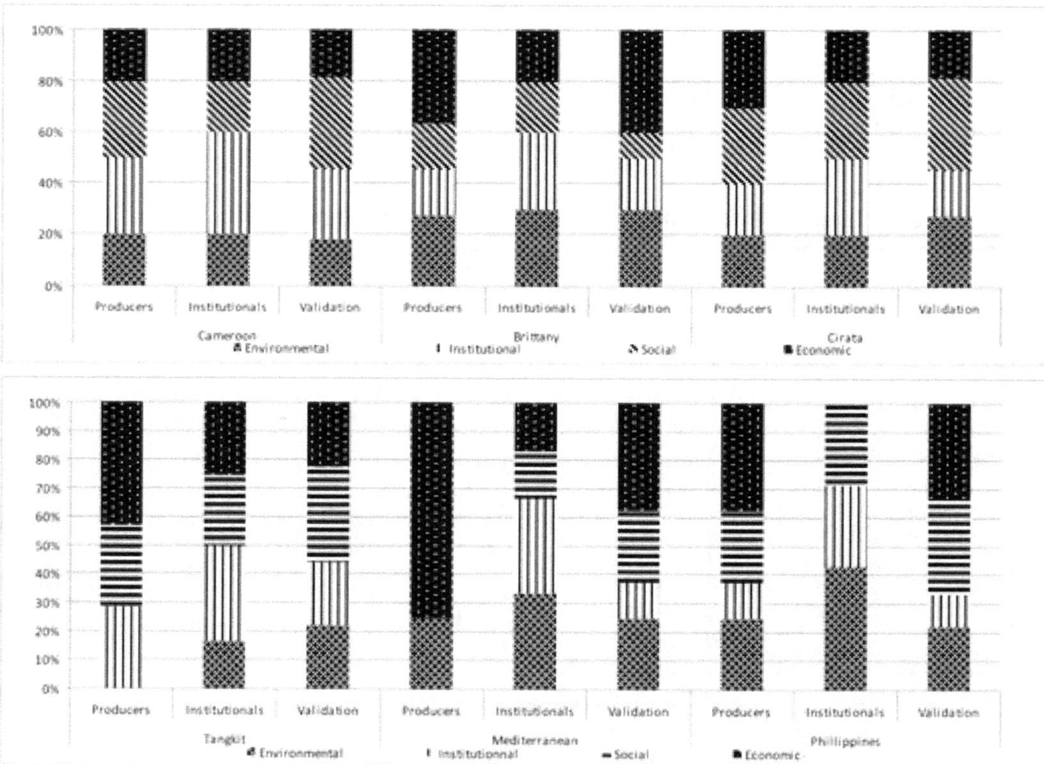

Fig. 4. Distribution of principles during the selection phase (choice of PCI by type of actors) and during the validation phase on the various study areas according to SD dimensions.

5. APPLYING LIFE CYCLE ASSESSMENTS TO THE AQUACULTURE SYSTEMS UNDER STUDY

The life cycle assessment (LCA) used in the project is a standardised method (ISO, 1997, 2000; Jolliet et al., 2005) used to establish environmental diagnoses of products or services. It is a method for the aggregation of knowledge. It is based on an inventory of all the resources used and of polluting emissions, including the extraction of raw materials, the development of the product, its use and its destruction (thrown away, recycled...). The functional unit selected is **1 tonne of aquaculture product** delivered to the first buyer. The allocation rules between flows are economic (environmental impacts are divided according to the value of the co-products). Calculations were based on the CML method (2001) modified in accordance with Papatryphon et al. (2004). Several categories of potential environmental impact were selected within the project framework as they were considered to be relevant for aquaculture. They are the following: 1) eutrophication (kg PO_4 eq) concerns the impacts on aquatic and terrestrial ecosystems associated with nitrogen and phosphorus enrichment; 2) acidification (kg SO_2 eq) assesses the potential acidification of ground and water due to the emission of acidifying molecules in the air, the ground or in water; 3) climatic change (kg CO_2 eq) assesses the production of greenhouse gases by the system; 4) the use of energy (MJ) concerns all the energy resources used; 5) the use of net primary production (kg C) represents the trophic level of farming from the quantity of carbon used and derived from primary production. For some sites, the following have been added: 6) the water dependency (m^3) defined as the amount of water flowing through the fish farm and required to produce fish; 7) the utilisation of the surface (m^2) which reflects the way the production system takes over the land, including the production of inputs (in particular the crops necessary for the manufacture of aquaculture feed). Work carried out on LCA within the framework of the EVAD project built on the experience of similar approaches already undertaken in aquaculture (Aubin and Van der Werf, 2009; Aubin et al., 2009; Papatryphon et al., 2004).

6. RESULTS

Over and above producing a guide for the co-construction of indicators for the sustainable development of aquaculture (Rey-Valette et al., 2008a, 2008b), the research undertaken made it possible to establish sustainability diagnoses for the different aquaculture systems studied, which, given their diversity, led to very instructive comparisons. The functionality of the approach suggested by the guide was tested when producing these diagnoses.

The diagnoses of the sustainability of aquaculture systems were first established for each area (territorial diagnoses §6.1) then at a global level by developing a synthesis of these diagnoses (into a meta-diagnosis §6.2). These diagnoses were undertaken at the criteria level, which is the most relevant analytical level to qualify the sustainability factors of these systems. The evaluation must be sufficiently detailed by theme to make the diagnosis intelligible. At this criteria level, even if the indicators which constitute these criteria are not all identical, it is possible to compare aquaculture systems. Then aggregating the results at the principles level facilitates comparisons even when selected indicators and criteria are not identical. One of the advantages of this type of approach is to make it possible to compare different aquaculture systems for which sustainability is measured with indicators adapted to the local characteristics and available information. Finally, these approaches based on simple criterion and indicator systems selected in each area were complemented by life cycle assessments for the environmental dimension (§ 6.3).

6.1 Territorial diagnoses of aquaculture system sustainability

Typologies carried out by area showed two differentiation factors common to all areas. These factors were the size of the farm and the nature of the capital or the ownership system associated with other factors, which vary according to the sites and are related to manpower, funding or to marketing methods (Lazard et al., 2009, 2010). These typologies reveal quite a large diversity in production and regulatory systems. Leaving aside the Tangkit site (Indonesia) where aquaculture systems are very homogeneous, three to four differentiated farm types were identified in each area, regardless of whether there was a large number of farms or not. This diversity in sustainability profile can even be found in aquaculture systems where the number of farms is low.

This article will not go into the details of the diagnoses undertaken at each area level. These diagnoses do make it possible to describe the situation of aquaculture systems in detail for each of the criteria in relation to the local context and issues. In order to present a global overview, there are several ways in which to aggregate these criteria. It is possible and important to present diagnoses for the two levels which characterise the approach, i.e. the sustainability of aquaculture farms and the contribution of aquaculture to the sustainability of territories in which each of the systems is located. Criteria can also be broken down according to the four dimensions of sustainability in order to identify in particular, the weaknesses and the strengths by sustainable development pillar. Finally, at a more subtle level, the sustainability profiles can also be considered in terms of farm types. The diversity of strategies goes hand in hand with the contrasting situations concerning sustainability. Hence, the 46 farms in Brittany and the 150 farms in Cameroon may be grouped into 4 sustainability profiles whilst the 18 farms in the Mediterranean (France and Cyprus) are distributed into three profiles. On the other hand, the two areas in Indonesia and the one in the Philippines, which comprise respectively 4,010 and 1,771 production units, are covered by a single sustainability profile.

The global overviews of the sustainability of the various aquaculture systems are presented here (figures 5 and 6) at the principles level in order to facilitate comparison. Working at this level makes it possible to generate general diagnoses by area which highlight the strengths and the weaknesses of the relevant aquaculture system. The aggregation process for principles is mostly based on the average score obtained by the various criteria that make up the principle, in particular when the results correspond to homogeneous or weakly- differentiated sustainability classes. For particular cases, where the principles rest on a restricted number of criteria characterised by very different sustainability scores, we adopted a common arbitration process consisting of choosing some of the selected criteria as the most determinant in the functioning of aquaculture systems. Average values had to be avoided in order to help identify strengths and weaknesses and make the diagnosis as operational as possible to facilitate decision-making and support action plans for sustainable aquaculture.

6.2 Meta diagnosis of the aquaculture systems under study

The comparative analysis of the results obtained for each area establishes several types of finding as regards the sustainability profiles of aquaculture systems and the accompanying policies that need to be implemented. In addition to its qualities in terms of the integration of representations and issues and its capacity to convey comprehensively the various dimensions of sustainability, the PCI method has the advantage of facilitating comparisons between diversified situations. The nesting of sustainability evaluation levels provides several comparative scales for aquaculture systems starting from indicators which are not necessarily the same. Hence comparability constraints and the conditions of adaptation to local specificities can be reconciled. A database was built of the selections made by the actors from different countries. It comprises 13 principles (table 3), 64 criteria and 129 indicators (Rey-Valette et al., 2008a, 2008b). Despite the system diversity, 10 principles and 25 criteria are common to 4 of the 6 areas. The proportion of common indicators is significantly lower with only 30 indicators common to three areas. Although the technical systems under study in Indonesia are very differentiated as regards both farming systems (cages and ponds) and aquaculture operators (farmers and entrepreneurs), there are many criteria common to the two Indonesian areas of Tankgit and Cirata. This observation tends to show the importance of cultural and institutional aspects for sustainability. Inversely, Cameroon, where aquaculture is struggling to develop, is a particular case which stands out from other areas in terms of principle selection and prioritisation. This situation tends to indicate that the degree of maturity of the sector is also a determining factor for sustainability.

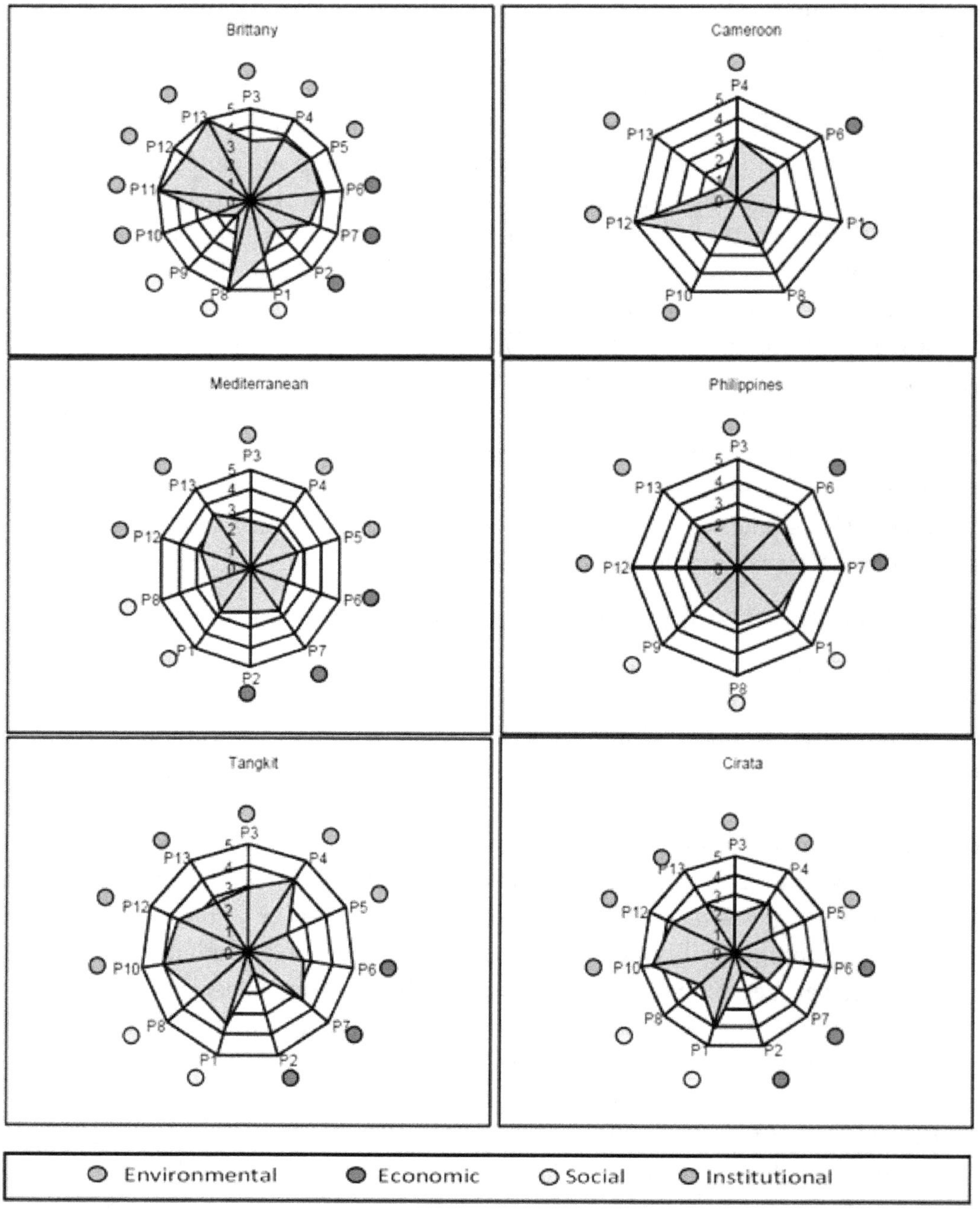

Fig. 5. Evaluation at principle level of the sustainability of aquaculture enterprises by country.

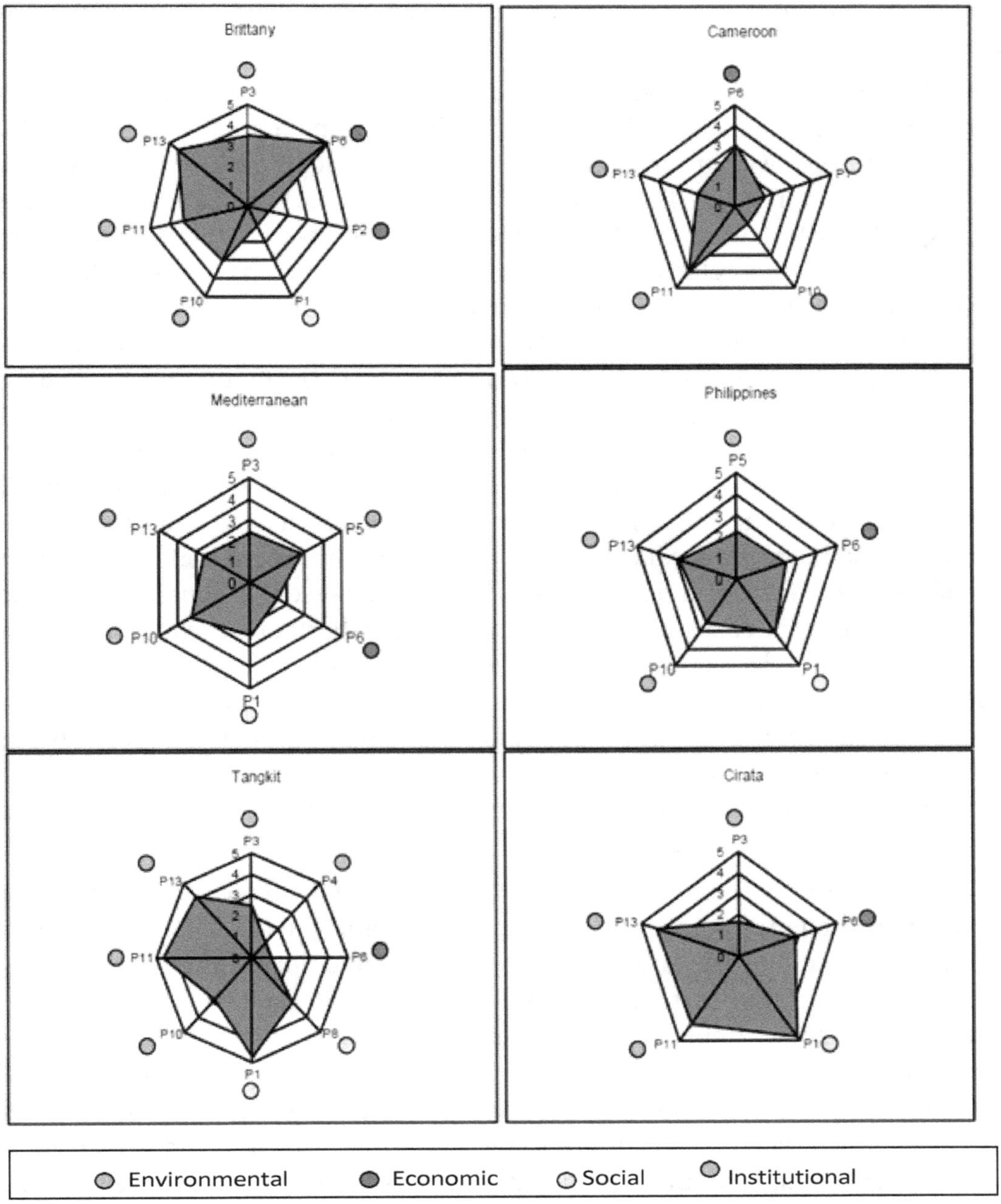

Fig. 6. Evaluation at principle level of the contribution of aquaculture enterprises to territorial sustainability by country.

Given the restricted number of areas studied, the main objective of the comparison process is not to compare the results in an attempt to obtain a universal diagnosis which could not be representative, but rather to study the structure of sustainability profiles and the types of criteria selected according to the areas. Table 3 presents the number of criteria selected by principle, distinguishing between those relating to farm sustainability and those relating to the evaluation of their contribution to territorial sustainability.

The analysis of the relative weights of the 13 principles highlights four principles which are little represented. These are: biodiversity and animal well-being (P5), social conditions within farms (P9), the role of aquaculture as a development factor for the territory (P10) and the capacity to participate in governance arrangements (P11). Overall, although these principles were of little importance, there appears to be a relative equilibrium between the dimensions of sustainable development. The analysis of the results of the selections made by the actors, depending on their status, shows that there are contrasting and specialised visions of sustainability in Tangkit, in the Mediterranean and in the Philippines depending on the type of actors, in particular between the producers who have a very sectoral vision of sustainability and institutional actors with a wider perception (figure 4). These contrasts decreased during the validation phase when actors altered their choice following discussions with other types of actors. Choosing a wide range of stakeholders appears therefore to be an essential condition to ensure an equilibrium between the dimensions, which is itself an essential condition in order to respect the spirit of sustainable development. Furthermore, the analysis of the types of criteria selected according to the area shows that actors tend to select criteria relating to aspects which seem to them to be problematic. This approach is therefore perceived by them as a management and programming tool to bring about progress in their aquaculture systems. This is a different process to labelling approaches or certification schemes which are often linked to marketing strategies and where the emphasis is on the strengths in order to build the image of the sector.

It would seem that the sectoral approach is dominant compared to using the territory as the entry point, as fish farm sustainability involves 46 criteria (60% of which are common to 4 areas) whereas the evaluation of the contributions of farms to territorial sustainability uses only 27 criteria (48% of which are common to 4 areas) (table 4). The share of territorial criteria is 29% for Brittany and 36% for Cameroon, the Mediterranean and the Philippines (table 5). The detailed analysis of the types of criteria selected according to the approach shows that the contribution of aquaculture to territorial sustainability concerns mainly the environmental and institutional dimensions of sustainable development. If we accept that the number of criteria selected is a kind of indicator of actors' awareness of sustainable development, the areas can be divided into three groups of decreasing appreciation: 1) Brittany, 2) the Mediterranean and Indonesia and finally 3) Cameroon and the Philippines.

The comparison of area sustainability profiles at the principle level produces a structural diagnosis of the kinds of strengths and handicaps concerning sustainability. For greater understanding, these results can be shown as "traffic lights", using green in the case of higher level sustainability classes (4 and 5) and red for lower classes (1 and 2), with class 3 corresponding to average scores remaining neutral (figure 7). The analysis of the results makes it possible to establish a typology of the areas in three classes, depending on the relative importance of their strengths and handicaps. Hence, Brittany is relatively well situated in terms of sustainability with, however, differentiated scores according to the various principles. On the contrary, the Mediterranean and the Philippines have more regular profiles which show some homogeneity in the results for all the principles with no outstanding strengths/constraints. Finally, Cameroon and Indonesia have, like Brittany, uneven profiles based on the principles but at a lower level of sustainability.

Table 4. Number of criteria selected by the 6 areas according to the principles and dimensions of sustainable development.

	Sector	Territorial		Sector	Territorial
Environmental			**Economical**		
P3. Ensure that natural resources and the environmental carrying capacity are respected	4	5	P6. Increase the capacity to cope with uncertainties and crises	6	5
P4. Improve the ecological yield of the activity	4	1	P7. Strengthen the long term future of exploitations	5	0
P5. Protect biodiversity and respect animal well-being	1	1	P2. Develop approaches which promote quality	2	1
Social			**Institutional**		
P1. Contribute to meet nutritional needs	5	3	P10. Strengthen the role of aquaculture in local development	2	4
P8. Strengthen sectoral organisation and identit	6	1	P11. Promote participation and governance	1	3
P9. Strengthen companies' social investment	1	0	P12. Strengthen research and sector-related information	5	0
			P13. Strengthen the role of the State and of public actors in putting sustainable development into place	4	3

Table 5. Distribution of the types of sectoral or territorial criteria selected by each area.

	Brittany (France)	Cameroon	Mediterranean	Tangkit (Indonesia)	Cirata (Indonesia)	The Philippines
Sector	42	18	25	33	32	18
Territory	17	10	14	18	14	10
Total	59	28	39	51	46	28
% territory	29%	36%	36%	35%	30%	36%

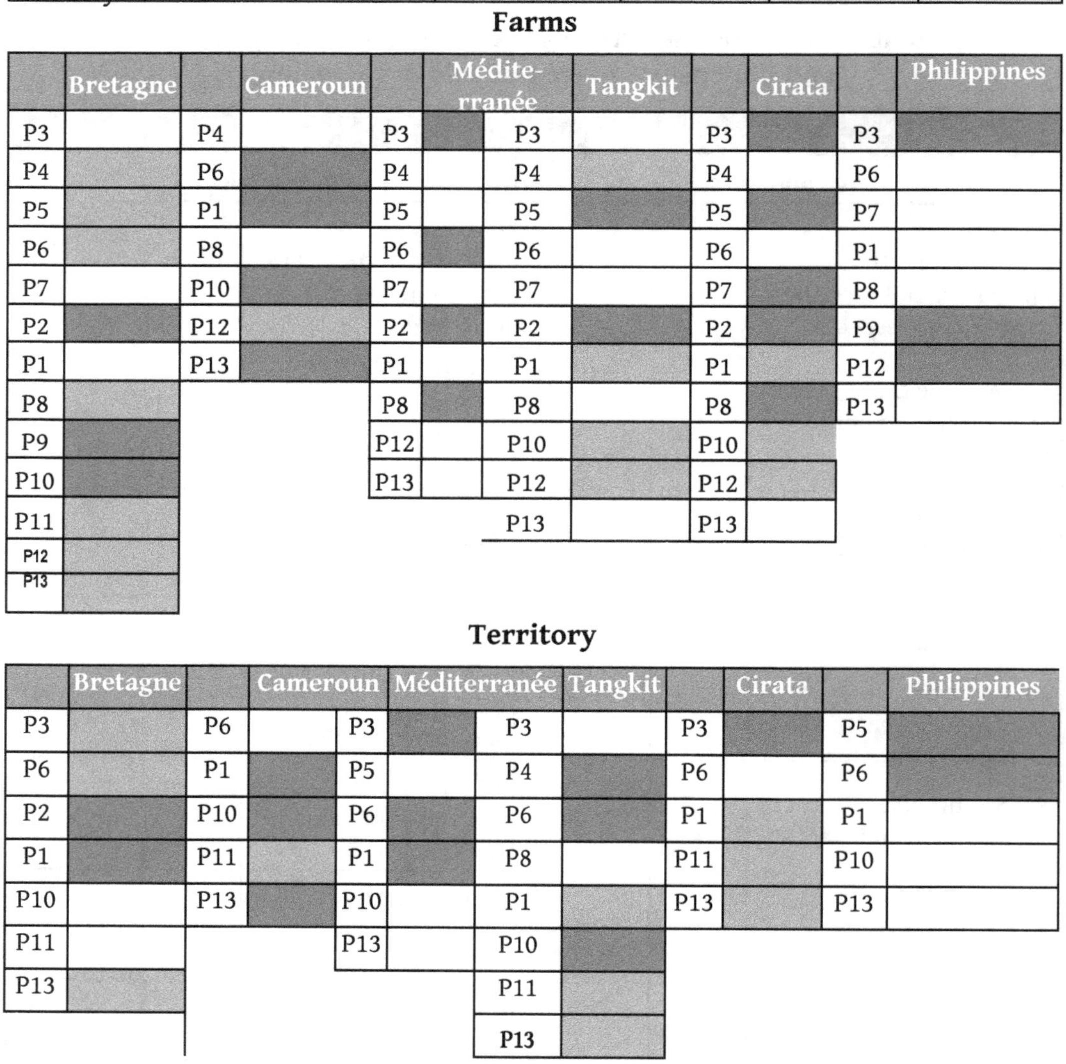

Fig. 7. Stylised presentation in terms of strengths (green) and constraints (red) of the results of the sustainability diagnosis by area in terms of sustainability principles.

This varying homogeneity in the scores is a fundamental result for the definition of sector- specific accompanying policies. Depending on the case, these policies will have to define the measures for regulation, incitation or raising awareness, focusing on a greater or lesser number of factors. This situation means that different types of public policies in terms of integration and progressiveness have to be designed.

6.3 Environmental diagnoses of aquaculture systems based on the LCA method

It will be recalled that this analysis was undertaken by studying 7 factors (eutrophication, acidification, contribution to climate change, and the utilisation of energy, water, surface area and net primary production) in all the areas. The results of the calculation of different impact categories obtained by LCA are presented as a relative evaluation of each of them for the various systems under study (figure 8).

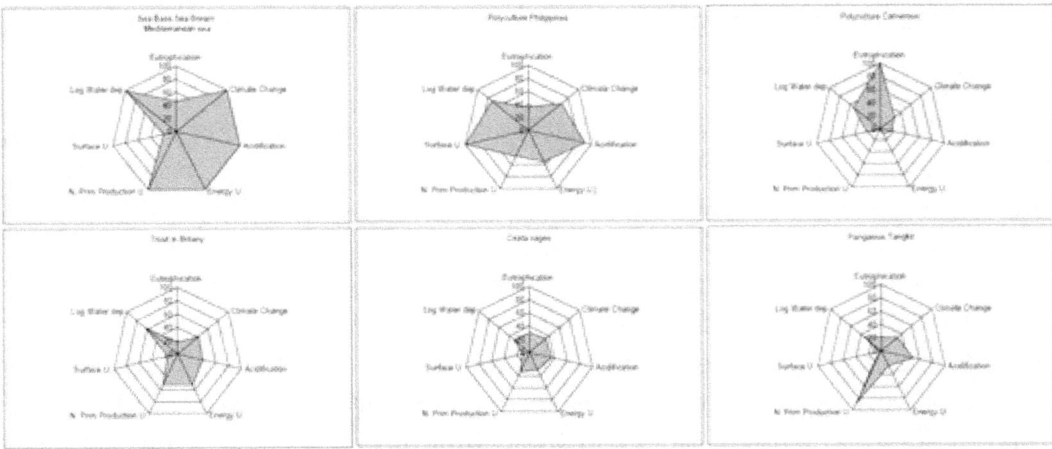

Fig. 8. Environmental profile of the 6 aquaculture systems under study in the EVAD project Radial graphs comparing the relative impact, for seven impact categories, of the 6 fish production systems. Points closer to the centre of the graph have less environmental impact. Values for Water Dependence have been log10-transformed.

For LCAs undertaken relative to a quantitative unit of production as they have been carried out here, it should be noted that there is no direct relationship between the level of intensification of the farming system and the level of impact. In particular, the Cirata fish farms in Indonesia (cages) and the bass and bream production in the Mediterranean, also in cages, are both very intensive, but show a very low level of impact for the former and a very high level for the latter. This might be explained by the species choice (predominantly plant- eating – omnivorous) and the goal of maximum productivity (by associating species: common carp and tilapia) in the first case and by the choice of carnivorous species (bass – bream) and a poor conversion index in the second case. In Brittany, trout has a profile which is similar to that of bass and bream but with markedly lower impact. This might be explained by the predominant effect of a high protein content feed in the two cases, which affects the level of impact significantly but the food conversion ratio is half in the case ofd trout. In the case of polyculture in Cameroon, only two impact categories show high levels. These are eutrophication and water dependency. This result shows the poor capacity of the system to utilise the nutrients (nitrogen and phosphorus) brought by the inputs (manuring, wheat bran...) combined with inadequate water management. Polyculture impacts are relatively high in the Philippines. They show the low productivity of the system, in particular due to the significant mortality of shrimp, which are the primary income source in the system. As a result, the quantity of inputs (fish fingerlings, shrimp post-larvae, molluscs for feed, energy...) does not produce sufficient output, and the same is true for land and water. In *Pangasius* fish farms in Tangkit, the predominant impact is the use of net primary production due to excess levels of fish meal (based on local species) incorporated into the feed.

7. CONCISE SUMMARY

Initially, it is crucial to acknowledge that the classification of areas of sustainability, derived from the multicriteria evaluation, aligns with the relative priority established by the outcomes of the life cycle assessment. Consequently, in both instances, Brittany (the technological model of intensive agriculture) has superior results, whereas more extensive systems, which may have been perceived as better aligned with natural systems in their environmental aspects and thus intuitively more "sustainable," receive significantly lower scores. This outcome indicates a correlation between sustainability results and the degree of control and delegated responsibilities, aligning with one of the definitions of sustainable development posited by Godard and Hubert (2002), who define sustainable development as "considering the repercussions of our actions."In general terms, this approach sits within the framework of companies taking sustainable development into account. Corporate social responsibility (CSR) may be defined as the fact that they "*assume the responsibility for the consequences of their actions and take pro-active measures to make their relationships with the rest of society and environment sustainable in the long term*" (Vivekanandan, 2008). According to this author, it leads to the notion of accountability defined as "*the responsibility of an actor to justify and account for his actions*" and supposes an adapted multi- level governance. However, in the case of enterprises, these strategies, as noted by Hommel and Godard (2008), do not just aim to integrate ethical issues but also to preserve the image and the capacity of enterprises to operate ("social licence to operate") and may pervert the collective dynamics which sustainable development seeks to initiate. These results, both from the multicriteria evaluation and from the LCA environmental approach, call into question the traditional classification between systems based on natural productivity management and "above ground" controlled systems and therefore most of the recommendations that are typically proposed, which, following in particular Naylor et al. (2000) and on the basis of partial approaches to sustainability, tend to promote more extensive systems.

It should also be noted that the criteria and the principles relating to the institutional dimension of the governance of aquaculture systems and territories were largely selected by stakeholders. This leads into the issue of the countries' political profiles (in particular the levels of decentralisation) and demonstrates the interaction between sectoral sustainable development policies and the political reforms of public action processes. It should be noted that theses aspects are little (or not) taken into account in standards and labels which are usually more focused and based on a sectoral or thematic approach, especially environmental or social. However, it is important to emphasise that these criteria concerning the institutional dimension are generally evaluated on the basis of expert opinion and should therefore be the subject of further research of a methodological nature. Finally, this institutional dimension, which is a governance issue, is increasingly defined as the fourth pillar of sustainable development (Goxe, 2007).

Lessons learnt from work carried out on the areas make it possible to put forward a number of more general conclusions which demonstrate the value of the method.

1. Combining a participatory and procedural approach with the integration of international reference frameworks has proved to be efficient. A fair level of learning and appropriation was achieved during the evaluation exercise yet worldwide the tendency is to establish norms and practices which are intended to apply generally, regardless of the zone or the scale, through "dialogues" and "good practices" originating from top-down work of NGOs and procurement centres (WWF, 2005, 2006, 2007, 2010).
2. The lessons learnt from this project – one proof is the diversity in the choice of indicators – confirms the idea that sustainable development cannot be fractal, i.e. the same content regardless of scale. One dimension which appears to be essential, although usually missing in the field of animal or vegetal productions, is that concerning the contribution of enterprises to the sustainable development of the territory in which they are located. The appropriation of this dimension strengthens local actors when they participate in integrated management arrangements for these territories where they can better defend the contribution of their activity to the territory.
3. This type of participatory approach clarifies that indicators can serve several functions: from simple measurement to the inventory of priorities and including the implementation of local rules. Given the little that most actors from the areas under study knew about the concept of sustainable

development, the indicators were clearly useful to give it some sense and bring it within general reach. However, indicators should only be developed once collective representations involving a diversity of stakeholders encompassing all the aspects of the activity have been identified.

4. Finally, between coercion, mimicry and professionalisation (Aggeri et al., 2005) which are different ways to adopt sustainable development, the production of the Co- construction guide (Rey-Valette et al., 2008a, 2008b), broadly disseminated following the project, clearly follows the third route. It emphasises the determinant importance of the choice of route to implement sustainable development for its adoption and the emergence of innovations within aquaculture systems.

This sustainable development method closely aligns with the ecosystem strategies proposed by the Millennium Ecosystem Assessment (2005). Nevertheless, considering this would have complicated the definition employed for the aquaculture system. Consequently, it would be beneficial, as a continuation of the study provided herein, to evaluate the relevance and social legitimacy of a methodology that interrogates the characteristics of aquaculture ecosystems and incorporates the concept of ecosystem services. (Chevassus-au-Louis, 2009; FAO, 2008).

8. REFERENCES

Aggeri, F., Pezet, E., Abrassart, C., Acquier, A., 2005. Organiser le développement durable. Expériences des entreprises pionnières et formation de règles d'action collective. Vuibert Ademe ed. Paris.

Aubin J., Van der Werf H.M.G., 2009. Pisciculture et environnement: apports de l'analyse du cycle de vie. Cah. Agric. 18 (2-3), 220-226.

Aubin, J., Papatryphon, E., Van der Werf, H.M.G., Chatzifotis, S., 2009. Assessment of the environmental impact of carnivorous finfish production systems using life cycle assessment. Journal of Cleaner Production 17 (3), 354-361.

Belton, B., Little, D., Grady, K., 2009. Is Responsible Aquaculture Sustainable Aquaculture? WWF and the Eco-Certification of Tilapia. Society and Natural Resources 22 (9), 840-855.

Billard, R., 1980. La pisciculture en étang. INRA, Paris.

Boulanger, P.-M., 2007. Political uses of social indicators: Overview and application to sustainable development indicators. International Journal of Sustainable Development 10 (1-2), 14-32.

Boyd, C.E., McNevin, A.A., Clay, J., Johnson H.M., 2005. Certification issues for some common aquaculture species. Rev. Fisher. Sci.13, 231-279.

Boyd, C.E., Lim, C., Queiroz, J., Salie, K., De Wet, L., McNevin, A., 2008. Best management practices for responsible aquaculture, in: USAID/Aquaculture Collaborative Research Support Program. Oregon State University, Corvallis, Oregon.

Bürgenmeier B., 2004. L'économie du Développement Durable. Question d'Economie et de Gestion. De Boeck, Bruxelles.

Caffey, R.H., Kazmierczak Jr., R.F., Avault, J.W., 2000. Developing Consensus Indicators of Sustainability for Southeastern United States Aquaculture. LSU AgCenter, Department of Agricultural Economics & Agribusiness Working Draft Bulletin No. 2000-01.Available at SSRN: http://ssrn.com/abstract=242312 or doi:10.2139/ssrn.242312

Carney, D. (Ed.), 1998. Sustainable rural livelihoods. What contribution canwe make? Papers presented at the DFID Natural Resources Advisers' Conference, July 1998. DFID, London.

Chevassus-au-Louis, B., Lazard, J., 2009. Situation et perspectives de la pisciculture dans le monde: consommation et production. Cah. Agric. 18 (2-3), 82-90.

Chevassus-au-Louis, B., Salles, J.M., Bielsa, S., Richard, D., Martin, G., Pujol, J.L., 2009. Approche économique de la biodiversité et des services liés aux écosystèmes. Contribution à la décision publique. Centre d'analyse stratégique (Premier Ministre), Rapports et documents, Paris. www.strategie.gouv.fr

Chia E., Rey-Valette H., Lazard J., Clément O., Mathé S., 2009. Evaluer la durabilité et la contribution au développement durable des systèmes aquacoles: propositions méthodologiques. Cah. Agric. 18 (2-3), 211-219.

Consensus, 2005. Defining Indicators for Sustainable Aquaculture Development in Europe. A multi-stakeholder workshop held in Ostende, Belgium, 21-23 November 2005.

Delgado, C., Wada, N., Rosegrant, M.W., Meijer, S., Ahmed, M. 2003. Fish to 2020. Supply and Demand in Changing Global Markets. International Food Policy Research Institute, Washington, DC.

Dey, M.M., Eknath, A.E., Sifa, L., Hussain, M.G., Tran Mai Thien, Nguyen Van Hao, Aypa, S., Pongthana, N., 2000. Performance and nature of genetically improved farmed tilapia: a bioeconomic analysis. Aquacult. Econ. And Manage. 4 (1-2), 5-11.

Droz, Y., Lavigne J.C., 2006. Éthique et développement durable. IUED Karthala Ed., Genève & Paris.

Edwards, P., 1999. Aquaculture and Poverty: Past, Present and Future Prospects of Impact. A Discussion Paper prepared for the Fifth Fisheries Development Donor Consultation, Rome, Italy (22-24 February 1999).

Edwards, P., 2000. Aquaculture, Poverty Impacts and Livelihoods. ODI/Natural Resource Perspectives 56. http://www.odi.org.uk/nrp/56.html

Edwards, P., Pullin, R.S.V., Gartner, J.A., 1988. Research and Education for the Development of Integrated Crop-Livestock-Fish Farming Systems in the Tropics. ICLARM Studies and Reviews 16. International Center for Living Aquatic Resources Management, Manila, Philippines.

FAO, 1995. Code of Conduct for Responsible Fisheries. FAO, Rome.

FAO, 2008. Building an ecosystem approach to aquaculture. FAO Fisheries and Aquaculture Proceedings 14, Rome.

FAO, 2009, La situation mondiale des pêches et de l'aquaculture. Département des Pêches et de l'Aquaculture de la FAO, Rome.

Fraser, E.D., Dougill, A.J., Mabee, W.E., Reed, M., McAlpine, P., 2006. Bottom up and top down: Analysis of participatory processes for sustainability indicator identification as a pathway to community empowerment and sustainable environmental management, Journal of Environmental Management, 78, 114-127.

GFCM (General Fisheries Commission for the Mediterranean), 2010. Indicators for sustainable development of aquaculture. Third coordinating meeting of the working groups, 24-26 February, 2010, 62-73. FAO, Rome.

Godard, O., Hubert B., 2002. Le développement et la recherche scientifique à l'INRA. Rapport intermédiaire de mission. INRA, Paris.

Goxe, A., 2007. Gouvernance territoriale et développement durable: implications théoriques et usages rhétoriques. in: Pasquier R., Simoulin, V., Weisbein, J. (Eds.), La gouvernance territoriale. Pratiques, discours et théories. Droit et Société 44. L.G.D.J. Ed., Paris, pp. 151-170.

Gupta, M.V., Acosta, B.O., 2004. From drawing board to dining table. The success story of the GIFT project. NAGA, WorldFish Center Quarterly 27 (3-4), 4-14.

Hatchuel, A., 2000. Intervention Research and the Production of Knowledge. in: LEARN Group (Ed.), Cow Up a Tree: Knowing and Learning for Change in Agriculture. Case Studies from Industrialised Countries. INRA Editions, Paris, pp. 55-68.

Hilden, M., Rosenström, U., 2008. The use of Indictors for Sustainable Development, Sustainable Development 16, 237-240.

Hommel T., Godard O., 2008. Que peut-on espérer des entreprises socialement responsables? in: Jacquet, P., Pachauri, R.K., Tubiana, L. (Eds.), Regards sur la Terre. La gouvernance du développement durable. Les presses de Sciences Po, Paris, pp. 167-178.

Irz, X., Stevenson, R.J., 2005. Aquaculture and poverty. A case study of five coastal communities in the Philippines. in: Aquaculture Compendium. CAB International, Wallingford, UK.

ISO (International Organization for Standardisation), 1997. International Standard 14040. Environmental management - Life cycle assessment - Principles and framework. International Organisation for Standardisation, Geneva.

ISO (International Organization for Standardisation), 2000. International Standard 14042. Environmental management - Life cycle assessment - Life cycle impact assessment. International Organisation for Standardisation, Geneva.

IUCN, 2005. Minutes of the workshops on aquaculture held within the framework of the World Conservation Forum, Bangkok, 17-25 November 2005. FAO, Rome.

Jodelet, D., 1989. Représentations sociales: un domaine en expansion. in: Jodelet, D. (Ed.), Les représentations sociales. Presses Universitaires de France, Coll. Sociologie d'aujourd'hui, Paris.

Jolliet, O., Saadé, M., Crettaz, P., 2005. Analyse du cycle de vie. Comprendre et réaliser un écobilan. Presses polytechniques et universitaires romandes/ Collection gérer l'environnement n 23, Lausanne.

Kautsky, N., Berg, H., Folke, C., Larsson, J., Troell, M., 1997. Ecological footprint for assessment of resource use and development limitations in shrimp and tilapia aquaculture. Aquaculture Research 28, 753-766.

Lazard, J., 1992. Réflexions sur la recherche en aquaculture tropicale: le tilapia d'Afrique et le tilapia d'Asie. La Jaune et la Rouge (Revue des anciens élèves de l'Ecole Polytechnique), 489, 23-26.

Lazard J., 2009. La pisciculture des tilapias. Cah. Agric. 18 (2-3), 174-192.

Lazard, J., Baruthio, A., Mathé, S. Rey-Valette, H., Chia, E., Clément, O., Morissens, P., Mikolasek, O., Legendre, M., Levang, P., Aubin, J., Blancheton, J.P., René, F., 2009. Diversité des systèmes aquacoles et développement durable: entre structures d'exploitation et représentations. Cah. Agric. 18 (2-3), 199-210.

Lazard, J., Baruthio, A., Mathé, S., Rey-Valette, H., Chia, E., Clément, O., Morissens, P., Mikolasek, O., Legendre, M., Levang, P., Aubin, J., Blancheton, J.P., René, F., 2010. Aquaculture system diversity and sustainable development: fish farms and their representation. Aquat. Living Resour. 23, 187-198.

Mathé, S., Brunel, O., Rey-Valette, H., Clément, O., 2006. Recensement des initiatives en faveur de la durabilité de l'aquaculture. CEP/UICN-Université de Montpellier 1, Montpellier, France.

Mickwitz, P., Melanen, M., Rosenström, U., Seppälä, J., 2006. Regional eco-efficiency indicators- a participatory approach. Journal of Cleaner Production 14 (18), 1603-1611.

Millennium Ecosystem Assessment, 2005. Ecosystems and Human Well-being: Synthesis. Island Press, Washington, DC.

Naylor, R.L., Goldberg, R.J., Primavera, J.H., Kautsky, N., Beveridge, M.C.M., Clay, J., Folke, C., Lubchenco, J., Mooney, H., Troell, M., 2000. Effects of aquaculture on world fish supplies. Nature 405, 1017-1024.

Papatryphon, E., Petit, J., Kaushik, S.J., Van der Werf, H.M.G,. 2004. Environmental impact assessment of salmonid feeds using Life Cycle Assessment. Ambio 33 (6), 316-323.

Prabu, R., Colfer, C., Dudley, R., 2000. Directives pour le développement, le test et la sélection de critères et indicateurs pour une gestion durable des forêts. Cirad Ed., Montpellier, France.

Rey-Valette, H., Clément, O., Mathé S., Chia E., Lazard J., 2007a. Le choix des principes, critères et indicateurs de développement durable de l'aquaculture: étapes et condition de l'appropriation du développement durable. Communication au colloque international « Instituer le développement durable. Appropriation, professionnalisation, standardisation », Lille 8-10 novembre 2007.

Rey-Valette, H., Laloë, F., Le Fur, J., 2007b. Introduction to the key issue concerning the use of sustainable development indicators. International Journal of Sustainable Development 10 (1-2), 4-13.

Rey-Valette H., Blancheton J.P., René F., Lazard J., Mathé S., Chia E., 2007c. Le développement durable: un défi pour l'aquaculture marine en Méditerranée. Cah. Agric. 16 (4), 1-10.

Rey-Valette, H., Clement, O., Aubin, J., Mathé, S., Chia, E., Legendre, M., Caruso, D., Mikolasek, O., Blancheton, J.P., Slembrouck, J., Baruthio, A., René, F., Levang, P., Morissens, P., Lazard, J., 2008a. Guide de co-construction d'indicateurs de développement durable en aquaculture. Cirad (Ed.), Montpellier, France.

Rey-Valette, H., Clement, O., Aubin, J., Mathé, S., Chia, E., Legendre, M., Caruso, D., Mikolasek, O., Blancheton, J.P., Slembrouck, J., Baruthio, A., René, F., Levang, P., Morrissens, P., Lazard, J., 2008b. Guide to the co-construction of sustainable development indicators in aquaculture. Cirad (Ed.), Montpellier, France.

Tacon A.G.J., Metian M., 2008, Global overview on the use of fish meal and fish oil in industrially compounded aquafeeds: Trends and future prospects. Aquaculture 285, 146-158.

Veverica, K.L., Molnar, J.J., 1997. Developing and extending aquaculture technology for producers. in: Egna, H.S., Boyd, C.E. (Eds.), Dynamics of Pond Aquaculture. CRC Press, Boca Raton, New York, pp. 397-413.

Vivekanandan, J., 2008. Vers la gouvernance multi niveau. in: Jacquet, P., Pachauri, R.K., Tubiana, L. (Eds.), Regards sur la Terre. La gouvernance du développement durable. Les presses de Sciences Po, Paris, pp. 127-137.

Wijkström, U., 2003. Short and long-term prospects for consumption of fish. Veterinary Research Communication 27 (suppl. 1), 461-468.

World Bank (The), 2007. Changing the Face of the Waters. The Promise and Challenge of Sustainable Aquaculture. The World Bank, Washington, DC.

WWF (World Wildlife Fund), 2005. Tilapia aquaculture dialogue. http://www.worldwildlife.org

WWF (World Wildlife Fund), 2006. Principles for responsible tilapia aquaculture. http://www.worldwildlife.org

WWF (World Wildlife Fund), 2007. Tilapia aquaculture dialogue meeting summary. http://www.worldwildlife.org/cci/pubs/2007summary.pdf

WWF (World Wildlife Fund), 2010. Pangasius aquaculture dialogue. http://www.worldwildlife.org

2 Chapter

GEOGRAPHIC INFORMATION SYSTEMS AS AN INTEGRATIVE TOOL FOR MARICULTURE MANAGEMENT IN PARATY, RIO DE JANEIRO, BRAZIL.

1. INTRODUCTION: AN OVERVIEW

The global depletion of marine life resources, caused by environmental degradation and overfishing, has prompted initiatives in aquaculture as an alternative for producing protein-rich meals. A 2004 FAO estimate indicated that of the 132 million tons of aquatic food produced in 2003, 31.7% originated from aquaculture, reflecting a 22.8% rise over the previous six years (25.8 % in 1998). Although marine aquaculture contributes with only 12.6 % of the whole 2003 production, it experienced an increased production of 24.4 % since 1998. Furthermore, marine fish capture has reached peak values in the year 2000 (86.8 million tons), but afterwards it showed a consistent decrease (81.3 million tons in 2003; Food and agricultural organization, 2004).

Although aquaculture has been seem as a major alternative for the threat of reducing stocks of fish worldwide, its insertion among the other coastal uses has shown to present significant environmental issues. Troell et al. (2003) analyzed peer-reviewed papers published in the literature of mariculture and identified a number of findings and techniques that provide new approaches for the sustainability of the activity. It is clear from Troell and colleagues' article that integration of the activity within the coastal zone management is the only way that can lead to a long term sustainability. On the other hand, Focardi and Corsi (2005) observed that the techniques to manage mariculture are still complicated and demand adaptation for each environmental condition, and is unaffordable for coastal populations in developing countries. The lack of sound management techniques drive fish-farmers to use chemicals that improve production, or to work with fed cultures and exotic species that impact the water quality, and affect ecosystems. Among the 10 future actions reported in Troell et al. (2003), the last one concerns the transfer of technology from researchers to producers, constituting the most challenging task for the sustainability of the activity.

Considering the complexity of uses in the coastal region and the possibility of conflicts that compromise the sustainability of the different activities, it is largely agreed that the application of geographic information system (GIS) modeling to determine the boundaries of the activities is helpful. However, when applying GIS, comprehensive strategies are necessary in order to identify the most important variables and parameters that will more reliably reduce the conflicts and warrant the sustainability. For instance, Mason Bay in Eastern Maine, US was studied by Congleton Jr et al. (1999) that gathered informations obtained from sediment characteristics, infra-red satellite images and hydrodynamic informations to establish suitable areas for mariculture of the soft-shell clam. In their work, focusing at the hydrodynamic characteristics, the authors were able to establish the parameters that would favor seeding juvenile shellfish. They also suggest that the system would be able to determine areas for suitable growth of the shellfish, but no GIS maps could be presented.

Geographic information system were also applied using algorithms that overlaid variables that were important for mariculture, like salinity, current velocity and wave height (Comert et al., 2008; Saitoh et al., 2011; Windupranata and Mayerle, 2009). When applied with special softwares, this methodology could generate comprehensible maps that rate suitability for mariculture. In a recent article Saitoh et al. (2011) develop a model that was coupled with IPCC (2007) previsions of climate warming presented rates ranging from 0 (totally unsuitable) to 8 (very suitable) for the cultivation of the japanese scallop in Funka Bay (Japan). The maps provided by the authors show the different scenarios according with the increase in global temperatures.

Mariculture has been developed in Brazil since the late 1970's, when the first experimental oyster cultivations were installed in Arraial do Cabo, Rio de Janeiro (Muniz and Jacob, 1986). Throughout the decades of the 1980's and the 1990's the development of the activity was still shy, for there was not a clear perception among investors of the restricted loading capacity of commercial and artisanal fisheries in the Brazilian coast (Magro et al., 2000; Rossi-Wongtschowski and Cergole, 2001). With the considerable contributions of the REVIZEE project (see some results on http://www.mma.gov.br/revizee/capa/menu.html) on the knowledge of the living resources of the Brazilian continental shelf, the need for alternative non-extrative procedures emerged. Later, in the early 2000's, the marine farms started to spread throughout the country and together with its development, sustainability problems, as well as conflicts with other uses in the coastal areas emerged and management solutions were demanded. Presently, the state of Santa Catarina is the fastest growing mariculture region in the country and in some areas, environmental problems are starting to emerge. Novaes et al. (2010) made a brief review of the activity in that State and showed how the application of a fairly complex GIS model could contribute to the management and location of the farms. In a participatory process, experts and local shellfish farmers, based on their knowledge and experience decided what were the most important variables allowing for the development of a reliable algorithm. The result of the work was a map presented in the article showing unsuitable areas in red and suitable areas in green.

This chapter delineates the approaches employed to administer mariculture in the Municipality of Paraty, Rio de Janeiro, Brazil, utilizing Geographic Information Systems (GIS) as a tool. The employed methodology was deemed the most effective means to organize the disseminated environmental, ecological, social, and political information. The methodology employed to Paraty involved the organization and presentation of information in a summary map, which encompassed the activities and topics influencing mariculture.

The chapter outlines a comprehensive framework of the planned GIS, detailing information sources and the proposed interface with society and municipal authorities. A set of metrics and variables is proposed as appropriate for future monitoring programs that will underpin long-term management initiatives.

2. METHODOLOGY

2.1 Study area

The Municipality of Paraty is located 236 km South-West of the city of Rio de Janeiro and is a tropical region within the geographic coordinates of 44.4° and 44.8° W, 22.9° and 23.4° S. The region is known for its outstanding natural scenic and cultural richness, underlined by the pristine environmental conditions. Most of its coast is located in the Ilha Grande Bay, a leaked embayment protected from waves by the Grande Island (Figure 1). The geology of the region was described by Muehe (1998), who included it among the Northern Crystalline Slopes. The steep relief characteristics of the region with heights of 1600 m is the result of the seaward drift of the Brazilian Coast Chain that dives into a very rugged coastline, with countless embayments and small islands. These characteristics are very favorable for mariculture, because waves and currents are not too strong, facilitating navigation and installation of fish-farm apparatus.

The coastal hydrography was classified as dendritic and the rivers follow the geologic alignements in the direction NE-SW and NW-SE (Figure 1). Due to the steep relief, the rivers are not very long and drainage basins are relatively small. On the other hand, the incidence of heavy rain events, mainly during winter periods yield a significant fresh-water input to the bay. In spite of thee significant fresh water inputs, the water of the bay is not very colored because of the thick vegetation cover observed in the continental portion (BRASFELS, 2005).

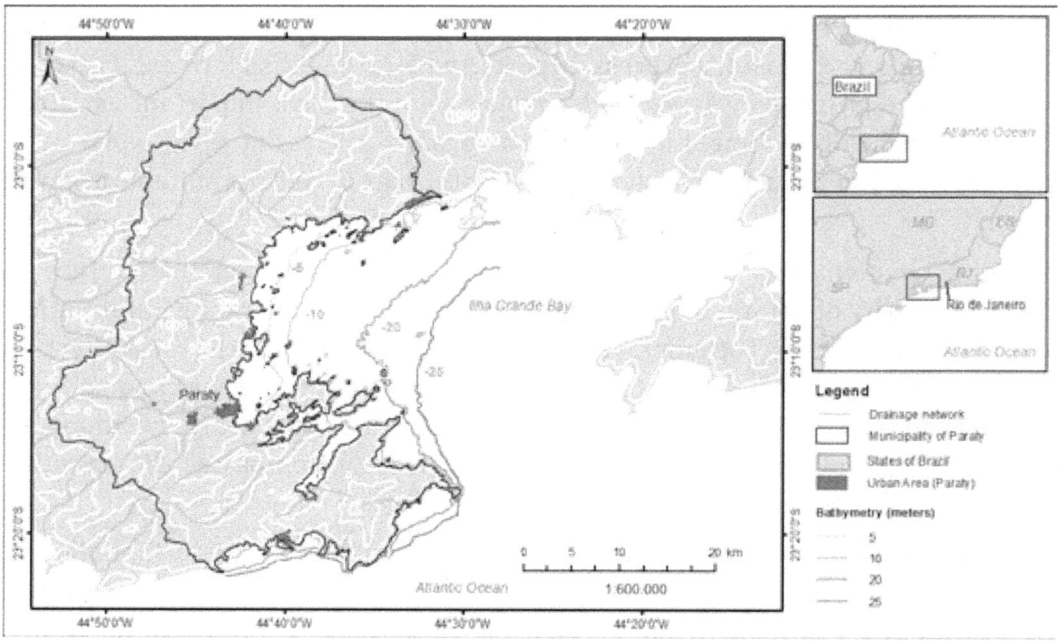

Fig. 1. Study area showing altimetry (100, 500, 1000 and 1500 meters), bathymetry (5, 10, 15, 20, 25 meters) and main drainage.

Most of the population of the Municipality is located in the small flat coastal plains, trapped between the sea and the steep vegetated mountains, a very favorable situation for touristic activities, generating potential conflicts with mariculture. Due to its pristine conditions, the region is still populated with traditional communities like "Caiçaras" (ancient fishermen groups of Portuguese and Indian origin), "Quilombolas" (black communities of ancient slavery refugees) and Indians. These groups have developed, through time, traditional and sustainable fishing techniques, but today, overfishing is threatening their way of life and mariculture may be one of their options of economic activity (Diegues, 2004).

The population of the Municipality of Paraty was estimated in 2006 to be circa 33.7 thousand inhabitants, which are mostly located in the rural areas (52% of the houses; IBGE, 2009). After the construction of the road BR-101, the connection with the great cities of Rio de Janeiro and São Paulo pulled the occupation and increased touristic predatory activities in the region. As established by Instituto Brasileiro de Geografia e Estatística (2000) the rate of occupation of the houses is quite low (72%) and among these, only 43% is perennially occupied. On the other hand, touristic rental of houses, house construction and services are the main economic activities of the Municipality GDP (TCE, 2003). The significant increase of the occupation without the due increase in sanitary infrastructure is generating eutrophication and uncontrolled pollution that also tend to conflict with fish-farms.

The maritime portion is characterized by a flat bathymetry, that reaches 25 meters in the edge of the Ilha Grande Bay (Figure 1). Wherever emerged areas are steep, the bathymetry is also locally steep and in these areas depths 20 meters may be observed 200 or 300 meters from the coastline. These areas are very suitable for mariculture because there is no depth limitation and the installations are very close to land.

2.2 Regulations for mariculture in Brazil

The Regulation number 06/2004 from the Brazilian Government established the driving lines for the ordainment of the mariculture and introduced the National and Local Plans for the Development of Mariculture (PNDM and PLDM respectivey), that constituted a reference document for the delimitation of mariculture parks and farms. Besides aiming to promote the organization of the mariculture initiatives, the document is intended to promote mariculture that offer an alternative income for artisanal fishermen, during periods of low productivity of the extractive fisheries. Nonetheless, the PNDM also aims to provide conditions and regulations for the commercial mariculture, wherever conflicts between artisanal communities and fishermen occurs. The Regulation number 06/2004 in its local approach (PLDM at the State or Municipality level) describes in detail the type and structure of the studies that are necessary to bind the mariculture areas. The approach presented in the document can be seen as a guideline for a generic Environmental Impact Assessment (EIA) for mariculture activities, including cultivation structures, transportation structures, processing and commercialization facilities.

The Local Plans for the Development of Mariculture (PLDM) should be carried out by expert interdisciplinary groups, which define the broadness of each PLDM as a function of the already installed farms and forthcoming activities associated. The work that was done in Paraty is part of the activities of the experts interdisciplinary group of the state of Rio de Janeiro.

2.3 GIS data

From the analysis of the available environmental data, it is clear that although scattered and scarce, oceanographic and biologic surveys are reliable for the construction of a local plan for the development of mariculture. Most of the works were obtained from university research programs (mostly master dissertations or doctorate theses) or constitute part of Environmental Impact Assessments and monitoring programs from logistics or industrial enterprises recently installed in the region. The objectives of these surveys were very distinct from those of mariculture and therefore the approach is frequently inadequate, but data is reliable and the information is suitable for the aims established in the present work. The spatial distribution of samples is normally focused on the studied problem, for instance, an EIA for a nuclear power plant from Ilha Grande Bay (Rio de Janeiro) is focused on the dispersion of pollutants in the case of accident. In this case, the hydrodynamic model (MRS, 2007) was constructed with a grid that is not adequate for the location of the farms. The simulation of the dispersion of hot water from the plant is also useless for the fish farmers. Furthermore, due to the fact that there is no articulation between these works, analytical methods, and units are sometimes incompatible, and it is difficult to correlate studies carried out in various periods. It is undeniable that an integrated program supported by the stakeholders and mainly the government is necessary, overcoming political, economic and social instabilities. Besides, the institutions responsible for this monitoring program should be strengthened, regardless its political colors.

For most mariculture areas, hydrodynamic data are scarce and circulation numeric models should be developed. Transport (dispersion) models may help to determine areas where water quality is more suitable for mariculture and may also help to determine loading capacity values for the activity (Alves and Wasserman, 2002). These hydrodynamic models may contribute to a better understanding of the oceanographic processes that dilute and disperse contaminants and therefore subsidizing the choice of better areas for mariculture. The dispersion models mentioned above are particularly interesting when evaluating the impact of coliform bacteria from domestic sewages on the cultivation stands.

In the state of Santa Catarina (Brazil) an expert group has developed a GIS system with a considerable number of variables that were thoroughly discussed by scientists and stakeholders (Novaes et al., 2010). Their system showed a reliable distribution of "suitability for mariculture" within the Baía Sul (Florianópolis), the determined variables and parameters cannot be applied elsewhere, heavily depending on the characteristics of the activity in each location.

The construction of a long term plan for the mariculture in the studied area depends on reliable monitoring programs, however, considering the present development of mariculture in Brazil, it is not possible to expect a broad and comprehensive set of data and the first management programs must be based on secondary information. Nonetheless, this management program must be the basis for a perennial monitoring of the environmental and social conditions, that will be constantly up-to-dated. This management program, in the form of a Geographic Information System will provide a dynamic process that follow the evolution of the knowledge, the evolution of the occupation of the spaces and the activity itself (Cicin-Sain et al., 1989).

In a generalist approach, Table 1 reports variables that should be considered as reference and should certainly be included in a monitoring program. However, for each location, some adaptations are necessary, including or excluding specific variables as a function of their own particular characteristics.

Table 1. Suggested variables for monitoring programs of mariculture areas.

1. Municipal limits and hydrographic basins
2. Detailed hydrography
3. Bathymetry
4. Water parameters (at least temperature and salinity)
5. Sediment contamination (metals, organic matter, hydrocarbons, pesticides, nutrients)
6. Classified coastal line (beaches, rocky coasts, mangroves, dumping areas, coastal vegetation)
7. Roads, railways and other accesses
8. Conservation units
9. Mariculture parks and areas
10. Fishermen and fish farmers associations, communities
11. Fish and shellfish processing units
12. Maritime uses (navigation paths, nautical tourism, nautical leisure, anchoring areas, marinas, harbours, fishing areas)
13. Reef areas, rocky and sandy banks,
14. Restricting areas (regulation)
15. Soil uses and cover of the continental adjacent area (with the LCC - Land Cover Classification of FAO)

Finally, an important point that has also to be discussed is the scale. When the parameter do not affect directly the activity like coastal vegetation, municipal limits, etc, smaller scales should be used (1:50,000). For parameters that directly influence mariculture like depth and sediment characteristics, bigger scales should be used. The problem is that sometimes available data grid is not tight enough to allow bigger scales. Another problem is that data variation is such that very tight sampling grid is not enough and a temporal series is necessary (that is the case for temperature or salinity). Even if the environmental parameters can only be represented in small scales (up to 1:50,000), the location of the marine farms must be precisely set (1:10,000) because overlapping of the areas should cause conflicts between farmers.

A broad range of information was obtained from academic studies, environmental impact assessment reports, management plans for the neighboring protected areas, municipal master plans and other sources. Furthermore, *in situ* observations, interviews and participatory mapping with stakeholders enriched the work. Cartographic information from the mainland were based on the topographic maps of the region (1:50,000), supplied by the Brazilian Institute of Geography and Statistics (IBGE) and the Directorate of Geographic Service (DSG – Brazilian Army). The cartographic information from the marine area was based on the bathymetric map (1:40,000) from the Directorate of Hydrography and Navigation of the Brazilian Navy (DHN).

The data was introduced in the software ArcView 9.0®, ESRI Software, and the approach was based on the concept of indicators that are variables recognized as significantly affecting the water quality or constituting elements of conflict with other activities. These variables were classified into production indicators and constraining indicator. Among the production indicators were temperature, salinity, depth, wind, currents, wave parameters, sediment characteristics, nutrient concentrations, chlorophyll-a concentrations, natural beds of seaweeds and eelgrass bank locations.

The constraining indicators surveyed were grouped into: 1) Restricted Areas (Law protected areas, leisure boating areas; anchorage distance from beaches – 200 m and from rocky shores – 50 m), 2) Activities with Potential Conflict (boat channels, diving areas, nautical tourism areas, and other activities related with Tourism, and existing marine farms) and 3) Pollution (areas of direct influence of rivers and streams, effluents from urban centers). This indicators, were also identified in other studies presented in the literature (Ferraz, 2006; IED-BIG, 2001, 2002; Perez-Sanchez and Muir, 2003; Scott, 1998) and were considered relevant for the study area.

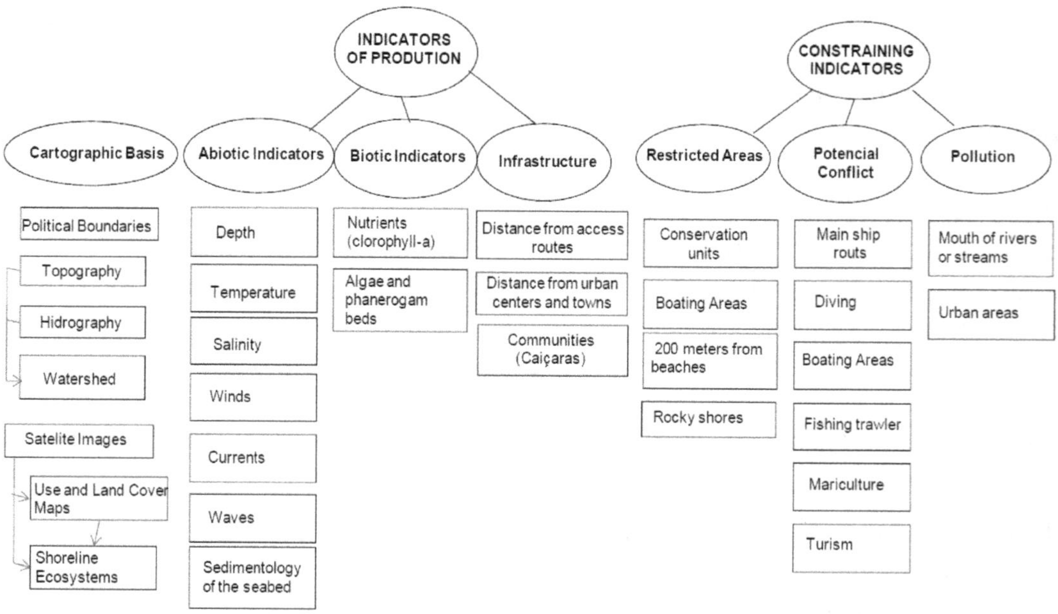

Fig. 2. Flowchart of the indicators used for the delineation of potential areas for mariculture in Paraty.

Furthermore, buffers considering distance from routes (streets and roads), distance from urban centers, distance from traditional communities (Caiçaras) and distance from marinas and airports were included. Although the evaluation of the reach of effluents from urban sewages is tentative, a buffer was included in the mouth of rivers and channels that were considered contaminated. The Figure 2 is a brief outline of mariculture production indicators and of constraining indicators.

3. RESULTS AND DISCUSSION

The study area was divided into areas of direct and indirect influence. The former (ADI) includes the aquatic marine area of the municipality of Paraty, located between the coastline and the isobath 25 meters. For practical reasons, these limits considered the distance from the coast, because the costs with diving apparatus, fuel, risk of theft and risk of navigation increase for farther operations. The area of indirect influence (AII) corresponded to the watersheds that contribute with freshwater and effluents from rivers and human activities into the ADI. (Figure 3)

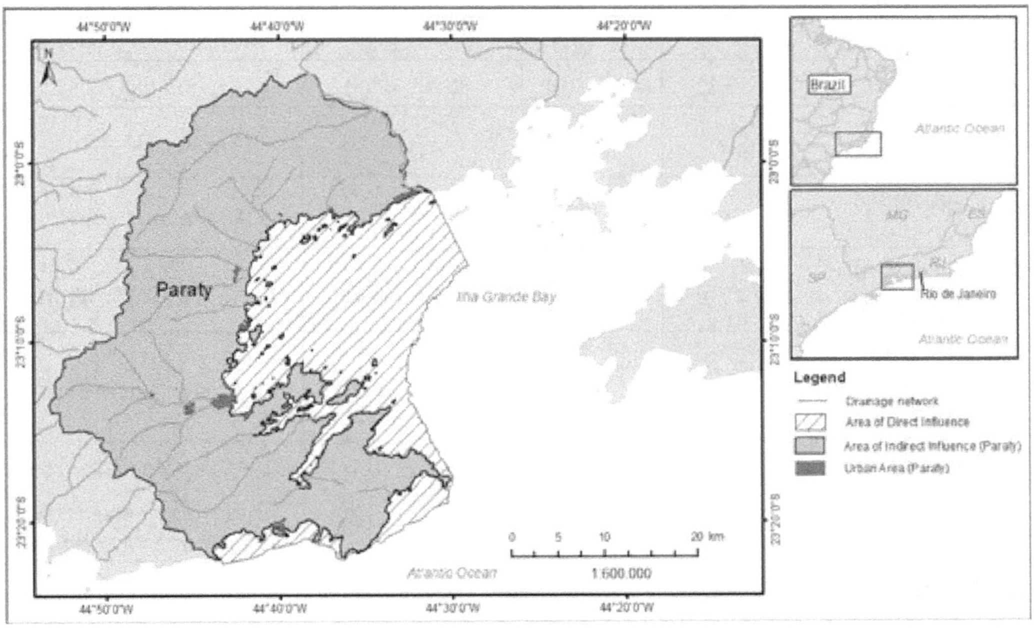

Fig. 3. Location of the Area of Direct Influence (ADI) and Area of Indirect Influence (AII).

The Areas of Direct Influence were classified into three categories regarding the potential for mariculture of scallops, that is the main mariculture activity in the area: ideal areas, appropriate areas or unsuitable areas. The classification considered the physiological characteristics of these organisms and ability to adapt to changes in the physical-chemical properties of the environment (IED-BIG, 2002; Scott, 1998). In Table 2 are shown the range of values for each indicator, classified as ideal or appropriate. Areas that had values outside these ranges were classified as unsuitable and were not represented on the map.

Table 2. Range of values of the indicators of production for the cultivation of scallops (*Nodipecten nodosus*) in the study area.

Indicators of production	Classes/ranges		References for the classification of ranges	Data source from the study area
	Ideal	Good		
Depth (m)	5 - 15*	16 - 25	(IED-BIG, 2002)	Bathymetric map (DHN)
Temperature (ºC)	18 - 22	15 - 17 and 23 - 25	(IED-BIG, 2002)	(Bormann, 2005)
Salinity	34 - 36		(IED-BIG, 2002)	(Bormann, 2005)
winds (m/s)	x	x	x	BRASFELS (2005)
Sea Current (cm/s)	10 - 60	60 - 70	(Scott, 1998)	BRASFELS (2005)
Waves (m)	< 0,25	0,25 - 0,4	(Scott, 1998)	BRASFELS (2005)
Sedimentology of the seabed	x	x	x	(DIAS et al., 1990))
Nutrients (chlorophyl-a)	x	x	x	(Bormann, 2005)
Distance from land access routes (km)	0 - 5	5 - 10	(Scott, 1998)	Participatory Mapping
Distance from the urban centers and towns (km)	0 - 4	4 - 10	(Scott, 1998)	Land cover and use maps
Distance from Traditional Communities (km)	0 - 2	2 - 5	(Scott, 1998)	Topographic maps from IBGE and *Management Plan from EPA Cairuçú*

3.1 Production indicators

According to FAO (1991) the ideal temperatures for growing scallops range between 20°C and 23°C. Temperature is among the main parameters to be monitored in the cultivation of scallops due to high sensitivity to small variations of these organisms In the study area, a more recent study has shown that the ideal temperatures for the growth of *Nodipencten nodosum* range from 18°C to 22°C, not quite different from what has been observed for other species worldwide (IED-BIG, 2002). When adult, these organisms have a regular growth in temperatures ranging between 16°C and 25°C and the marine water temperature in the study area varies between 22.5°C and 26°C in summer and between 21°C and 23ºC in winter (Bormann, 2005; IED-BIG, 2002). The optimum depth for scallops ranges between 5 and 15 m (IED-BIG, 2002), but depths of 16 and 25 m are still appropriate. The depth is closely related with the temperature, because in the calm waters of the study area, there is strong thermal gradients.

The adequate distance from the land routes is not farther than 5 km. Beyond this distance, the production becomes expensive and the risk of theft of equipments and of scallops become unacceptable. After reaching the coast, the access to the consumers market must also be considered, due to the need of low temperature containers. In this case, after reaching the routes, 4 km was considered as an ideal distance from consumers centers and 10 km is still acceptable (Scott, 1998). The ideal distance between shell-farmers residences and crops is 2 km, but the distance is still appropriate between 2 and 4 km. Beyond 4 km the daily transportation between their residences and the farm becomes expensive and navigation safety gets compromised (Scott, 1998).

Hydrodynamic indicators like winds, currents and waves were grouped together and represented as restrictions indicators in the southern portion of the study area that is located in the open sea and is exposed to stormy conditions. However, for these areas, farmers can consider the use of fixed structures that resist this extreme oceanographic conditions, as has been done in the neighboring Municipality of Angra dos Reis.

3.2 Constraining indicators

Due to its pristine state, the region has a great number of Conservation Units, that may or may not constitute a constraining factor, as a function of its type: the sustainable use units allows the installation of some few economic activities since it does not constitute any threat for the environment. In the case of the sustainable conservation units of Paraty, the installation of marine farms is allowed. On the other hand, the Integral Protection Units has as main objective the preservation of untouched environments. In many of these conservation units even scientists are not allowed, unless an special authorization is granted by the Brazilian Environmental Agency (IBAMA) and mariculture of any kind is not allowed. Data concerning Federal and State Conservation Units were obtained from Foundation CIDE (2005) while Municipal conservation units were obtained from the Municipality of Paraty (Macrozonation and the Master Plan of the Municipality).

The nautical maps of the Brazilian Navy number 1633 and 1634 were used to determine the coastal limits that after Marinha do Brasil (2003) restrict any aquatic activities within 200 meters from sandy beaches or 50 meters from rocky shorelines. The activities developed in the marine environment, which may constitute potential conflicts with mariculture have also been plotted: navigation channels, diving areas, leisure boating areas, trawlers' fishing and tourism areas. In addition to areas where mariculture is already being developed.

The identification of the main navigation routes was carried out using the method of "participatory mapping" (Tuan, 1975, 1983), which consisted of collecting information from community members. In the fieldwork, the main routes were set with the aid of the Secretary of Fisheries and Aquaculture of Paraty. Information on that matter was also obtained from the Brazilian Navy Agency at the Port of Paraty. It has to be underlined that the navigation routes obtained and plotted are preferred areas, but there is no rigid limits.

The areas for diving and tourism were identified following consultation with operators and tourist guides of Paraty. The points of nautical tourism, as well as the classification of areas according to the degree of potential for tourism, were obtained from a Zonation Map of the Touristic Potentials, part of the Master Plan for Tourism Development by the Municipality of Paraty. The degree of potential for tourism ranged from 1 to areas with lower potential to 5, for those with greatest potential.

The locations of the areas where fish-trawling is developed were obtained from the Management Plan of the EPA Cairuçu. This Environmental Protection Areas comprises the whole southern coast and all of the Island of the Municipality of Paraty and therefore the document includes the whole coastal region. Furthermore, the areas of installed marine farms, including those of scallop, algae and fish were obtained from the Secretariat of Aquaculture and Fisheries of the Presidency of the Brazilian Republic (SEAP-PR), which has a register of areas required for the activity.

In Paraty, the main source of contamination is domestic sewages, that frequently are directly dumped in the drainage and rivers. Except for a few small shipyard (leisure boats), no possible source of contamination from industrial activity is present in the drainage basin. Agricultural activities are also scarce in the region (Rocha, 2005) because of the steepness of the terrains and therefore very little nutrient inputs from fertilizers can be identified (MRS, 2006). Based on the above statements and order to access the impact and extent of the domestic pollution in shellfish farm areas, a buffer of 500 meters or 1000 meters radii from the mouth of the rivers and streams was established, beyond which, mariculture is adequate or ideal (respectively). These limits were established without any consistent ground, but were corroborated by IED-BIG (2002). For a better establishment of these buffers a hydrodynamic transport numeric model has to be developed, in order to simulate conservative and non-conservative contaminants.

According to Instituto Brasileiro de Geografia e Estatística (2000), 85% of households don't have sewages treatment that may constitute a serious restriction to mariculture. Contamination with coliforms or with nutrients may cause harmful algae blooms (HAB) constituting serious threats to consumers of sea food. After an outbreak of HAB, the recovery of the activity is very difficult and demands a lot of new investments. The importance of the evaluation of the impact of these contaminants was recently demonstrated in the outbreak of what looked like a shellfish disease in the Ilha Grande Bay (Côrtes et al., 2009). The disease colored the kidney of scallops and as scientists tried to identify what was responsible for the anomaly, no side environmental data could support any of the hypotheses. Furthermore, the studies initially carried out were not integrated and did not constituted reliable evaluations. Considering the risks in cultivation of mollusks and fishes, including contamination with harmful algae blooms, it is advisable that a technical and scientific team should be ready for action in the case of outbreak of diseases.

Finally, the local shellfish farmers were listened in order to determine the areas that their experience indicate as more suitable for the cultivation of mussels, utilizing the participatory mapping method (Tuan, 1975, 1983). Rough outlines were established together with the stakeholders, who identified in nautical charts, in the scale 1:40,000, the most viable areas for cultivation, following what is called "cognitive knowledge".

3.3 Data analysis

The identification of potential areas for mariculture was made according to the results obtained in synthesis maps via the Boolean intersection method (Burrough and McDonnell, 1997). This method of analysis does not require quantitative rating of the production indicators and indicators that showed significant restriction for mariculture.

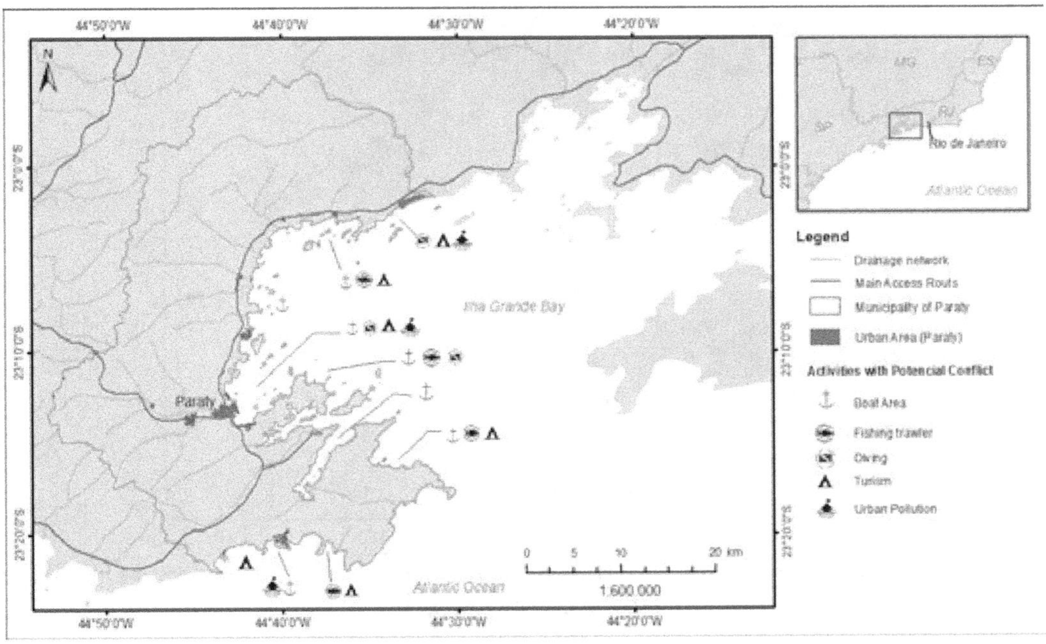

Fig. 4. Potentially conflicting activities in the coastal region of Paraty.

Data on potentially conflicting activities were presented in a separate map (Figure 4), considering the large amount of information being displayed. The areas of conflict are not considered as areas of exclusion of mariculture, but constitute multiple use territories, where acquaintance of use has to be achieved from agreements between local managers and other stakeholders.

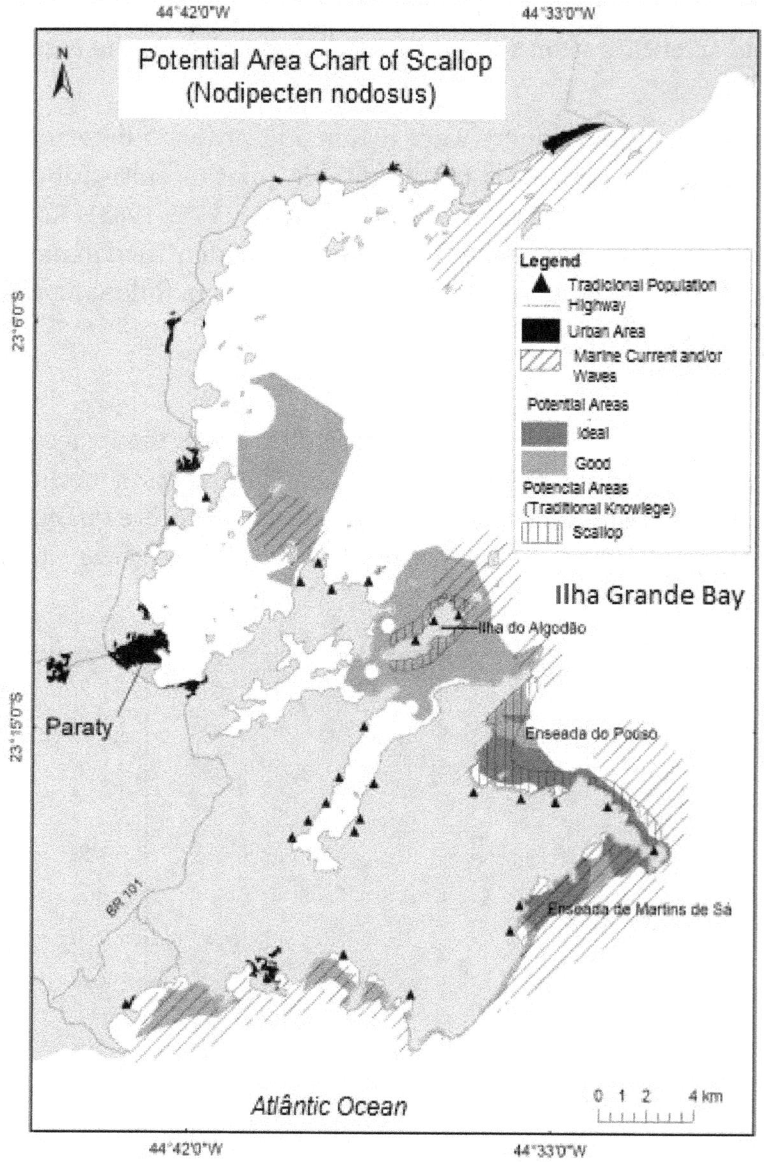

Fig. 5. Summary map, showing the suitable areas for cultivation of scallops (*Nodipecten nodosus*).

It is interesting to note that the areas established in the participatory process overlap those established with the indicators of production and constriction. The analysis of Figure 5 indicate that the northern portion of Paraty is not quite suitable for the cultivation of the scallop, because during summer water can present very high temperatures. This higher temperatures are attributed to the reduced hydrodynamics and shallowness. In this northern portion, the more significant inputs of fresh or contaminated waters, higher levels of suspended matter and other factors are also responsible for the reduced suitability for scallop cultivation.

From Figure 5, the best (ideal) areas for mariculture of scallops are located near the Ilha do Algodão and Enseada do Pouso (Pouso Cove) in the semi-sheltered and barely occupied southern region. Furthermore, the settlement of traditional Caiçara populations favors the installation of shellfish farms that may constitute a sustainable source of incomes for these poor communities. Farther South, the Enseada Martins de Sá (Martin de Sá Cove) is also considered as ideal for the cultivation of scallops, however special structures should be installed in order to stand the destructive potential of the currents and waves expected during storms in the area.

In the surroundings of the city of Paraty, mariculture is not advisable due to heavy boat traffic and the influence of contaminants generated mainly from domestic sewage and fuel oil from leisure boats. In the northern portion of the study area, the presence of Conservation Units prevents the installation of mariculture parks in the surroundings of the islands. However, the proper management of the water quality in this portion of the study area, would allow the installation of the activity in the unrestricted areas. The knowledge of the load capacity of each environment has to be developed, permitting to establish a safer dimension of the activity. Conflicts also must be managed with a strong participatory process, establishing strategies for the multiple use of the territory.

4. CONCISE SUMMARY

In conclusion, in regions where mariculture remains limited, the regulatory processes are relatively straightforward and lack formal farm management; conversely, in areas with established practices, the procedures are considerably more intricate, leading to disputes that necessitate advanced systems. It is evident that, over time and with the advancement of mariculture, the strategic planning of marine territory utilization is a prudent approach that will prevent issues similar to those experienced in Northeast Brazil, when entire fields of Malaysian shrimp were devastated.

In Paraty, several factors were identified for assessment to prevent usage conflicts, overexploitation of natural resources (endangering sustainability), and to ascertain the most appropriate locations for the activity. The objects were mapped in a GIS system, serving as an effective tool for identifying locations for mariculture parks, where farm installations are favored. The arrangement of operations in mariculture parks, where conditions are more conducive, may provide a solution for the anticipated expansion of the sector, mitigating conflicts with other activities and environmental concerns.

The study's findings indicate that the production indices influencing optimal mariculture locations in Paraty include: the proximity of marine farms to traditional communities; annual depth and temperature fluctuations; and the influx of freshwater and pollutants. In storm-prone regions, cultivation is recommended; nevertheless, substantial investments in equipment are necessary to prevent production losses during inclement weather. An economic assessment must be conducted to assess the profitability of these initiatives for local communities.

The adoption of the GIS approach in various domains may require changes to accommodate local features; yet, it can significantly enhance the management and long-term sustainability of mariculture.

5. REFERENCES

Alves, A.R., Wasserman, J.C., 2002. Determinação do tempo de renovação em sistemas lagunares. Mundo Vida Alter. Est. Amb. 3, 48-53.

Bormann, R.S., 2005. Atlas Geotemático das Baías de Angra dos Reis e Ilha Grande, Desenvolvidos num Sistema de Informações Geográfica, Instituto de Ciências Biológicas Universidade Santa Úrsula, Rio de Janeiro, p. 45.

BRASFELS, 2005. Environmental Impact Assessment of Dredging Services for the deepening and manutention of the access for the Brasfels Shipyard. Ecology Brazil, Rio de Janeiro, RJ.

Burrough, P.A., McDonnell, R.A., 1997. Principles of Geographical Information Systems. Oxford University Press, Oxford.

Cicin-Sain, B., Knecht, R.W., Jang, D., Fisk, G.W., 1989. Integrated Coastal and Ocean Management: Concepts and Practices. Island Press, New York.

CIDE, 2005. Anuário Estatístico do Rio Janeiro. Estado do Rio de Janeiro - Fundação Centro de Informações e Dados do Rio de Janeiro, Rio de Janeiro, pp. CD-ROM.

Comert, Ç., Akinci, H., Åžabin, N., Bahar, Ö., 2008. The value of marine spatial data infrastructure for integrated coastal zone management. Fresenius Environmental Bulletin 17, 2240-2249.

Congleton Jr, W.R., Pearce, B.R., Parker, M.R., Beal, B.F., 1999. Mariculture siting: A GIS description of intertidal areas. Ecol. Model. 116, 63-75.

Côrtes, M.B.V., WASSERMAN, J.C., AVELAR, J.C.L., 2009. Gestão da Qualidade Sanitária de Moluscos Bivalves para Cultivos das Baías da Ilha Grande (Paraty, Angra dos Reis e Mangaratiba) e da Guanabara (Niterói), V Congresso Nacional de Excelência em Gestão. LATEC - UFF, Niterói, RJ, Brazil, pp. 1-11.

DIAS, G.T.M., PEREIRA, M.A.A., DIAS, I.M., 1990. Mapa geológico - geomorfológico da Baía da Ilha Grande e Zona Costeira adjacente, esc. 1:80000 - Relatório Interno LAGEMAR,Universidade Federal Fluminense, Niterói.

Diegues, A.C.A., 2004. Fishing, constructing societies. NUPAUB - University of São Paulo, São Paulo.

FAO, 1991. Training manual on breeding and culture of scallop and sea cucumber in China. Yellow Sea Fisheries Research Institute in Qingdao, People's Republic of China, Regional Seafarming Development and Demonstration Project (RAS/90/002), Qingdao, China.

Ferraz, G.M., 2006. Aspectos socio-ambientais de áreas costeiras com potencialidade aqüícola no Município de Niterói - RJ, M.Sc. dissertation in Environmental Science,. Universidade Federal Fluminense, Niterói, p. 103.

Focardi, S., Corsi, I., 2005. Safety issues and sustainable development of European aquaculture: new tools for environmentally sound aquaculture. Aquaculture International 13, 3-17.

Food and agricultural organization, 2004. State of world fisheries and aquaculture (SOFIA). FAO Fisheries Department., Rome, p. 153.

IBGE, 2009. Anuário Estatístico do Brasil. Fundação Instituto Brasileiro de Geografia e Estatística (IBGE). Rio de Janeiro, RJ.

IED-BIG, 2001. Curso Internacional de Biotecnologia de Cultivo de Ostra Nativa (Crassotrea rhizophorae) e Mexilhão (Perna perna). Instituto de Ecodesenvolvimento da Baía da Ilha Grande - Laboratório de Larvicultura de Moluscos, Angra dos Reis, RJ.

IED-BIG, 2002. Curso Internacional de Biotecnologia de Cultivo de "Coquille de Saint Jacques" (Nodipecten nodosus). Instituto de Ecodesenvolvimento da Baía da Ilha Grande - Laboratório de Larvicultura de Moluscos, Angra dos Reis, RJ.

IBGE, 2000. Brazilian Census of 2000. Fundação Instituto Brasileiro de Geografia e Estatística (IBGE) -Centro de Documentação e Disseminação de Informações, Rio de Janeiro, pp. CD-ROM.

IPCC, 2007. Summary for policymakers, in: Pachauri, R.K., Reisinger, A. (Eds.), Climate Change 2007:the Physical Science Basis. Contribution of Working Group I to the Fourth Assessment Report of the Intergovernmental Panel of Climate Change. Cambridge University Press, Cambridge, UK, pp. 1-18.

Magro, M., Cergole, M.C., Rossi-Wongtschowski, C.L.D.B., 2000. Síntese de conhecimentos dos principais recursos pesqueiros costeiros potencialmente explotáveis na costa Sudeste-Sul do Brasil: Peixes. Programa REVIZEE. MMA - Ministério do Meio Ambiente, dos Recursos Hídricos e da Amazônia Legal;

CIRM - Comissão Interministerial para os Recursos do Mar, Brasília - DF, p. 145.

Marinha do Brasil, 2003. Normas da Autoridade Marítima para Amadores, Embarcações de Esporte e/ou Recreio epara Cadastramento e Funcionamento das Marinas, Clubes e Entidades Desportivas Náuticas - NORMAN 03/DPC, in: Coasts, D.o.P.a. (Ed.), Norman 03. Marinha do Brasil, Rio de Janeiro.

MRS, 2006. Relatório de Impacto Ambiental da Usina Nuclear de Angra III. MRS Estudos Ambientais Ltda/IBAMA Brasília, DF, p. 211.

MRS, 2007. Estudo de Impacto Ambiental da Unidade 3 da Central Nuclear Almirante Alberto Álvaro. Eletronuclear - MRS Estudos Ambientais Ltda., Rio de Janeiro.

Muehe, D., 1998. O litoral brasileiro e sua compartimentação, in: Cunha, S.B.G., A.J.T. (Ed.), Geomorfologia do Brasil. Editora Bertrand Brasil S.A, Rio de Janeiro, RJ, p. Capítulo 7.

Muniz, E.C., Jacob, S.A., 1986. Condition Index, Meat Yield And Biochemical-Composition Of Crassostrea-Brasiliana And Crassostrea-Gigas Grown In Cabo-Frio,. Aquaculture International 59, 235-250.

Novaes, A.L.T., Vianna, L.F.d.N., Santos, A.A.d., Silva, F.M., Souza, R.V.d., 2010. Planos Locais de Desenvolvimento da Maricultura de Santa Catarina. Panorama da Aqüicultura 122, 52-58.

Perez-Sanchez, E., Muir, J.F., 2003. Fishermen perception on resources management and aquaculture development in the Mecoacan estuary, Tabasco, Mexico. Ocean Coastal Manage. 46, 681-700.

Rocha, S.P.d., 2005. Análise espacço temporal do uso e cobertura da terra no entorno da BR- 101 - Trecho Angra dos Reis e Parati/RJ, Departamento de Geografia. Universidade do Estado do Rio de Janeiro, Rio de Janeiro, p. 94.

Rossi-Wongtschowski, C.L.D.B., Cergole, M.C., 2001. A área de dinâmica de populações e avaliação de estoques pesqueiros na região Sudeste-Sul, in: M. C. Cergole, Carneiro, M.H. (Eds.), Análise das principais pescarias comerciais do Sudeste-Sul do Brasil: Dinâmica das frotas pesqueiras. MMA - Ministério do Meio Ambiente, dos Recursos Hídricos e da Amazônia Legal; CIRM - Comissão Interministerial para os Recursos do Mar, Brasília - DF, pp. 4-9.

Saitoh, S.I., Mugo, R., Radiarta, I.N., Asaga, S., Takahashi, F., Hirawake, T., Ishikawa, Y., Awaji, T., In, T., Shima, S., 2011. Some operational uses of satellite remote sensing and marine GIS for sustainable fisheries and aquaculture. ICES Journal of Marine Science 68, 687-695.

Scott, P.C., 1998. Considerações Sobre o Uso da Baía de Sepetiba - RJ para Maricultura Apoiada num Sistema de Informação Geográfica (SIG). Instituto de Ciências Biológicas e Ambientais. Universidade Santa Úrsula, Rio de Janeiro, p. 45.

TCE, 2003. Economic Study 2003 Paraty. Secretaria-Geral de Planejamento do Estado do Rio de Janeiro - Tribunal de Contas do Estado do Rio de Janeiro, Rio de Janeiro, RJ, p. 109.

Troell, M., Halling, C., Neori, A., Chopin, T., Buschmann, A.H., Kautsky, N., Yarish, C., 2003. Integrated mariculture: asking the right questions. Aquaculture 226, 69-90.

Tuan, Y., 1975. Images and Mental Maps. Annals of the American Association of Geographers 65, 205-213.

Tuan, Y., 1983. Espaço e Lugar: A perspectiva da Experiência. Editora Difel, São Paulo.

Windupranata, W., Mayerle, R., 2009. Decision support system for selection of suitable mariculture site in the western part of Java Sea, Indonesia. ITB Journal of Engineering Science 41 B, 77-97.

Chapter 3

NUTRIENT UPTAKE BY ORGANISMS ON ARTIFICIAL REEFS IN AQUACULTURE

1. INTRODUCTION: AN OVERVIEW

An artificial reef (AR) consists of one or more objects, either natural or anthropogenic, intentionally placed on the seafloor to affect physical, ecological, or economical processes associated with living marine resources (Seaman & Jensen, 2000). Historically, the primary application of ARs has been to alleviate the depletion of fishery stocks caused by overfishing and habitat degradation (Pitcher & Seaman, 2000). Additional objectives of AR implementation are shoreline conservation, harbor stabilization, recreational surfing, aquaculture, and habitat protection and restoration. (Bombace, 1997; Fabi & Fiorentini 1997; Pickering et al., 1998).

Reef sessile animals, mainly comprising polychaetes, bivalves and gastropods, barnacles, sponges, tunicates and corals can take up and accumulate particles from the surrounding water column through their feeding behavior (Fernández et al., 2004; Gao et al., 2006). During particle capture by reef communities from overlying waters, the metabolic processes of the organisms drive the flux of nutrients between reef-water interfaces. Despite the well- known capacity of reef communities to take up, ingest and digest particulate matter suspended in the ambient water column (Wotton, 1990; Ribes et al., 1998, 1999), studies on nutrient budgets of faunal organisms in reef communities are scarce (Hearn et al., 2001; Ribes et al., 2003), especially for artificial reefs. Faunal recruitment generally differs between natural and artificial reefs, due to the modifications of the physical environment which result from the reef structure, as well as the settlement preference of the organisms to various substratum materials (Glasby & Connell, 1999; Smith & Rule, 2002). As a result, community feeding and physiological processes do not show the necessary consistency between artificial and natural reefs, even for very small temporal and/or spatial scales. Heterogeneous assemblages of epifaunal organisms create a wide range of microcosms and, subsequently, modified physiological processes that cannot be simulated in studies with individual organisms (Sebens et al., 1996). Functional processes of artificial reef communities and concurring nutrient budgets in ecosystems are rarely documented despite numerous reports on the colonization and succession of AR epifauna to mimic the natural community structure (reviewed by Svane & Petersen, 2001).

In fish culture zones (FCZ), the foraging of enhanced faunal biomass through the deliberate, calculated installment of AR systems can function as biofilters to reduce organic pollutants from fish rafts. Numerous studies have shown that the rapidly expanding industry of marine fish farming over the last decades has resulted in nutrient and organic enrichment, owing to the release of organic and inorganic waste from uneaten food, feces and dissolved excretory products (see Wu, 1995; Pearson & Black, 2001; Islam, 2005 for review). In trout farms, for example, only 19 – 28% of the total nitrogen (N) supply could be harvested as the end-product of commercial fish (Foy & Rosell, 1991; Hall et al., 1992). A recent study (Tsutsumi et al., 2006) using sediment traps reported that annual mean organic fluxes to the sea floor from fish rafts were 2.11 g C m^{-2} d^{-1} in total organic carbon (TOC) and 0.26 g N m^{-2} d^{-1} in total nitrogen (TN), 2.5 and 2.2 times higher for TOC and TN fluxes, respectively, than those for natural organic flux outside the fish farm. The conceptual model developed by Islam (2005), based on the published data, showed that 132.5 kg nitrogen and 80.0 kg phosphorus (P) are released into the environment for each ton of fish production and the annual global N and P releases from cage fish culture are 1,325 – 1,387.5 tons and 240.0 – 250.0 tons, respectively.

In the subtropical waters of Hong Kong, Leung et al. (1999) reported that only 8.6% of the total N input to the fish cages was harvested in the form of fish in areolated grouper *Epinephelus areolatus*. Gao et al. (2005) conducted a one-year study which investigated the effects of nutrient release from fish farming in Hong Kong. Their results showed that, on average, total organic carbon, total Kjeldahl nitrogen and total phosphorus levels of the sediment just beneath the fish cages were 82.8, 128.5 and 1315.7%, respectively; this was higher compared to those of the control seabed areas, which were far away from the fish cages, without the effects of farming activities. The elevated nutrient levels derived from the farming activities stimulate the occurrence of nuisance and toxic algal blooms. The final decay of excess plant production consumes considerable dissolved oxygen, leading to the mass mortality of the cultured fish (Wu, 1995). As a result of the changes in environmental conditions due to the impacts of farming waste, subsequent alterations in infaunal (Gao et al., 2005) and epifaunal (Cook et al., 2006; Sarà et al., 2007) community structures occurred in the fish culture, as well as in the adjacent waters.

In Hong Kong, cage culture has been widely practiced for the rearing of high-priced carnivorous fishes such as grouper *Epinephelus awoara*, snapper *Lutjanus russellii* and seabream *Acanthopagrus latus*. Unlike the conventional land-based culture of carp, tilapia and milkfish, cage farming of carnivorous species in the open system requires a large amount of high-protein trash fish, e.g., pilchards and anchovies, as feed. The relatively higher food conversion rate (FCR) of trash fish feed (Leung et al., 1999; Pearson & Black 2001), and the inevitable direct effluent of farming waste to the farming waters, have raised the concern of the adverse impacts of cage aquaculture on open and coastal waters, as well as on the culture farms themselves.

Deployment of ARs, especially in the soft-bottom subtidal zone where hard substrata are unavailable, can provide additional habitats to fulfill species life-history requirements, and eventually increase faunal assemblages. The increased animal assemblages on AR complexes can enhance the environmental cleansing capacity, due to elevated particle removal efficiency by the epibionts, leading to improved environmental conditions (Antsulevich, 1994). Use of filter-feeding organisms on ARs as biofilters has been proven to be one of the more efficient measures in removing farming waste in the fish culture zone without contributing to secondary pollution. Angel et al. (2002), for example, reported that the loading of planktonic microalgae, in terms of chlorophyll concentration, was reduced to a level 15-35% lower than the ambient concentrations when farming water traversed the artificial reefs which were deployed under finfish cages. The nutrient removal effectiveness of epifaunal communities on both natural and manmade reefs, whether under FCZ or under natural conditions, has not been quantified in Hong Kong or globally to date. This study involved the collection of sessile organisms on artificial reefs (ARs) deployed in the Kau Sai fish culture zone in Hong Kong on a bimonthly schedule. The research quantified their nutrient acquisitions and expenditures, including feeding, absorption, carbon respiration, and nitrogen and phosphorus excretion, to evaluate the efficiency of nutrient removal by AR epifaunal organisms.

2. MATERIALS AND METHODS

2.1 Experimental site and AR design

The study area is a marine fish culture zone located in Kau Sai Bay, which is a semi-enclosed embayment in the eastern waters of Hong Kong (22°21'N and 114°19'E). The fish raft area, which is confined within the inner part of the bay, is approximately 4.6 ha with water depth ranging from 11-16 m. Each fish cage is approximately 4 m × 4 m × 4 m. The total fish stock is ~500 t with an average density of 4.5 kg m^{-3}. The cultured species are mainly grouper (*Epinephelus awoara*), snapper (*Lutjanus russellii*) and seabream (*Acanthopagrus latus*). Small trash fish (mainly anchovies, *Thryssa* spp.) are used as fish feed. The daily feed supply is 3 – 5% of total stock, i.e., 15 – 25 t d^{-1}.

In April 2002, 16 AR sets made of cement concrete, with dimensions of 3 m (length) × 3 m (width) × 4 m (height) and total surface area of 250 m^2, were deployed around the fish culture zone boundary in Kau Sai Bay. Twelve removable settlement plates were fixed on two sides of each AR for future sampling.

2.2 Sample collection and preparation

From August 2003, about one year after the AR deployment, AR plates were collected by SCUBA divers at bimonthly intervals. For each field cruise, five pieces of the removable settlement plates were randomly selected and retrieved. To reduce the bias from to the effects of water depth on the development of epifaunal communities on the ARs, all plates were collected from water depths between 10-13 m. On board the vessel, each plate was labeled individually and placed in a plastic tray. After field retrieval, the sampled AR plates were delivered to the laboratory. The plates were cut into small pieces and washed with seawater to remove excess mud. After the laboratory treatment, the AR plates were transported back to Kau Sai FCZ immediately and cultured *in situ* in water depths of 10-13 m by hanging them on a fish raft for at least 24 hours prior to biofiltration determination so that the AR organisms would recover to their original status before sample collection and treatment.

2.3 Determination of feeding processes of epifaunal organisms

Five replicates of AR plates, approximately 200 cm², were separately cultured in static flow systems as described by Smaal & Widdows (1994) at the FCZ (Fig. 1). An extra AR plate of similar area (control plate), which has been defaunated by means of refrigeration at -20 °C for 48 hours, was placed in an additional culture setup as a control.

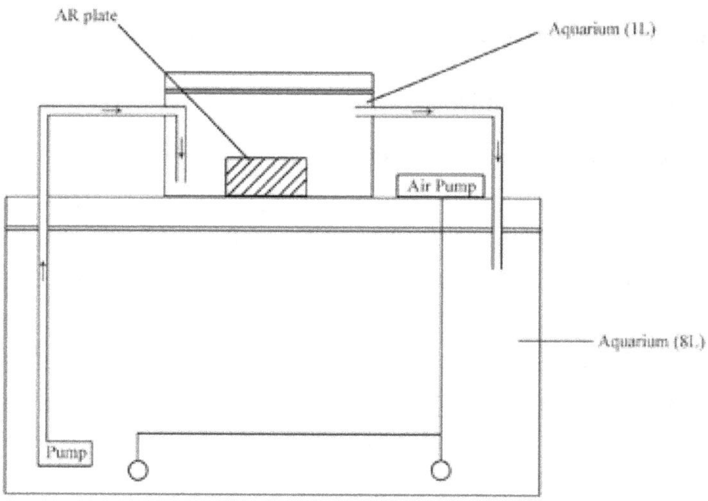

Fig. 1. A water-circulating static system used to determine the clearance rate of the epifaunal organisms on AR plates (adopted from Smaal & Widdows, 1994).

The AR plates were cultured for one to three hours, depending on the biomass on the plates and the production of fecal pellets, with seawater pumped from the 10-13 m water layer. The experimental time was adjusted so that the maximum reduction in particle concentration was 30% (Smaal & Vonck, 1997; Wong &

Cheung, 2001). After cultivation, three replicates of 500 ml water samples and all produced fecal pellets were collected from the six sets of systems and filtered through pre-weighed ashless 47 mm GF/C filters and rinsed with double distilled deionized water to remove salt adsorbed on the surface of particles or filter paper. The filter papers were dried in an oven at 60°C and weighed to the nearest 0.1 mg. The concentration of total particulate matter (TPM: mg l^{-1}) for each water sample was measured and fecal egestion rate (ER: mg h^{-1}) was calculated with the equation:

$$ER = DW_{fp}/t \quad (1)$$

where DW_{fp} is the dry weight of the produced fecal pellets for each AR plate and t is the culture time. Fecal samples were kept frozen at -80°C for future organic carbon, nitrogen and phosphorus analyses (see below for details).

Clearance rate (CR: l h^{-1}) is defined as the volume of water cleared per unit time and calculated following the exponential decrease of the seston concentration with the equation (Coughlan, 1969; Karlsson et al., 2003):

$$CR = V/t \ln (C_0/C_t) \quad (2)$$

where V = water volume in the experimental aquarium, t = time that the experiment lasted, and C_0 and Ct = seston concentrations in the control aquarium (representing the initial seston concentration) and AR aquarium (representing the final seston concentration), respectively.

Filtration rate (FR: mg h^{-1}) is defined as the food mass filtered by the AR epibiota per unit time and computed as the CR times particle concentration in weight unit (Hawkins et al., 1998). The absorption rate (ABR: mg h^{-1}), representing the food content that was retained in the digestive gut after egestion, was calculated as FR-ER.

2.4 Determinations of oxygen consumption and nutrient excretion

To determine oxygen consumption (V_{O2}: μg O h^{-1}), each of the experimental and control plates was placed in a separate sealed 800 ml glass chamber. The sealed chambers were fully filled with seawater pumped from the 10-13 m water column and bathed in a plastic tank supplied with seawater which was continuously pumped from the same water depth to reduce the effects of temperature fluctuations on the respiratory activities of the animals. After 30-60 minutes, depending on the area occupied by the organisms, the DO level of the chambers was measured with a YSI DO meter. V_{O2} was calculated with the equation:

$$V_{O2} = (DO_c - DO_e) \times V/t \quad (3)$$

where DO_c and DO_e are the DO levels of control and experimental chambers, respectively, representing initial and final DO levels, V the volume of chamber and t the test time.

To test the ammonium (V_N: µg N h^{-1}) and phosphate (V_P: µg P h^{-1}) excretion rates, experimental and control AR plates were maintained in separate glass beakers filled with 750 ml seawater for 30 to 60 minutes. Water samples collected from the experimental and control beakers were stored in an icebox and taken back to the laboratory for nitrogen and phosphorus determination. Ammonium-nitrogen and phosphate-phosphorus concentrations of the seawater samples were colorimetrically determined using the phenolhypochlorite and phospho-molybdate methods, respectively, with a Flow Injection Analyzer (FIA, Lachat, QuikChem 8000) (Strickland & Parsons, 1977). V_N and V_P were calculated with an equation similar to that used for the oxygen consumption test.

2.5 Seston and faeces analysis

Triplicate 1-litre seawater samples were filtered on pre-combusted and weighed glass-fiber filters and dried at 60° C to the nearest 0.1 mg, then combusted in a muffle furnace at 450° C for six hours and reweighed to determine the total particulate matter (TPM: mg l^{-1}), particulate organic matter (POM: mg l^{-1}), particulate inorganic matter (PIM: mg l^{-1}) and organic content (f = POM/TPM). Another aliquot of suspended particulate matter on filter paper was dried but not combusted and kept in a freezer at -80º C for future carbon, nitrogen and phosphorus measurements. Particulate organic carbon (POC: µg mg^{-1}) and particulate organic nitrogen (PON: µg mg^{-1}) of seston and faecal pellets on filter papers were measured with a CHNS/O Analyzer (PerkinElmer, PE2400 Series II) (Ribes et al., 2003). For POC measurement, carbonates were removed by fuming the seston on the filters over concentrated HCl (37%) for three days. Samples were then re-dried and re-weighed. The percentage of carbon was corrected for the weight change due to carbonate removal (Kristensen & Andersen, 1987). Particulate organic phosphorus (POP) was determined following the wet digestion method. The filters retaining a known weight of suspended particulate matter or faecal pellet samples were digested with concentrated sulfuric acid using a digestion block at 200º C for 30 minutes for the digestion of detritus and 370º C for two hours for the digestion of other phosphorus-containing constituents. Copper sulphate ($CuSO_4$) was used as a catalyst and potassium sulfate (K_2SO_4) was added to raise the boiling point of the digesting acid. The concentrations of total phosphate were determined with a Flow Injection Analyzer (Lachat, QuikChem 8000).

2.6 Nutrient element assimilation

The assimilation rates of the nutrient elements (C, N and P) were measured as scope for growth (SFG: µg h^{-1}) for each element, which is defined as the difference between acquisition and expenditure (Warren & Davis, 1967; Smaal & Widdows, 1994). SFG of C, N or P might be calculated as:

$$\text{SFG } (\mu g\ h^{-1}) = \text{ABR} - \text{Respiration or Excretion} \quad (4)$$

Oxygen consumption was converted to C excretion based on a mean respiratory quotient (the ratio between carbon dioxide formed and oxygen used) of 0.85:1 µg O_2 ≡ 0.32 µg C. For C, excretion is zero, while for N and P, respiration is zero (Hawkins & Bayne, 1985; Smaal & Vonck, 1997). The assimilation of the nutrients by the organisms during each bimonthly test interval was estimated as:

$$\text{Assimilation (g)} = [\text{SFG } (\mu g\ h^{-1}) \times 24 \text{ hs} \times 60 \text{ d}]/10^6 \quad (5)$$

Water conditions, including temperature, salinity and DO level, during the field experiments were measured every day. Since water stratification frequently occurs in semi-enclosed Kau Sai FCZ in summer, due to the heating of surface water and weak water movement the temperature, salinity and DO levels in August were measured for each 1 m of water layer (Gao et al., 2007).

2.7 Measurement of colonization of epifaunal organisms

After field biofiltration determination, the AR pieces were preserved in 5% borax-buffered formalin for measurements of the coverage area by the epifaunal organisms following the methods of Nelson et al. (1994). The area of the AR plate surface which was covered as a projection of the total individuals of each species was measured with a 10 × 10 cm^2 transparent plastic board, of which the surface was evenly divided into 0.5 × 0.5 cm^2 grids. Since the filtration efficiency of animals is proportional to their biomass, the fouling organisms were removed carefully from the AR plates with a cutter, and the wet biomass was weighed to the nearest 0.1 mg with an electronic balance after blotting away the surface water on filter paper.

2.8 Data analysis

The relationships of physiological processes of AR epifauna, including feeding, respiration, nitrogen and phosphorus excretion, to wet biomass were analyzed with linear regression. Once the regressive relationships between physiological parameters and biomass were statistically significant (significance level $p < 0.05$), the physiological parameters were standardized to the equivalent values of 1 g biomass. Regression analyses were applied to reveal how environmental factors, such as food conditions, temperature, DO level and salinity, affected the physiological processes by simple linear or nonlinear procedures, depending on the most appropriate function to be fitted in each case (Zar, 1999). Multiple regression analysis was conducted when physiological parameters were correlated with more than one environmental condition, and the collinearity between the independents was tested using the collinearity statistics of SPSS (Belsley et al., 2004). Highly correlated independents were eliminated from the independent list and the model was reconstructed until all the intercorrelation between the independent variables was removed from the regressive model. Residuals were also analyzed to check normality and constant variance. Data were transformed, if necessary, to meet the requirements of data normality and homogeneity of variances. All statistical procedures were performed with the SPSS software for Windows, Release 14.0.

3. RESULTS

3.1 Hydrography, food conditions and nutrient contents

Bimonthly changes in the bottom hydrographic parameters at the experimental site, such as temperature (T), dissolved oxygen (DO), salinity (S) and food conditions (in terms of total particulate matter (TPM), particulate organic matter (POM), particulate inorganic matter (PIM) and organic content (*f*) and nutrient contents, including POC, PON and POP of the suspended particulate matter) have been reported in Gao et al. (2007). In brief, water stratification, with a thermocline and a halocline recorded at 4-10 m, occurred in August and exerted considerable effects on bottom conditions. The bottom temperature underwent seasonal variations with high values (~26°C) in October, decreasing gradually to the lowest (~15°C) in February. Bottom temperatures during the summer (August) were even lower than in October because of the stratification, despite August having the highest surface value (~31°C). Bottom DO levels were quite constant over the study period (5.7 – 7.2 mg l^{-1}) except the extremely low level of 1.7 mg l^{-1} in August due to the presence of stratification in summer. Bottom salinity ranged from ~30 – 33 PSU throughout the year. Mean TPM and PIM concentrations in winter (TPM 12.62-15.15 mg l^{-1}; PIM 7.65-11.29 mg l^{-1}) were higher than in summer (TPM 10.16-10.68 mg l^{-1}; PIM 4.51-5.75 mg l^{-1}) due to the wave-driven re- suspension of sediment from the seabed, leading to *f* values (0.25-0.39%) in winter (December to February) relative to those in June and August (0.43-0.58%). Mean POM, POC and PON showed higher values during the warmer months (POM 4.41-6.17 mg l^{-1}; POC 129.73-135.56 μg l^{-1}; PON 22.04-24.56 μg l^{-1}) and lower values in colder months (POM 3.18-4.97 mg l^{-1}; POC 70.33-81.64 μg l^{-1}; PON 11.25-15.60 μg l^{-1}). A seasonal pattern of fluctuations in POP levels (1.56-2.29 μg l^{-1}) was not obvious.

3.2 Colonization and succession of AR epifauna

Variations in coverage area and biomass of the epifaunal organisms on the AR plates during the biofiltration measurements from August 2003 to June 2004 are illustrated in Fig. 2. Coverage area by epifaunal organisms on AR plates gradually dropped to the lowest value (mean 19%) from August to October, owing to the adverse bottom conditions during the summer season. The epibiota, however, resumed recolonization on the ARs after the summer, and the coverage area and biomass peaked in April 2004 (mean 43%) due to the settlement of juvenile organisms after reproduction in the spring. High mortality of the juveniles after settlement resulted in a reduction in coverage area and biomass of the AR plates sampled in late spring (June). This could be verified through direct observations of a considerable number of empty shells, especially the barnacle *Balanus trigonus*, on the plates in June 2004. The barnacle *Balanus trigonus*, the bivalve *Pseudochama retroversa* and the tunicate *Styela* species were continuously dominant during the experimental period. Other relatively dominant species included bivalves *Bentharca* sp. And *Dendostrea folium*, polychaete *Spirobranchus tetraceros*, bryozoans *Rhynchozoon paratridenticulata*, *Schizoporella erratoidea* and coral *Heterocyathus japonicus*. Other polychaete tubeworms such as *Hydriodes* sp. frequently occurred, but never dominantly occupied the AR surface.

3.3 Feeding of epifaunal organisms

As illustrated in Fig. 3, feeding processes of epifaunal communities on AR plates, in terms of clearance rate (CR: mg l^{-1} m^{-2}), filtration rate (FR: mg h^{-1} m^{-2}), egestion rate (ER: mg h^{-1} m^{-2}) and absorption rate (ABR: mg h^{-1} m^{-2}) of each m^2 AR plate, were linearly proportional to their biomass (BM: g m^{-2}) according to following equations:

$$CR = 18.41 + 20.24 \times BM\ (r^2 = 0.91,\ t_{28} = 17.05,\ p < 0.01)$$

$$FR = 334.85 + 233.17 \times BM\ (r^2 = 0.84,\ t_{28} = 12.32,\ p < 0.01)$$

$$ER = 246.55 + 134.57 \times BM\ (r^2 = 0.76,\ t_{28} = 9.41,\ p < 0.01)$$

$$ABR = 88.30 + 98.60 \times BM\ (r^2 = 0.69,\ t_{28} = 7.97,\ p < 0.01)$$

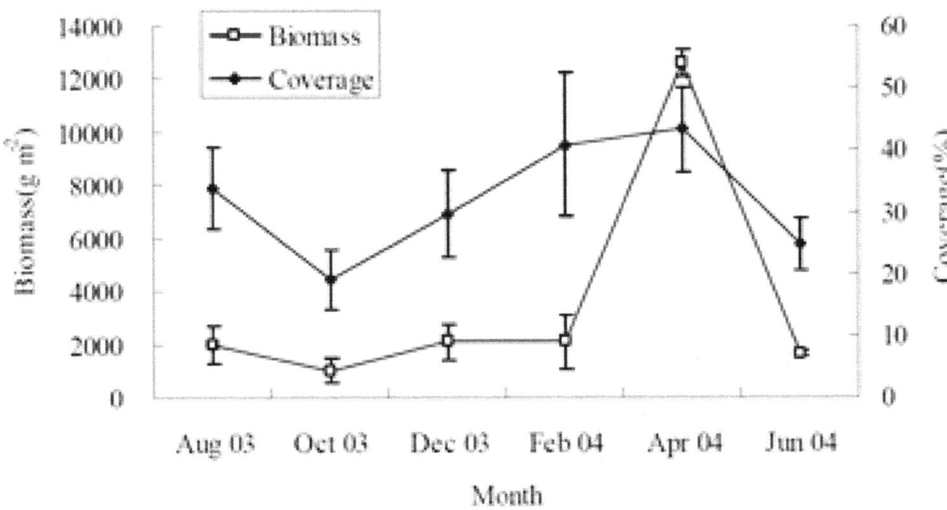

Fig. 2. Bimonthly variations in the communities of AR epifanual organisms in terms of biomass (g m^{-2}) and coverage area (%) during the biofiltration measurements from August 2003 to June 2004 (values were presented as mean ± 1SD, n = 5) (from Shin et al., 2011).

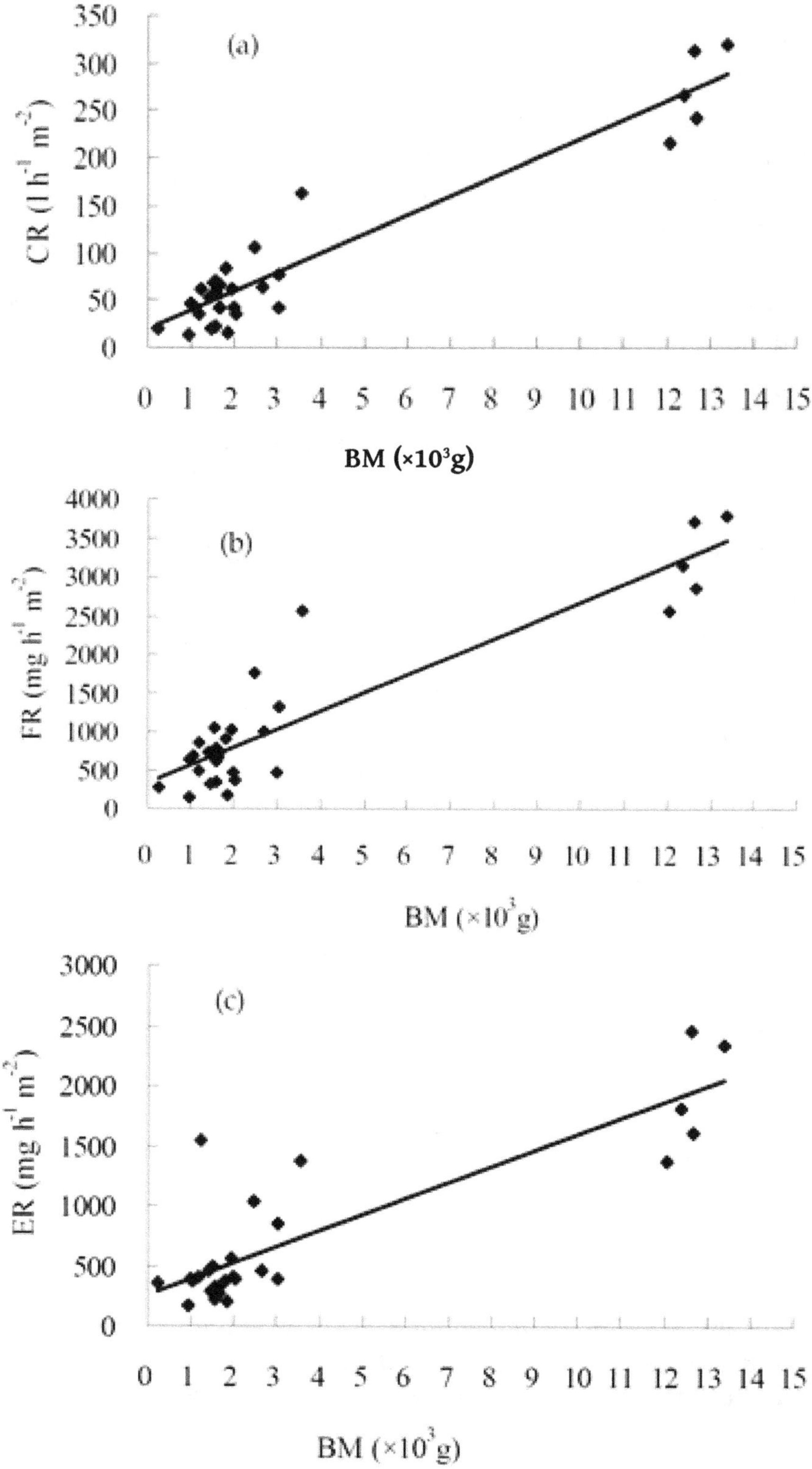
(a)
CR (l h⁻¹ m⁻²)
BM (×10³g)
(b)
FR (mg h⁻¹ m⁻²)
BM (×10³g)
(c)
ER (mg h⁻¹ m⁻²)
BM (×10³g)

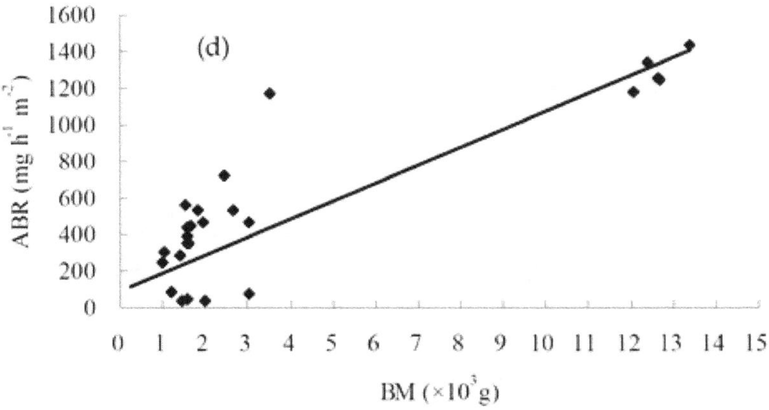

Fig. 3. Relationships of (a) clearance rate (CR: l h^{-1} m^{-2}), (b) filtration rate (FR: mg h^{-1} m^{-2}), (c) egestion rate (ER: mg h^{-1} m^{-2}) and (d) absorption rate (ABR: mg h^{-1} m^{-2}) of epifaunal organisms on AR plates to their biomass (BM: g m^{-2}) (from Shin et al., 2011).

The standardized CR (SCR), FR (SFR), ER (SER) and ABR (SABR) of 1 g biomass (wet weight) for each month, which were calculated based on the linearly proportional relationships between the feeding processes and biomass, are summarized in Table 1. The relationship of SCRs to the fluctuations in TPM is shown in Fig. 4. At the start, SCR increased with increasing TPM and achieved a peak mean value of ~0.06 l h^{-1} when TPM had an intermediary value of ~14 mg l^{-1} in October 2003. After this peak value, SCR gradually decreased with increasing TPM. The relationship of SCR to TPM could be described as the quadratic polynomial equation,

$$SCR = -0.36 + 0.059 \times TPM - 0.0021 \times TPM^2 \ (r^2 = 0.22, F_{2, 26} = 2.73, p < 0.05)$$

Table 1. The standardized clearance rate (SCR: l h^{-1} g^{-1}), filtration rate (SFR: mg h^{-1} g^{-1}), egestion rate (SER: mg h^{-1} g^{-1}) and absorption rate (SABR: mg h^{-1} g^{-1}) of 1 g biomass (wet weight) of epifaunal organisms on ARs for each bimonthly determination (mean ± 1SD, n = 5) (from Shin et al., 2011).

Month	SCR (l h^{-1} g^{-1})	SFR (mg h^{-1} g^{-1})	SER (mg h^{-1} g^{-1})	SABR (mg h^{-1} g^{-1})
Aug 03	0.015 ± 0.0047	0.14 ± 0.051	0.13 ± 0.042	0.010 ± 0.021
Oct 03	0.055 ± 0.021	0.67 ± 0.29	0.62 ± 0.52	0.049 ± 0.026
Dec 03	0.025 ± 0.013	0.55 ± 0.21	0.40 ± 0.095	0.15 ± 0.081
Feb 04	0.036 ± 0.011	0.63 ± 0.16	0.35 ± 0.097	0.28 ± 0.073
Apr 04	0.021 ± 0.003	0.25 ± 0.035	0.15 ± 0.033	0.10 ± 0.0053
Jun 04	0.041 ± 0.0046	0.44 ± 0.050	0.18 ± 0.029	0.26 ± 0.027

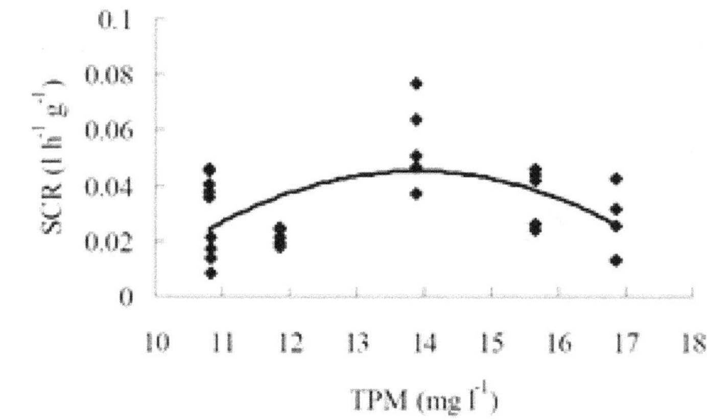

Fig. 4. Relationship of standardized clearance rate (SCR: l h-1 g-1) to concentration of total particulate matter (TPM: mg l-1) (from Shin et al., 2011).

SFR was independent of any food characteristics throughout the experimental period. If the lowest summer filtration rate in August 2003 is excluded, SFR showed a positive relationship to the food quantity in terms of POM, which could be summarized in the following power equation (Fig. 5),

$$SFR = 0.028 \times POM^{1.90} \ (r2 = 0.38, t_{23} = 3.74, p < 0.01)$$

In addition to food conditions, SCR, SFR and SABR were significantly correlated with bottom DO levels (for SCR, Pearson correlation coefficient $r = 0.48, p < 0.01$; for SFR, $r = 0.64, p < 0.01$; for SABR, $r = 0.59, p < 0.01$). The positive relationship of SABR to DO could be expressed as an exponential equation,

$$SABR = 0.0083 \times e^{0.47DO} \ (r2 = 0.57, t_{23} = 5.54, p < 0.01)$$

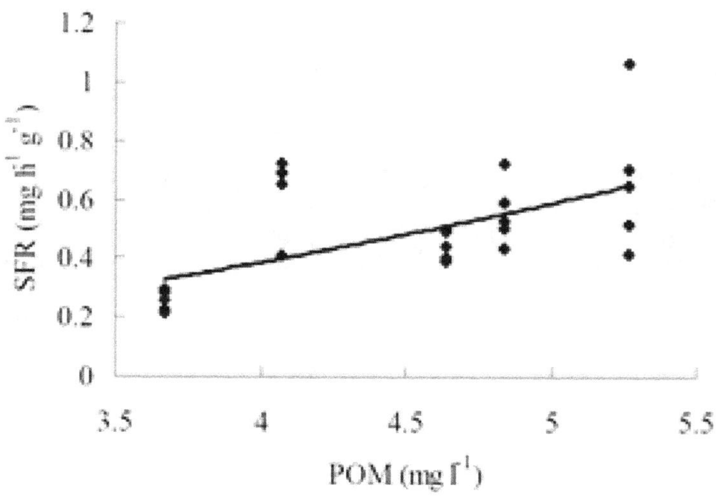

Fig. 5. Relationship of standardized filtration rate (SFR: mg h-1 g-1) to particulate organic matter (mg l-1) (from Shin et al., 2011).

Integrating the effects of food conditions and bottom dissolved oxygen levels on the feeding behavior of AR epifaunal communities, relationships of SCR and SFR to food quantity and DO could be described by the following multiple regression equations,

$SCR = -0.37 + 0.047 \times TPM - 0.0017 \times TPM^{2} + 0.069 \times DO^{0.16}$ ($r^2 = 0.38$, $F_{5, 25} = 36.0$, $p < 0.01$ $SFR = 2.61 - 4.80 \times POM^{-0.28} + 0.31 \times DO^{0.68}$ ($r^2 = 0.62$, $F_{5, 25} = 60.7$, $p < 0.01$)

3.4 Nutrient budgets

Relationships of oxygen consumption rate (V_{O2}: mg h^{-1} m^{-2}), nitrogen (V_N: mg h^{-1} m^{-2}) and phosphorus (V_P: mg h^{-1} m^{-2}) excretion rate to biomass (BM: g m^{-2}) are shown in Fig. 6. V_{O2}, V_N and V_P all showed linear relationships to biomass as following equations:

$$V_{O2} = -25.46 + 68.51 \times BM\ (r^2 = 0.97,\ t_{28} = 31.20,\ p < 0.01)$$

$$V_N = 0.89 + 1.29 \times BM\ (r^2 = 0.80,\ t_{28} = 10.63,\ p < 0.01)$$

$$V_P = -0.27 + 0.42 \times BM\ (r^2 = 0.88,\ t_{28} = 14.02,\ p < 0.01)$$

Table 2. Standardized rates of oxygen consumption (SV_{O2}: µg h^{-1} g^{-1}), carbon respiration (SV_C: µg h^{-1} g^{-1}, 1 µg $O_2 \equiv$ 0.32 µg C), and excretion of nitrogen (SV_N: µg h^{-1} g^{-1}) and phosphorus (SV_P: µg h^{-1} g^{-1}) of 1 g epifaunal biomass (mean ± 1SD, *n* =5) (from Shin et al., 2011).

Month	SV_{O2} (µg h^{-1} g^{-1})	SV_C (µg h^{-1} g^{-1})	SV_N (µg h^{-1} g^{-1})	SV_P (µg h^{-1} g^{-1})
Aug 03	21.38 ± 4.99	6.84 ± 1.60	3.00 ± 0.99	0.68 ± 0.17
Oct 03	73.50 ± 14.40	23.52 ± 4.61	1.75 ± 0.57	0.44 ± 0.15
Dec 03	61.84 ± 13.06	19.79 ± 4.18	1.45 ± 0.20	0.44 ± 0.13
Feb 04	71.45 ± 4.57	22.87 ± 1.46	1.03 ± 0.13	0.51 ± 0.21
Apr 04	66.49 ± 6.12	21.28 ± 1.96	1.36 ± 0.37	0.41 ± 0.57
Jun 04	47.45 ± 4.22	15.18 ± 1.36	1.67 ± 0.45	0.48 ± 0.28

Based on these linear models, the standardized V_{O2} (SV_{O2}: µg h^{-1} g^{-1}), carbon respiration (SV_C: µg h^{-1} g^{-1}, 1 µg $O_2 \equiv$ 0.32 µg C), V_N (SV_N: µg h^{-1} g^{-1}) and V_P (SV_P: µg h^{-1} g^{-1}) of 1 g biomass are summarized in Table 2. SV_{O2} and SV_C showed the lowest rates in August 2003. In contrast, SV_N value in August 2003 was much higher than other months and SV_P appeared relatively constant throughout the experimental period.

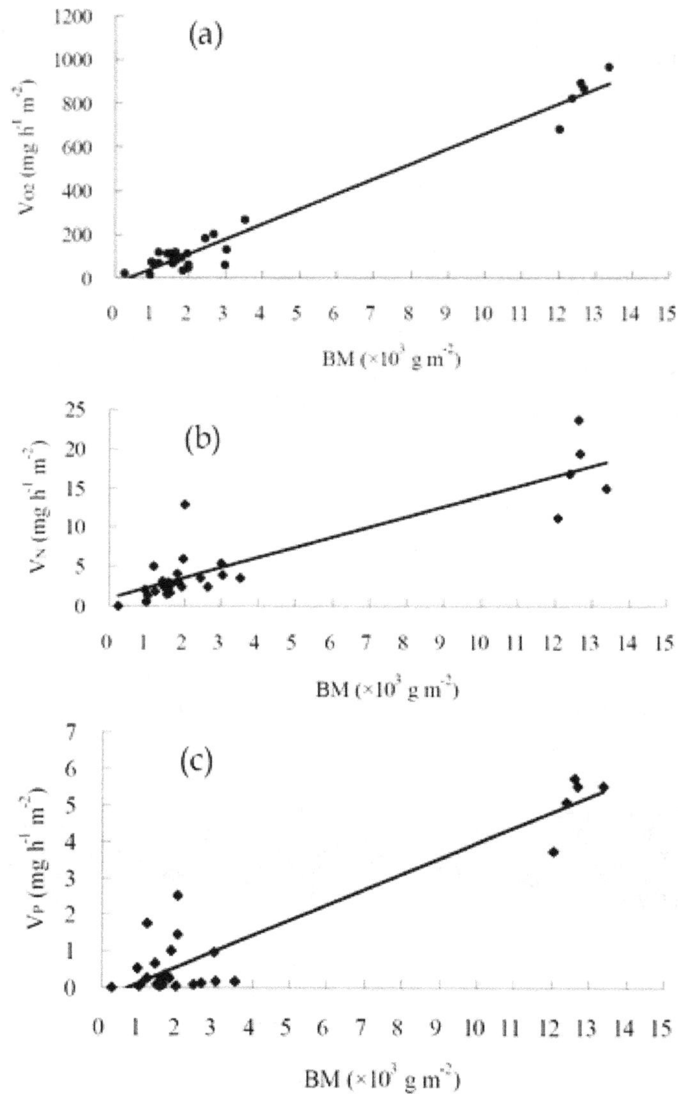

Fig. 6. Relationships of (a) oxygen consumption ($V_{O2:}$ mg h-1 m-2), (b) excretion rate of nitrogen (V_N: mg h-1 m-2) and (c) phosphorus (V_P: mg h-1 m-2) to biomass of AR epifaunal organisms (BM: g m-2) (from Shin et al., 2011).

As a result of the balance between nutrient absorption via food uptake and nutrient consumption due to carbon respiration or nitrogen and phosphorus excretion, nutrient budgets in terms of carbon, nitrogen and phosphorus SFGs by 1 m² AR plate in October and December of 2003, and February and June of 2004 were positive, indicating that nutrient absorption via food uptake was larger than their expenditure due to carbon respiration and nitrogen and phosphorus excretion in these four months. In contrast, nutrient SFGs in August 2003 and April 2004 were negative, showing that AR epifaunal organisms consumed more nutrients than they absorbed in these two months. As a result, the annual mass values of carbon, nitrogen and phosphorus that were assimilated by the epifaunal organisms on each 1 m^2 AR plate were 588.12 g, 156.06 g, and 25.71 g, respectively (Table 3).

Table 3. Bimonthly changes in SFGs and total assimilation of carbon, nitrogen and phosphorus for 1 m^2 AR plate (mean ± SD, *n* = 5) (from Shin et al., 2011).

Month	SFG (µg h-1 m-2)			Assimilation (g m-2)		
	C	N	P	C	N	P
Aug 03	-30022±7971	-13684±2043	-1015±299	-43.23	-19.71	-1.46
Oct 03	58141±13574	2939±423	104±31	83.72	4.23	0.15
Dec 03	43484±13818	26318±5846	6000±865	62.62	37.90	8.64
Feb 04	180821±54910	63018±9419	10402±1426	260.38	90.75	14.98
Apr 04	-145966±36402	-13258±1387	-2566±630	-210.19	-19.08	-3.70
Jun 04	301955±13392	43036±7750	4929±192	434.82	61.97	7.10
Yearly budget	–	–	–	588.12	156.06	25.71

4. DISCUSSION

The findings of this study indicated that artificial reefs (ARs) deployed under fish rafts can function as biofilters to reduce particulate nutrient enrichments in fish culture zone and adjacent waters. This is accomplished via food uptake by the epifaunal organisms on the AR surface, despite the net release of nutrients to the environments in two of the six experimental months (a the result of the metabolic budget between nutrient absorption and consumption).

For nutrient acquisition (feeding and absorption) and expenditure (respiration and excretion), high goodness-of-fit (r^2 values) of the regressive models between physiological rates and biomass on unit areas of AR plates revealed that the nutrient removal efficiencies of AR systems were primarily controlled by the changes in community development due to the colonization and succession of epifauna on the AR surface. Community structure of sessile organisms on the surface of hard habitats is controlled by both biotic factors, such as reproductive cycle, larval settlement, growth and life span of each component species and abiotic factors, such as salinity, temperature, pH, dissolved oxygen and the texture of the substratum itself (Moran and Grant, 1993; Kocak & Kucuksezgin, 2000; Qian et al., 2000). In temperate and subtropical regions such as Hong Kong, most marine organisms show bimodal reproductive cycles peaking in spring and autumn (Morton & Morton, 1983). The summer bottom hypoxic condition at the study site, caused by the presence of stratification, led to a high number of mortalities in the epibiotic communities from the summer to early autumn months (Gao et al, 2007). After the summer hypoxia, settlement and growth of the epifaunal communities resumed from autumn to late spring, owing to continuous larval recruitment and juvenile and adult growth under such relatively suitable hydrographic conditions. The epifaunal Communities fully developed in late spring (April), in terms of coverage area, and biomass peaked at this time due to the extensive larval settlement during the spring reproductive stage. The slow growth of epifaunal communities at the beginning of summer (June) could be attributed to termination of larval settlement and collapse of the populations of some dominant species, e.g., barnacles, ascidians and sponges, caused by the high mortality of juvenile individuals that settled in the early spring months.

The standardized feeding rates, i.e., nutrient acquisition efficiencies, of the epifauna for 1 g biomass were significantly related to the environmental factors, such as TPM, POM and DO level. Numerous research studies revealed that bivalves, ascidians and other filter-feeding species possess regulating functions to optimize energy acquisition in the presence of various seston concentrations and composition (Robbins, 1984; Riisgård, 1988; Petersen & Riisgård, 1992; Bayne, 1993; Petersen et al., 1995). In the present study, fluctuations in standardized clearance rate (SCR) in response to the changing food quantity, in terms of TPM, indicated the presence of a pre-ingestive regulating mechanism in filter-feeding aquatic organisms (Fig. 4). At the relatively low TPM concentration (< 14 mg l^{-1}), SCR increased with increasing TPM; when TPM was larger than the threshold concentration of 14 mg l^{-1}, SCR showed down-regulating pattern, i.e., decreased with increasing TPM. Generally, the filtration rate of epifauna was elevated in response to increased POM (Fig. 5), as a higher supply of suspended particles increased the opportunity for the epifauna to take up the particulate matter (Ribes et al., 2003).

The concentration of dissolved oxygen is another factor determining the metabolism of aquatic animals (Hummel et al., 2000; Huang & Newell, 2002). During the experimental period, DO levels underwent considerable fluctuations and significantly affected the metabolism of the component species and, consequently, the entire community structure. The hypoxic conditions (< 2 mg l^{-1}) in summer months significantly inhibited feeding, absorption and nutrient consumption by the epifauna. As a result, all standardized metabolic rates of unit epifaunal biomass, including clearance rate, filtration rate, absorption rate, carbon respiration and nitrogen and phosphorus excretion, dropped to the lowest levels under such extremely adverse summer conditions (Table 2). Owing to the adverse effects of hypoxia on nutrient acquisition and consumption of epibiotic organisms, carbon, nitrogen and phosphorus budgets showed negative values, indicating the net expenditure of body reserves under environmental stresses.

Studies on the use of artificial reefs as biofilters to reduce pollution under natural or farming conditions have rarely been reported. Angel et al. (2002) observed that the depletion of the chlorophyll *a* level might be detected in the water traversing the artificial reefs installed under fish farms compared to that in the upstream water, indicating the absorption of organic matter from fish farms by the organisms on ARs. Bugrov (1994) and Laihonen et al. (1996) also suggested that artificial reef-fish cage complexes may have the potential to remove particulate and dissolved matter from fish farm effluents.

Assimilation of carbon, nitrogen and phosphorus by epifaunal organisms on ARs led to a net accumulation of these nutrients from ambient waters to the biomass, suggesting the bioremediation function of AR epifaunal organisms as biofilters to reduce nutrient pollution from fish farming. By taking the total AR surface area of 250 m^2 AR^{-1} × 16 ARs = 4,000 m^2, the total filtration capacity was estimated at 8,736 m^3 d^{-1}, and removal of carbon was 2,352 kg, nitrogen 624 kg and phosphorus 103 kg per year, or 6.4, 1.7 and 0.3 kg d^{-1}, respectively. According to a local study by Leung et al. (1999), the nitrogen loss rate from fish farms to surrounding waters is 321 kg N t^{-1} fish production. Using this figure, the yearly nitrogen removal by all ARs in the present study is thus equivalent to nitrogen loss from about two tons of fish production. Such filtration capacity and efficiency are clearly beneficial to reducing the excessive nutrients generated from current fish farming practices.

Currently, the greatest limitation for the application of artificial reefs to remove farming waste is the fate of the accumulated nutrients in the forms of biomass on the ARs. To solve this problem, Antsulevich (1994) suggested an improved AR design with removable plates and the facilities for collecting matter which drops from the ARs. With such a specialized design for cleansing waters, the biomass with absorbed nutrients could be harvested conveniently from the farming waters. On the other hand, ARs provide additional habitat and shelter for the pelagic fish. The aggregates of the carnivorous fish thus function as another group of consumers to take up the sessile organisms on the ARs, as well as the farming wastes (Relini et al., 2002).

5. CONCISE SUMMARY

Research examining nutrient transfer between diverse biotic and abiotic components in aquaculture is crucial for understanding nutrient flow mechanisms in organically enriched aquatic systems within a broader biogeochemical framework. In light of the global concern regarding the escalating degradation of coastal reef habitats due to seashore overexploitation (Hilton & Manning, 1995; Wilkinson, 1998), this analytical process may facilitate a deeper comprehension of eutrophication in shallow coastal marine ecosystems (Nixon, 1995; Kemp et al., 2005). This study demonstrated that filter-feeding sessile organisms on artificial reefs can effectively assimilate nutrients in situ, revealing that the faunal communities within reef systems play a crucial role in the removal of eutrophic nutrients from adjacent waters, thereby highlighting the significance of reef habitats in nutrient dynamics within marine ecosystems.

6. REFERENCES

Angel, D.L.; Eden, N.; Breitstein, S.; Yurman, A.; Katz, T. & Spanier, E. (2002) In situ biofiltration: a means to limit the dispersal of effluents from marine finfish cage aquaculture. *Hydrobiologia*, Vol.469, pp. 1-10, ISSN 0018-8158

Antsulevich, A.E. (1994) Artificial reefs project for improvement of water quality and environmental enhancement of Neva Bay (ST.-Petersburg County Region). *Bulletin of Marine Science*, Vol.55, pp. 1189-1192, ISSN 0007-4977

Bayne, B.L. (1993) Feeding physiology of bivalves: time-dependence and compensation for changes in food availability. In: *Bivalve Filter Feeders in Estuarine and Coastal Water Eecosystem Processes*, R.F. Dame, (Ed.), 1-24, Springer-Verlag, ISBN 978-038756952, Berlin/Heidelberg, Germany

Belsley, D.A.; Kuh, E. & Welsch, R.E. (2004) *Regression Diagnostics: Identifying Influential Data and Sources of Collinearity*. Wiley-Interscience, ISBN 0471691178, New York

Bombace, G. (1997) Protection of Biological Habitats by Artificial Reefs, *European Artificial Reef Research: Proceedings of the first EARRN Conference*, pp. 1-15, ISBN 0-904175-28-6, Ancona, Italy, March, 1996

Bugrov, L.Y. (1994) Fish-farming and artificial reefs: complex for waster technology. *Bulletin of Marine Science*, Vol.55, p. 1332, ISSN 0007-4977

Cook, E.J.; Black, K.D.; Sayer, M.D.J; Cromey, C.J.; Angel, D.L.; Spanier, E.; Tsemel, A.; Katz, T. & Eden, N. (2006) The influence of caged mariculture on the early development of sublittoral fouling: a pan-European study. *ICES Journal of Marine Science*, Vol.63, pp. 637-649, ISSN 1054-3139

Coughlan, J. (1969) The estimation of filtering rate from the clearance of suspensions. *Marine Biology*, Vol.2, pp. 356-358, ISSN 0025-3162

Fabi, G. & Fiorrentini, L. (1997) Molluscan Aquaculture on Reefs, *European Artificial Reef Research: Proceedings of the first EARRN Conference*, pp. 123-140, ISBN 0-904175-28-6, Ancona, Italy, March, 1996

Fernández, D.; López-Urrutia, Á; Fernández, A; Acuña, J.L. & Harris, R. (2004) Retention efficiency of 0.2 to 6 μm particles by the appendicularians *Oikopleura dioica* and *Fritillaria borealis*. *Marine Ecology - Progress Series*, Vol.266, pp. 89-101, ISSN 0171- 8630

Foy, R.H. & Rosell, R. (1991) Loadings of nitrogen and phosphorus from a Northern Ireland fish farm. *Aquaculture*, Vol.96, pp. 17-30, ISSN 0967-6120

Gao, Q.F.; Cheung, K.L.; Cheung, S.G. & Shin, P.K.S. (2005) Effects of nutrient enrichment derived from fish farming activities on macroinvertebrate assemblages in a subtropical region of Hong Kong. *Marine Pollution Bulletin*, Vol.51, pp. 994-1002, ISSN 0025-326X

Gao, Q.F.; Shin, P.K.S.; Lin, G.H.; Chen, S.P. & Cheung, S.G. (2006) Stable isotopic and fatty acid evidence for uptake of organic wastes from fish farming through green-lipped mussels (*Perna viridis*) in a polyculture system. *Marine Ecology - Progress Series*, Vol.317, pp. 273-283, ISSN 0171-8630

Gao, Q.F.; Xu, W.Z.; Liu, X.S.; Cheung, S.G. & Shin, P.K.S. (2008) Seasonal changes in C, N and P budgets of green-lipped mussels (*Perna viridis*) and their application as biofilters to remove nutrients from fish farming in Hong Kong. *Marine Ecology - Progress Series*, Vol.353, pp. 137-146, ISSN 0171-8630

Glasby, T.M. & Connell, S.D. (1999) Urban structures as marine habitats. *Ambio*, Vol.28, pp. 595-598, ISSN 0301-0325

Hall, P.O.J.; Holby, O.; Kollberg, S. & Samuelsson, M.O. (1992) Chemical fluxes and mass balances in a marine fish cage farm. IV. Nitrogen. *Marine Ecology - Progress Series*, Vol.89, pp. 81-91, ISSN 0171-8630

Hawkins, A.J.S. & Bayne, B.L. (1985) Seasonal variation in the relative utilization of carbon and nitrogen by the mussel *Mytilus edulis*: budgets, conversion efficiencies and maintenance requirements. *Marine Ecology - Progress Series*, Vol.25, pp. 181-188, ISSN 0171-8630

Hawkins, A.J.S.; Bayn, B.L.; Bougrier, S; Héral, M; Iglesias, J.I.P.; Navarro, E.; Smith, R.F.M. & Urrutia, M.B. (1998) Some general relationships in comparing the feed physiology of suspension-feeding bivalve molluscs. *Journal of Experimental and Marine Biology and Ecology*, Vol.219, pp. 87-103, ISSN 0022-0981

Hearn, C.J.; Atkinson, M.J. & Falter, J.L. (2001) A physical derivation of nutrient-uptake rates in coral reefs: effects of roughness and waves. *Coral Reefs*, Vol.20, pp. 347-356, ISSN 0722-4028

Hilton, M.J. & Manning, S.S. (1995) Conversion of coastal habitats in Singapore: Indications of unsustainable development. *Environmental Conservation*, Vol.22, pp. 307-322, ISSN 0376-8929

Huang, S.C. & Newell, R.I.E. (2002) Seasonal variations in the rates of aquatic and aerial respiration and ammonium excretion of the ribbed mussel, *Geukensia demissa* (Dillwyn). *Journal of Experimental and Marine Biology and Ecology*, Vol.270, pp. 241- 255, ISSN 0022-0981

Hummel, H.; Bogaards, R.H.; Bachelet, G.; Caron, F.; Sola, J.C. & Amiard-Triquet, C. (2000) The respiratory performance and survival of the bivalve *Macoma balthica* (L.) at the southern limit of its distribution area: a translocation experiment. *Journal of Experimental and Marine Biology and Ecology*, Vol.251, pp. 85-102, ISSN 0022-0981

Islam, Md. S. (2005) Nitrogen and phosphorus budget in coastal and marine cage aquaculture and impacts of effluent loading on ecosystem: review and analysis towards model development. *Marine Pollution Bulletin*, Vol.50, pp. 48-61, ISSN 0025- 326X

Karlsson, Ö.; Jonsson, P.R. & Larsson, A.I. (2003) Do large seston particles contribute to the diet of the bivalve *Cerastoderma edule*? *Marine Ecology - Progress Series*, Vol.261, pp. 161-173, ISSN 0171-8630

Kemp, W.M. & other 17 authors (2005) Eutrophication of Chesapeake Bay: historical trends and ecological interactions. *Marine Ecology - Progress Series*, Vol.303, pp. 1-29, ISSN 0171-8630

Kocak, F. & Kucuksezgin, F. (2000) Sessile fouling organisms and environmental parameters in the marinas of the Turkish Aegean coast. *Indian Journal of Marine Sciences*, Vol.29, pp. 149-157, ISSN 0379-513

Kristensen, E. & Andersen, F.Ø. (1987) Determination of organic carbon in marine sediments: a comparison of two CHN-analyzer methods. *Journal of Experimental and Marine Biology and Ecology*, Vol.109, pp. 15-23, ISSN 0022-0981

Laihonen, P.; Hännunen, J.; Chojnacki, J. & Vuorinen, I. (1996) Some Prospects of Nutrient Removal with Artificial Reefs, *European Artificial Reef Research: Proceedings of the first EARRN Conference*, pp. 85-96, ISBN 0-904175-28-6, Ancona, Italy, March, 1996

Leung, K.M.Y.; Chu, J.C.W. & Wu, R.S.S. (1999) Nitrogen budgets for the areolated grouper, *Epinephelus areolatus*, cultured under laboratory conditions and in open-sea cages. *Marine Ecology - Progress Series*, Vol.186, pp. 271-281, ISSN 0171-8630

Moran, P.J. & Grant, T.R. (1993) Larval settlement of marine fouling organisms in polluted water from Port Kembla Harbour, Australia. *Marine Pollution Bulletin*, Vol.15, pp. 512-514, ISSN 0025-326X

Morton, B. & Morton, J. (1983) *The Sea Shore Ecology of Hong Kong*, Hong Kong University Press, ISBN 962-209-027-3, Hong Kong

Nelson, W.G.; Savercool, D.M.; Neth, T.E. & Rodda, J.R. (1994) A comparison of the fouling community development on stabilized oil-ash and concrete reefs. *Bulletin of Marine Science*, Vol.55, pp. 1303-1315, ISSN 0007-4977

Nixon, S.W. (1995) Coastal marine eutrophication: A definition, social causes, and future concerns. *Ophelia*, Vol.41, pp. 199-220, ISSN 0078-5326

Pearson, T.H. & Black, K.D. (2001) The environmental impacts of marine fish cage culture. In: *Environmental Impacts of Aquaculture*, K.D. Black, (Ed.), 1-31, Sheffield Academic Press, ISBN 1–84127–041–5, Sheffield, UK

Petersen, J.K. & Riisgård, H.U. (1992) Filtration capacity of the ascidian *Ciona intestinalis* and its grazing impact in a shallow fjord. *Marine Ecology - Progress Series*, Vol.88, pp. 9- 17, ISSN 0171-8630

Petersen, J.K.; Schou, O. & Thor, P. (1995) Growth and energetics in the ascidian *Ciona intestinalis. Marine Ecology - Progress Series*, Vol.120, pp. 175-184, ISSN 0171-8630

Pickering, H.; Whitmarsh, D. & Jensen, A. (1998) Artificial reefs as a tool to aid rehabilitation of coastal ecosystems: Investigating the potential. *Marine Pollution Bulletin*, Vol.37, pp. 505-514, ISSN 0025-326X

Pitcher, T.J. & Seaman, W. (2000) Petrarch's principles: How protected human-made reefs can help the reconstruction of fisheries and marine ecosystems. *Fish and Fisheries* Vol.1, pp. 73-81, ISSN 1467-2979

Qian, P.Y.; Rittschof, D. & Sreedhar, B. (2000) Macrofouling in unidirectional flow: miniature pipes as experimental models for studying the interaction of flow and surface characteristics on the attachment of barnacle, bryozoan and polychaete larvae. *Marine Ecology - Progress Series*, Vol.207, pp.109-121, ISSN 0171-8630

Relini, G.; Relini. M.; Rorchia, G. & De Angelis, G. (2002) Trophic relationships between fishes and an artificial reefs. *ICES Journal of Marine Science*, Vol.59, pp. S36-S42, ISSN 1054-3139

Ribes M, Coma R, Gili JM (1998) Seasonal variation of in situ feeding rates by the temperate ascidian *Halocynthia papillosa. Marine Ecology - Progress Series*, Vol.175, pp. 201-213, ISSN 0171-8630

Ribes, M.; Coma, R. & Gili, J.M. (1999) Natural diet and grazing rate of the temperate sponge *Dysidea avara* (Demospongia, Dendroceratida) through an annual cycle. *Marine Ecology - Progress Series*, Vol.176, pp. 179-190, ISSN 0171-8630

Ribes, M.; Coma, R. Atkinson, M.J. & Kinzie, R.A. (2003) Particle removal by coral reef communities: Picoplankton is a major source of nitrogen. *Marine Ecology - Progress Series*, Vol.257, pp. 13-23, ISSN 0171-8630

Riisgård, H.U. (1988) The ascidian pump-properties and energy-cost. *Marine Ecology - Progress Series*, Vol.47, pp. 129-134, ISSN 0171-8630

Robbins, I.J. (1984) Regulation of ingestion rate, at high suspended particulate concentrations, by some phleobranchiate ascidians. *Journal of Experimental and Marine Biology and Ecology*, Vol.82, pp. 1-10, ISSN 0022-0981

Sarà, G.; Lo Martire, M.; Buffa, G.; Mannino, A.M. & Baalamenti, F. (2007) The fouling community as an indicator of fish farming impact in Mediterranean. *Aquaculture Research*, Vol.38, pp. 66-75, ISSN 1355-557X

Seaman, W. & Jensen, A.C. (2000) Purpose and Practices of Artificial Reef Evaluation. In: *Artificial Reef Evaluation: With Application to Natural Marine Habitats*, W. Seaman (Ed.), 1-20, CRC Press, ISBN 084-939-061-3, Boca Raton, Florida, USA

Sebens, K.P.; Vandersall, K.S.; Savina, L.A. & Graham, K.R. (1996) Zooplankton capture by two scleractinian corals, *Madracis mirabilis* and *Monastrea cavernosa*, in a field enclosure. *Marine Biology*, Vol.127, pp. 303-318, ISSN 0025-3162

Shin, P.K.S., Gao, Q.F. & Cheung, S.G. (2011) Nutrient assimilation by organisms on artificial reefs in a fish culture (example of Hong Kong). In: *Fish Farms*, F. Aral (Ed.), InTech Publisher, ISBN 978-953-307-663-8, Croatia

Smaal, A.C. & Vonck, A.P.M.A. (1997) Seasonal variation in C, N and P budgets and tissue composition of the mussel *Mytilus edulis*. *Marine Ecology - Progress Series*, Vol.153, pp. 167-179, ISSN 0171-8630

Smaal, A.C. & Widdows, J. (1994) The Scope for Growth of Bivalves as an Integrated Response Parameter in Biological Monitoring. In: *Biomonitoring of Coastal Waters and Estuaries*, K. Kramer (Ed.), 247-268, CRC Press, ISBN 084-934-895-1, Boca Raton, Florida, USA

Smith, S.D.A. & Rule, M.J. (2002) Artificial substrata in a shallow sublittoral habitat: Do they adequately represent natural habitats or the local species pool? *Journal of Experimental and Marine Biology and Ecology*, Vol.277, pp. 25-41, ISSN 0022-0981

Strickland, J.D.H. & Parsons, T.R. (1977) *A Practical Handbook of Seawater Analysis*. Canadian Government Publishing Center, ISBN 066-011-596-4, Ottawa, Canada

Svane, I.B. & Petersen, J.K. (2001) On the problems of epibioses, fouling and artificial reefs, a review. *Marine Ecology*, Vol.22, pp. 169-188, ISSN 0173-9565

Tsutsumi, H.; Srithongouthai, S.; Inoue, A.; Sato, A. & Hama, D. (2006) Seasonal fluctuations in the flux of particulate organic matter discharged from net pens for fish farming. Fisheries Science, Vol.72, pp. 119-127, ISSN 0919-9268

Warren, C.E. & Davis, G.E. (1967) Laboratory Studies on the Feeding Bioenergetics and Growth in Fish. In: *The Biological Basis of Freshwater Fish Production*, S.D. Gerking (Ed.), 175-214, Blackwell Scientific Publications, ISBN 063-200-256-5, Oxford, UK

Wilkinson, C. (1998) *Status of Coral Reefs of the World.* Australian Institute of Marine Science, ISBN 0 642 32218 X, Townsville, Australia

Wong, W.H. & Cheung, S.G. (2001) Feeding rates and scope for growth of Green mussels, *Perna viridis* (L.) and their relationship with food availability in Kat O, Hong Kong. *Aquaculture*, Vol.193, pp. 123-137, ISSN 0044-8486

Wotton, R.S. (1990) Particulate and Dissolved Organic Matter as Food. In: *The Biology of Particles in Aquatic Systems*, R.S. Wotton (Ed.), 213-261, CRC Press, ISBN 0-8493- 5450-1, Boca Raton, Florida, USA

Wu, R.S.S. (1995) The environmental impact of marine fish culture: Towards a sustainable future. *Marine Pollution Bulletin*, Vol.31, pp.159-166, ISSN 0025-326X

Zar, J.H. (1999) *Biostatiscal Analysis* (4th Ed.), Prentice-Hall, ISBN 817-758-582-7, New Jersey, USA

4 Chapter

EJACULATE DISTRIBUTION AND SPERM RIVALRY IN ALTERNATIVE REPRODUCTIVE STRATEGIES OF SALMON AND TROUT

1. INTRODUCTION: AN OVERVIEW

Aquaculture output has risen worldwide, and currently, the majority of salmon consumed in Sweden is sourced from hatcheries. Aquaculture is anticipated to yield greater food production for human consumption than catch fisheries (Anon 2009). Freshwater aquaculture accounts for 48 percent of global value, while mariculture constitutes 36 percent. Norway and Chile are the foremost countries in the production of farmed salmonids, with 33 and 31 percent of aquacultural output, respectively (Anon 2009). A prevalent method of salmonid production in hatcheries entails the amalgamation of milt from multiple adult males to fertilize the eggs of one or several females. This technique consistently entails sperm rivalry among the milt of multiple males to fertilize the eggs of individual females.

Atlantic salmon (Salmo salar) and brown trout (Salmo trutta) are stenothermic fish species inhabiting cold waters, exhibiting significant population variations in gonadal maturation timing and life history strategies. Ecological factors and genetic predisposition drive seasonal fluctuations in reproductive cycle dynamics. This cycle is defined by a feedback mechanism: the gonadal development in smolt (sub-adults migrating from the nascent river to the sea) dictates the duration of the marine/lacustrine phase prior to the fish's return to freshwater spawning habitats. Northern, cold-climate populations may achieve precocial maturity as parr in their second year of life (Dalley et al. 1983; Myers 1984), while southern salmonid populations have been seen to mature precocially in their first year of life (Bagliniere & Maisse 1999).

Sperm competition refers to the contest among sperm from many males for the fertilization of a female's ova within a single reproductive cycle (Parker 1970). Due to their ability to mate repeatedly at a significantly higher frequency than females, males have developed adaptations to inhibit rival males from ejaculating with the same female, so ensuring paternity. Competition for reproductive success has led to the development of different reproductive strategies, which are grounded in genetic life cycle allocation and behavioral norms that influence how an individual distributes its reproduction throughout its lifetime (Brommer 2000). Atlantic salmon and brown trout display divergent male maturation phenotypes: anadromous men and precociously developed parr males, which are frequently observed.

Participate in sperm competition at spawning sites. Alternative reproductive techniques are phenotypes that reflect a life history strategy aimed at optimizing individual reproductive success, even at the expense of lower survival. In numerous fish species, different reproductive strategies are marked by a significant disparity in age and size at maturity, as well as variations in the relative investment in gonadal and somatic tissues (Taborsky 1994). The disparity in age and size at sexual maturation has resulted in divergent spawning behavioral strategies that serve as mechanisms for achieving reproductive success, including "guarder" behavior exhibited by dominant males and "sneaker" behavior displayed by subordinates. This paper examines energy allocation methods of different mating phenotypes in Atlantic salmon and brown trout, situating this selective pressure within the context of interbreeding between escaped farmed fish and their wild counterparts. The aim is to examine the immediate mechanics of sperm competition and its evolutionary consequences in two sympatric salmonid species that display distinct male maturation strategies, and to relate these findings to enhanced aquacultural output in contemporary contexts. This chapter will examine the potential impacts of these processes on the genetic population structure due to the escape of farmed fish from aquaculture facilities.

2. SPERM CELL

In the salmonid spermatozoon, the following parts are morphologically distinct:

1. sperm plasma membrane which is the mediator of the signals for sperm motility,
2. sperm head with a nucleus containing the haploid paternal genetic material,
3. sperm mid-piece with a circular mitochondrion, where glycogene, phospholipids and phosphocreatine are the substrates for ATP production that provides energy for sperm motility,
4. sperm flagellum with a central bundle of microtubules with a 9+2 organisational pattern, the axoneme, which is a locomotory component of the sperm cell.

2.1 Sperm membrane

Sperm plasma membrane is the semipermeable barrier that defines sperm body, about 10 nm thick (Baccetti 1985). Water surrounding the cell membrane tends to enter the cell, which would eventually cause it to burst. Freshwater fishes have evolved mechanisms to keep water outside the cytoplasm, in order to maintain cellular stability. Sperm plasma membrane functions as the main receptor of the environmental signals for motility, such as the hypotonicity in freshwater after ejaculation, which initiates sperm motility of freshwater teleosts (Morisawa and Suzuki 1980). High potassium concentrations in the seminal plasma of salmonid fishes are responsible for the inhibition of sperm motility (Stoss 1983). In contrast, changes in the external divalent cation concentrations and in osmolality initiate sperm motility concomitantly (Morisawa and Suzuki 1980). After ejaculation, potassium leaves sperm cell through ion channels hyperpolarizing thereby the cell membrane. This membrane hyperpolarization event is the trigger for the initiation of sperm motility (Morisawa 1994). Simultaneous increase in intracellular calcium levels causes activation of the enzyme adenylyl cyclase, which catalyzes the synthesis of cyclic AMP (cAMP) from ATP (Morisawa and Okuno 1982). cAMP is an intracellular signal, which activates the enzyme protein kinase, the enzyme activating tyrosine kinase with the function to phosphorylate a 15K protein (Morisawa and Hayashi 1985). Change in intracellular pH is not a primary factor for regulation of sperm motility in freshwater fishes (Morisawa et al 1999). Signal transmission traversing the sperm plasma membrane results in the cascade of events with an ultimate function to maintain the communication between a mature sperm cell and its environment.

2.2 Sperm head

Ellipsoid trout sperm head is 2,5 μm long and 1,5-2 μm in diameter, containing the cell nucleus (Billard 1983). Teleost fish spermatozoa have no acrosome at the anterior of the sperm head, a structure containing the enzymes that hydrolyse the egg envelopes, which is coupled with the presence of an orifice, the micropyle, on the teleostean egg (Ginsburg 1972). Sperm nucleus is transcriptionally inactive. The nucleoplasm of fish spermatozoa consists of nucleoprotamines, which after fertilization, when the hereditary material is activated, are substituted for histones (Figure 1).

Sperm movement ensues as the result of the viscous interactions of sperm flagella with the surrounding medium (Taylor 1951). Gray and Hancock (1955) calculated that the viscous drag of the sperm head was small relative to the viscous drag of the flagellum itself. Thus, sperm head has only a negligible effect on the sperm cell locomotion (Gray and Hancock 1955; Humphries at al. 2008).

2.3 Mid-piece

The salmon sperm middle piece is 0.30- 0.95 µm long (Vladić et al 2002). Sperm movement commences from the base of the flagellum, and is performed by sliding movements between flagellar proteins. A single ring-shaped mitochondrion surrounds the midpiece in salmonid fishes (Jamieson 1991; Figure 1). The mitochondrial function is to synthesize ATP by the process of oxidative phosphorylation from endogenous phospholipids, glycolipids and glycogene (Stoss 1983). ATP produced in the mitochondrial oxidative phosphorylation prior to ejaculation is the main energetic source for sperm motility. It is hydrolysed by the molecular motor, dynein ATPases, in the course of motlilty. An ATP molecule contains two phosphoanhydride bonds, which liberate free energy when hydrolysed to ADP or AMP. This ATP/ADP cycle is the fundamental mode of energy conversion in living systems. Atkinson (1968) proposed that the energy charge:

$$EC = \frac{2(ATP)+(ADP)}{2[(ATP)+(ADP)+(AMP)]} \qquad (1)$$

regulates the energy metabolism in all living systems. In salmon, sperm energy charge increased with sperm tail length (Vladić et al 2002), which agrees with the finding of greater ADP concentrations in shorter sperm cells (Vladić 2001).

In the process of energy transduction, the free energy of respiration is the driving force for ATP production by the F_1F_0 ATP synthase (Harold 1986; Kinosita et al 2000). It was suggested that trout mitochondria have a low oxidative phosphorylation capacity, as ATP stores are quickly depleted in the course of sperm motility due to hydrolysis of ATP by dynein ATPase (Christen et al. 1987). Thus, a rate of mitochondrial respiration in fish spermatozoa is insufficient to maintain endogenous ATP reserves for prolonged motility (Cosson at al. 1999). In the Atlantic salmon, length of the sperm mid-piece was positively associated with the sperm ATP concentrations confirming thereby mitochondrial origin of ATP (Vladić et al 2002).

2.4 Sperm flagellum

The salmon sperm flagellum is 35- 45 µm long (Vladić et al 2002). Sperm flagella contain a uniform microtubular structure of the axoneme, comprised of nine peripheral and two central microtubules, surrounded by the plasma membrane (Afzelius 1959). The two central tubules are comprised of single microtubules, while the nine peripheral tubules are comprised of a complete A-tubule and an incomplete B-tubule. The molecular motor enzyme complex, adenosine triphosphatase (ATPase) activity involves a "dynein", which is an ATPase protein that drives the sliding of the outer doublet microtubules in sperm flagella (Harrison & King 2000). Some dynein molecules are assembled together with other proteins into macromolecular complexes called dynein arms. The peripheral microtubules have two rows of dynein arms along the length of the principle part of the flagellum. Guanosine nucleotides instead of adenosine nucleotides as in actin were detected in an additional isolated microtubule protein, which was named "tubulin" (Mohri and Ogawa 1975). This dynein-tubulin system is the molecular motor, which drives the flagellar movement of spermatozoa (Woolley 2000). Dynein ATPase proteins are bound to the A-tubule. The dynein arms of one peripheral doublet will walk on its neighbour to produce force for the flagellar movement, under hydrolysis of ATP. Thus, the beating of sperm flagella is the result of an active sliding between adjacent doublets of the axoneme powered by the ATP-driven mechanochemical cycle (Omoto 1991; Harrison & King 2000).

A thin filament devoid of dynein arms is present at the terminal end of the sperm flagellum (Vladić et al 2002). This cell structure is an universal feature of animal spermatozoa (Retzius 1904, Franzén 1956). Length of this flagellar portion in salmon spermatozoa was 2- 4 µm and contains only the two central microtubular pair (Vladić et al 2002). Since this region of the sperm flagellum experiences an increased viscous drag relative to the rest of the sperm flagellum an improvement in sperm propulsive effectiveness as an adaptation to viscous resistances has been proposed (Omoto and Brokaw 1982). Importantly, sperm fertility was positively related to the size of the thin sperm tail end piece fillament, suggesting adaptation to the viscosity of the aquatic medium (Vladić et al 2002; Figure 2).

3. SPERM COMPETITION IN SALMONIDS

Success in sperm competition depends on behaviour of the competing males and of the contested female, as well as on the frequencies with which different behavioural tactics are played in the population: therefore sperm competition models are analysed in the framework of the evolutionary game theory (Maynard Smith 1982). These games are designed to find an "unbeatable" or evolutionary stable strategy (ESS), strategy which, after adopted by all individuals in the population, cannot be invaded by a mutant playing an alternative strategy. The game-theoretic models are phenotypic and do not include genetics of the participants (Maynard Smith 1982).

The assumptions of the sperm competition games are (sensu Parker 1998):

1. Males in competition have specified information about their physiological state, and a tactical decision is made about the behaviour that yields the highest pay-off for that individual. Indeed, phenotypic plasticity may be viewed as an individual response to its physiological state (McNamara and Houston 1996);
2. There is a range of possible ejaculation decisions that are dependent on his state, which includes resource-holding potential, fighting ability, age and size of the individual;

3. Several male ejaculates must compete for fertilization of a single egg clutch;

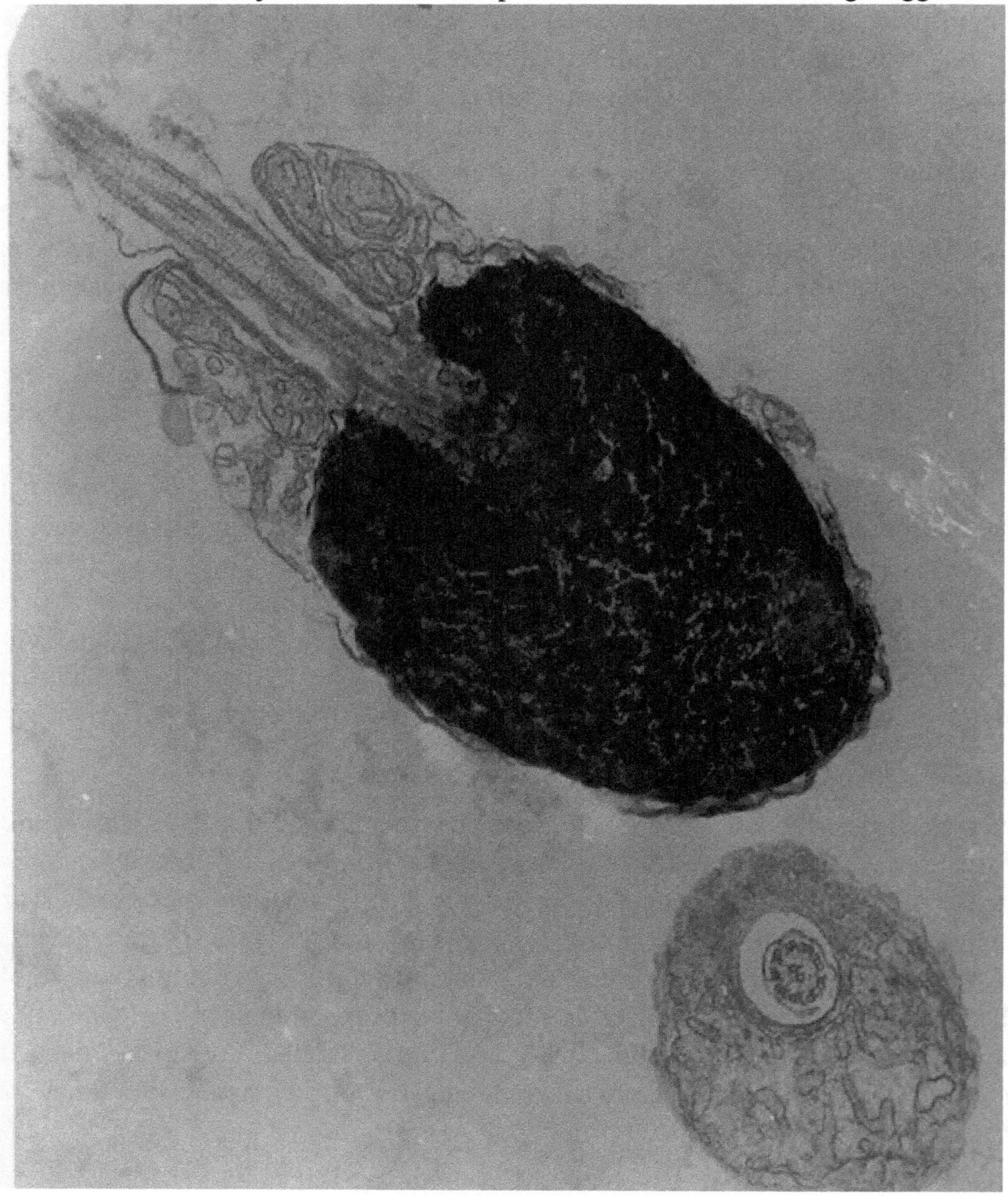

Fig. 1. Electron micrograph showing a longitudinal section through sperm head and middle piece of a salmon. Sperm nucleus contains densely packed chromatine, which is transcriptionally inactive. Below, a transverse section of the middle piece. Note the cell membrane around the sperm head and a single circular mitochondrion in the middle piece. On the transverse section of the middle piece the axoneme with a typical 9+2 microtubular complement is apparent. Magnification x 40 000. Courtesy by Björn Afzelius (reprinted from Vladić 2001, with permission).

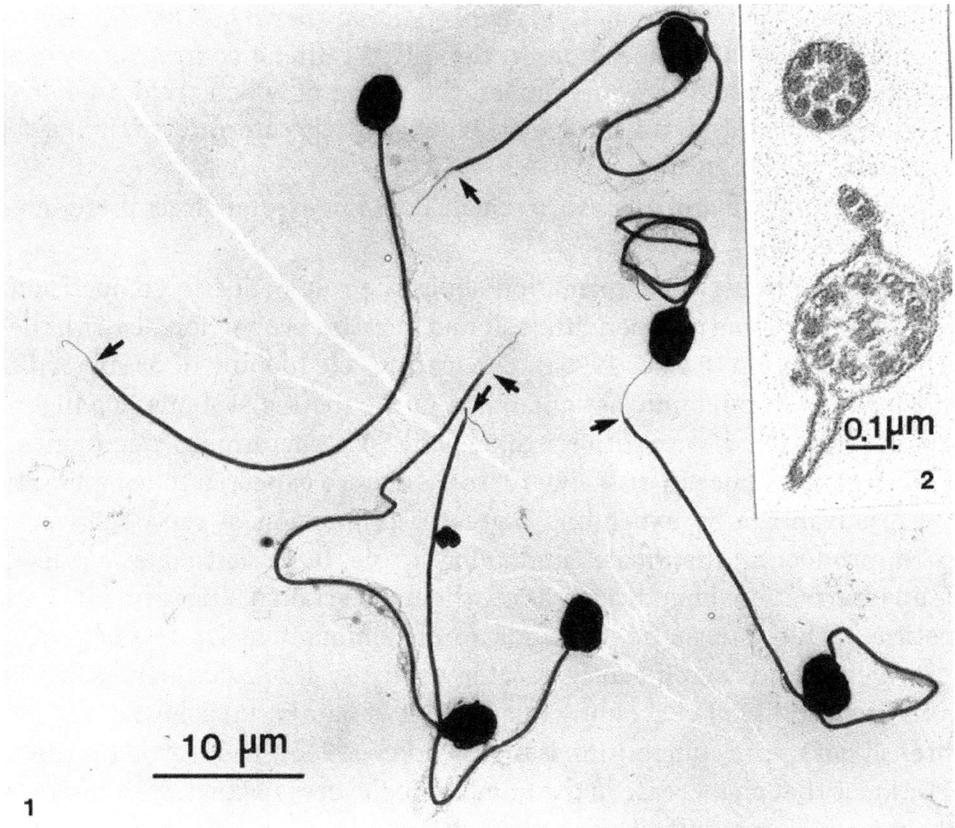

Fig. 2. 1) Whole mount of spermatozoa from a salmon parr. X2200. At the end of flagella, tail tip, containing only two centrally positioned, inner microtubules are indicated (arrows).

2) Sections through sperm tails from an anadromous salmon. The transected main piece (bottom) has nine microtubular doublets carrying inner and outer dynein arms and the two inner microtubules; the cell membrane forms two side fins. The end piece near the upper side fin contains only the two inner microtubules. The upper sperm transect contains 8 + 2 singlets; it is probably close to the transition region. X80 000. (from Vladić, Afzelius & Bronnikov, Biology of Reproduction, 66: 98-105).

4. Sperm may be used either randomly for fertilization depending on their densities in ejaculates (i.e. fair raffle), or spermatozoa originating from competing ejaculates may differ in some features determining mating order or ejaculate quality, leading to different propensities for a success (i.e. loaded raffle). In the *fair* raffle, all sperm compete on equal terms, that is competing ejaculates are physiologically and energetically equal. Thus, only an increased number of spermatozoa relative to competitor's spermatozoa in the competition will yield a greater chance for success in the fair raffle situation. When the physiological quality of spermatozoa differs

between competing males, it is expected that sperm quality will determine the outcome of competition. In the *loaded* raffle, a compensatory mechanism is predicted by which males, the sperm of which are devalued, compensate for this disadvantage by expending a greater proportion of reproductive effort on sperm (Parker 1990a,b).

5. In externally fertilizing fish, a female does not exert a direct preference for the competing ejaculates.

If males have an imperfect information about the role in sperm competition (eg. mating hierarchy), sperm expenditure should increase proportionally with the risk of sperm competition (Parker 1990a). The mating role (dominant or sub-ordinate) doesn´t have to be randomly assigned, like in the mating systems in which male characters correlated with fitness, like body size, determine dominance and access to females. Males in disfavoured role are here expected to compensate for mating disadvantage by expending a greater proportion of reproductive effort on sperm production than males in dominant role. Dominant males are likely to suffer an informational handicap making them uncertain of engagement in sperm competition. This informational handicap of dominant males favours sneaking tactic, which attains significantly greater paternity than dominant adult tactic when the probability of cuckolding (mating out of pair bond) is low, but not when the intensity of sperm competition is high (Parker 1990b). The reason for the latter expectation is that an increase in the number of players (ejaculates) in competition results in reduced pay- offs when the number of players is greater than 2 (Parker 1998). Therefore, disfavoured males ought to be selected to expand more energy in sperm production and/or quality than dominant males (Parker 1998).

In all salmonid species, there are at least two distinct life histories in males (Jones, 1959; Fleming 1996): one, with dominant anadromous males with variable degrees of fighting ability that have developed linear dominance hierarchy, and second, with small precociously mature males- parr that do not migrate to sea to acquire food for prolonged growth, but stay in the stream of their hatching or "grilse", who return to the spawning ground after single season in the sea. Because of the smaller amounts of food in the river, and different ecological conditions of the freshwater habitat (reviewed by Gibson, 1993) these males are miniature in size relative to dominant males (Gage et al 1995) (Figure 3). In salmonids, genes are propagated into future generations by means of alternative life history strategies. Thus, younger, precociously mature males are using "sneaking" tactic in the vicinity of the spawning salmon pair - after it have swam unnoticed into the spawning territory where dominant male have already courted a female, parr is trying to "sneak" into the sperm cloud of dominant male ejaculates, when it will also ejaculate spermatozoa in what is to be a trial to fertilize females eggs in the red (Figures 4 and 5). This strategy can be very costly for pre-cocious parr if discovered by dominant male, because dominant will not hesitate to retaliate. Such retaliatory behaviour can sometimes incure injuries, or even death of precocious salmon parr (Hutchings & Myers, 1987). Reproductive strategies can be defined as genetically based behavioural programmes influencing individual allocation decisions to reproductive effort between alternative tactics within a sex (Gross 1996). Phenotypes that became as coevolved responses of life history traits to ecological problems are alternative life history tactics (Stearns 1976). The alternative strategy involves a genetically based life history program, which has evolved under environmental, usually frequency-dependent selective pressure (Gross 1996). When average individuals have reduced fitness compared to individuals on extremes of phenotypic distribution, disruptive selection might select for extreme individuals (Rueffler et al 2006). Variability in response to environmental pressure among individuals within salmonid populations is shaped by differences in survival and reproductive success.

Atlantic salmon (Salmo salar) is an anadromous species, which spawn in freshwater, a feature characteristic for all salmonids. During its life history, two ecological environments are inhabited, a freshwater environment, in which salmon hatch and spawn, and a marine environment, where fast growth is achieved. The seaward migration is preceded by one to eight years. Before returning to the river of hatching for spawning the anadromous males and females stay between one and five years in the sea. Immature fish in the river (parr), become smolt in the spring of the second, third or fourth year. The process of smoltification involves various morphological, behavioural and physiological changes, as adaptations to marine environment. Salmon that return to the spawning river after only a single year are called 'grilse' (Mills 1971). Salmon do not feed during spawning migration, when males develop conspicuous lower hook and red bodily coloration, viz. secondary sexual characters (Figure 3). In northern latitudes, spawning may last from October to February, depending on duration of returning time to the nascent river. Males contribute no nest guarding or offspring tending to the female, but only their spermatozoa. Although after the spawning most of the males (called 'kelts') die due to high energetic costs that are paid in terms of intra-sexual fights for female acquisition and metabolic demands of sexual maturation (Jonsson et al 1991), relatively small proportion of females return to spawn in the following season (Mills 1971). Sea trout (*Salmo trutta*) males pursue shorter migratory routes and have greater iteroparity (i.e. mating in several consecutive seasons) than Atlatic salmon males, which undertake long migrations to the feeding grounds far off the coast and have an increased mortality rate after single spawning season (i.e. semelparity) (Belding 1934, Jones 1959, Mills 1989). Alternative male sexual maturation strategies are apparent both in salmon and brown trout, with adult, anadromous males which shed sperm simultaneously with precociously mature parr, a situation which results in sperm competition (Fig. 3). Number of males that may compete over fertilization of a single female eggs can vary between one and ten males (Hutchings 1986, Petersson and Järvi 1997). Gonad maturation was proposed to be determined by a genetical threshold (Thorpe 1986), which in concert with environmental control of maturation (shortening photoperiods; Lundqvist 1980; higher-than-average temperatures; Adams and Thorpe 1989; and food in excess enabling good growth, Alm 1959) determines male maturation pattern. Males that have a good growth rate tend to mature precociously (Alm 1959); they are maturing as precocious parr (Jones 1959, Mills 1989). A genetic component in male reproductive strategies was found in the Atlantic salmon (Glebe and Saunders 1986; Garant et al 2003). Individual precociously mature Atlantic salmon parr males have very variable fertilization success and may fertilize up to 65% of female eggs in the redd (Hutchings and Myers 1988, Thomaz et al 1997; Garcia- Vazquez et al 2002).

Precocious parr invest relatively more into gonadal tissue for their body mass than anadromous adults which invest more into secondary sexual traits which are frequently used in aggressive intrasexual interactions (Vladić & Järvi 2001). Also, higher metabolic demands of the parr reproductive strategy may be the reason that the relative heart weight in precocious parr is greater than this in immature fish (Armstrong and West 1994). This life history strategy is also associated with the impaired sea-water adaptability and reduced smoltification (Myers 1984, Lundqvist et al 1989). Male success is more dependent on social environment than is female success, which is dependent on the allometric relation between body size and gonad mass; therefore no single optimal life history is expected (Thorpe et al. 1998). Several decisions about the number of winters in the sea before returning to spawning may be exhibited within a single cohort. This results in the variable proportions of precociously mature parr, "grilse" and anadromous males that have spent a varying number of years in sea (Thorpe et al. 1998).

In reproductive biology, male quality equals individual reproductive success. Sperm movement in externally fertlizing fish is dependent on the cellular energy, produced in the mitochondria located in the sperm mid-piece. The synthesis of ATP is coupled to respiratory electron transport requiring the expression of mitochondrial genes. As the sperm cytoskeletal microtubular assembly, the axoneme, is extending throughout the sperm flagellum, sperm size is mainly related to sperm tail length (reviews in Gibbons 1981, Witman 1990). Reduction in sperm size with the increase in time between ejaculation and fertilization of the egg was predicted when the sperm tail size is positively correlated with sperm velocity at the expense of sperm longevity (Parker 1993). This prediction is applicable typically to internally fertilizing species. The logic is that it is difficult to adjust sperm size before given mating, since sperm have matured in the reproductive tract before the information about the role in competition could influence male ejaculation tactic. In externally fertilizing species, like salmon, it was found that sperm size decreases with sperm competition across fish species (Stockley et al 1997). In the Atlantic salmon, positive associations between different sperm length parameters and sperm energy charge, ATP concentrations and fertilization ability were found (Vladić et al 2002). In addition, salmonid sperm show adaptation to natural spawning temperatures (ie 3-4 ºC), whereas trout eggs exhibit higher thermotolerance than salmon eggs, possibly reflecting the southern origin of trout (Vladić & Järvi 1997). Sperm density was higher in both brown trout and salmon precocious parr, whereas salmon sperm are containing greater ATP concentrations than trout sperm (Vladić 2001). These features may be connected to the greater semelparity of salmon as compared to trout (Vladić 2000). Jonsson and Jonsson (2005) discuss greater energy allocation in reproduction of the precocially mature parr of Atlantic salmon than this in precocially mature trout parr in connection to conspicuous body size difference between the species and relatively longer migration distances to the feeding areas at sea in the Atlantic salmon as compared to the brown trout. Female eggs were found to be fertile after 512 s in water, significantly longer than sperm were mobile, i.e. 100-300 s at 2-4 Cº (Vladić & Järvi 1997). Salmon parr have greater sperm vigour (percentage of motile cells in ejaculates) (Vladić & Järvi 2001) and trade-off between sperm velocity and longevity after one-third of time since sperm activation (Vladić 2001), the result in agreement with the result published by Levitan (2000) (see Rosengrave et al 2009 for the discussion of effect of ovarian fluid on sperm behaviour). Therefore, studies on sperm traits should not imply contention of individual male quality unless these traits are tested in fertilization experiments. Besides sperm ATP content, sperm velocity was found to be the most important determinant of success in sperm competition (Gage et al 2004; Burness et al 2004; Yeates et al 2007). In addition, salmon parr were found to produce more ATP per sperm cell and are beter in fertilizing eggs both in the non-competitive situation (Vladić & Järvi 2001) and in sperm competition (Vladić et al 2010) confirming thereby the loaded model of sperm competition (Parker 1998).

Sperm density in the competition is high; therefore all eggs in externally fertilizing fish are expected to be fertilized instantaneously. At ESS, there will be a natural level of egg loss due to sperm death rate (Ball and Parker 1996). This "adaptive infertility" is opposed by an increase in sperm competition intensity; it benefits females to tolerate group spawning promoting thereby conflict between males and thereby sperm competition (Ball and Parker 1997). In addition, trade-off between offspring quality and quantity might be expected. This emanates from the fact that in structured populations, natural selection does not maximise short-term individual reproductive success (quality) but rather long- term value associated with genotype distribution in the population (McNamara et al 2011).

4. HUMAN IMPACT ON SALMONID EJACULATE ALLOCATION AND HERITABILITY OF PHENOTYPIC PLASTICITY

Human impact on wild habitats has proven devastating in many instances, as a consequence of the extensive hydroelectric power plants dam construction in most Swedish rivers. Occurrence of interspecific hybrids between the salmon and trout, which are called "laxing" in Sweden, was attributed to the shrinkage of the spawning area and destruction of natural spawning habitats caused by new hydroelectric plants (Jansson and Öst 1997). Such inter-specific hybrids are sterile. In addition, wild-farmed Atlantic salmon hybrids were found to have lowered fitness in comparison to wild fish, cautioning that frequent escapes from hatchery facilities may potentially reduce fitness for wild populations (Fleming et al 2000; McGinnity et al 2003; Araki et al 2007). Therefore, hatchery rearing for compensatory purposes has created new demographic pressures on endangered wild populations of these fish. Some traits are artificially selected for in hatchery environment, like genes for pathogen resistance (reviewed by Fjaelstad et al 1993), for instance genes of the Major Histocomapatibility Complex (MHC) that confer resistence to disease in vertebrates (Reusch et al 2001). However, considering the fact that local populations contain long-term adapted genomes to local environments, gene flow between farmed and wild fishes is likely to erode local adaptation and possibly lead to their extinction (McGinnity et al 2003). Hereditary basis for age at maturity in salmonids cautions that the compensatory breeding programs could have a significant demographic influence on wild Atlantic salmon populations (see Garant et al 2003), since the proportion of early maturing males may be substantial in hatcheries from where the supplementary fish are recruited, especially under favourable conditions of culture (i.e. higher-than-average temperature, food in excess) (reviewed by Jonsson & Jonsson 2006). Therefore a knowledge about ejaculate quality from alternative male morphotypes commonly engaging in sperm competition at spawning grounds has comprehensive areas of application, from those related to basic evolutionary questions to application of the findings in species management context.

Fig. 3. Differences in size between andromous and preciously mature parr of sea trout (above) and salmon from river Dalälven (below), showing the asymmetry in roles that these male morphotypes experience during sperm competition. Below, at the top, a female (81 cm long, 5.2 kg total weight), beneath a male (92 cm long, 6.7 kg total weight) with a large kype at the lower jaw, elongated nose and big adipose fin. The three smaller salmon are mature parr. Note differences in body morphology between parr males and anadromous adults, and differences in the expression of secondary sexual traits between the anadromous male, the female and precocious parr. Photo by Erik Petersson and Anna Löf. (Lower photo reprinted by permission from Vladić 2001, with permission).

Fig. 4. Sea trout precocious parr assumes an advantageous "sneaking" position beneath the female genital vent. Photo: Tomislav Vladić.

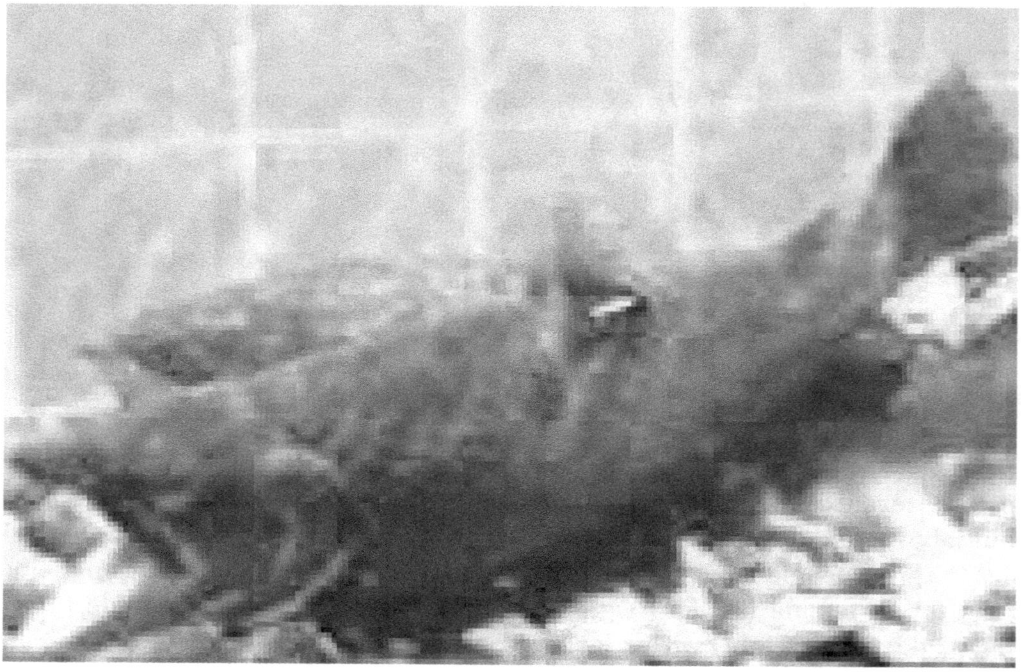

Fig. 5. Orgasm and ejaculation of an anadromous trout male. The female that expulsed eggs is behind the ejaculating male. Note the precocious parr in the sperm cloud beneath the anadromous male; it released sperm in sperm competition. Photo: Tomislav Vladić

Since behaviourally sub-ordinate males can acquire fitness benefits by exploiting sexual investment of behaviourally dominant males, male-biased operational sex ratio selects for alternative reproductive tactics, possibly through frequency-dependent selection. Conditional variation (variation depending on environmental influence) may produce different phenotypes from single genotype (Gross 1996). Reaction norms are genotype responses describing the manner individuals respond to environmental change within a population (Woltereck 1909). In salmonid fishes, growth rates influence choice of male mating tactic; males in good condition tend to mature early whereas males in poor condition tend to postpone reproduction in favour of prolonged body growth (Hutchings & Myers 1994; Thorpe et al 1998). This choice is a "threshold trait", where a liability toward maturation-age decision depends on the phenotype's position relative to some physiological (or environmental, see below) threshold value. It is evident that human impact exercised through high food ratios supplemented to hatchery fish aimed at compensatory releases changes male maturation pattern in the population, since conditional variation produces different phenotypes from the same genotype over environmental gradient (i.e. phenotypic plasticity). Therefore, decrease in egg size is expected when the food supplementation is in excess in hatcheries (Heath et al 2003), as offspring from larger eggs could have an advantage early in life (Einum & Fleming 1999). Interestingly, brown trout sea-running males were found to preferentially fertilize eggs of intermediate sizes, on the contrary to precocially maturing parr, which fertilized all eggs sizes indiscriminately; this mechanism was proposed as an expression of cryptic male choice (Vladić 2006). Recently, phenotypic plasticity in sperm production as a response to sperm competition risk was emphasized (Rudolfsen et al 2006; Cornwallis & Birkhead 2007; Pizzari et al 2007; Ota et al 2010).

Importantly for conditional strategies to evolve, environmental cue affecting gonad maturation must be reliable, whereas finesses of the alternative maturation life histories are not necessarily equal (Tomkins & Hazel 2008). Each reaction norm can be understood as different conditional strategy. A cue property of the genetic response in reaction norms as a result of selection, G, is that the selection differential before and after the episode of selection measured by mean mortality in different environments, S_k, and heritability of the plastic maturation trait, h^2, will vary in the function of environmental cue distribution, e, (e.g. feeding, temperature, photoperiod) within a given generation k:

$$\Delta G_k = \frac{S_k h^2}{e} \tag{2}$$

Evolutionary changes arise through change in frequency distribution of the environmental cue, each side of this distribution favoring different maturation conditional tactic (eg. early versus late maturing phenotypes) (Tomkins & Hazel 2008). Above equation implies that stronger additive genetic variation for the maturation trait, the stronger response to selection should be over an environmental gradient.

Heritability of phenotypic plasticity depends on the extent of genotype *x* environment interaction and strength of sexual selection. A genotype *x* environment reaction exists in the population if slopes of alternative fitness functions over environmental gradient cross. If slopes of fitness functions do not cross, male tactic may depend on individual male condition, which is commonly influenced by developmental constraints (Tomkins & Hazel 2008). These conditional reproductive tactics are central in sexual selection. Thus, genetic variance in phenotypic plasticity in the population, δ^2_{PL}, can be defined by partitioning total variance,

$$\delta 2_{PL} = \delta 2_{E} + \delta 2_{GxE} + \delta 2_{S} \quad (3)$$

where δ^2_E is environmental variance, δ^2_{GxE} is genotype *x* environment variance and δ^2_S is conditional variance due to sexual selection. Reduced phenotypic plasticity due to interbreeding between escaped farmed and wild fish may reduce capacity of population to cope with environmental change and thereby disrupt population dynamics.

Genetic diversity is crucial in small, isolated wild populations with increased levels of genetic drift, whereby chance events may cause random fixation of deleterious or invasive alleles. Maintaining genetic diversity and minimizing potential bottlenecks due to genetic drift requires minimum relatedness among individuals (Ohta 1982). In the generation *k*, genetic diversity can be defined as

$$\mathrm{GD}k = 1 - \frac{1}{2Nef} \quad (4)$$

where $2N_{ef}$ is the effective allele number (Crow & Kimura 1970; Caballero & Toro 2000). Common procedure employed in hatcheries is to mix ova from several females with sperm from several males in a single batch for supportive breeding purposes. This procedure induces sperm competition leading to reduction in the effective sample size of breeders and to increase in genetic variance in the population (Withler 1988; Withler & Beacham 1994). Although sperm potency leads to increased individual fertilization success, it may also lead to decreased genetic variation within populations when sperm traits are heritable due to increased variance in fertilization success between competing males and consequently decreased effective population size. Therefore equalizing milt volume in hatcheries should not necessarily reduce loss of genetic variation even at expense of favouring younger males in competition (Vladić et al 2010), because this practice reduces opportunity for natural selection to cleanse locally maladapted genetic contribution (Wedenkind et al 2007).

Effective number of breeders is defined as (Ridley 1993):

$$Ne = \frac{4NmNf}{Nm + Nf} \qquad (5)$$

where *Ne* is the effective number of breeders contributing to the gene pool in the following generation, Nm and Nf are numbers of male and female breeders, respectively. N*e* is the effective size of population with Mendelian segregation of genes (ideal population: equal sex ratio, constant population size and random probability of survival to adulthood) derived from the probability that two alleles at a locus are derived from the same grandparent. Means and variances are derived from the numbers of offspring surviving to maturity rather than from individual reproductive success (Campton 2004). Numbers of offspring surviving to maturity are affected by life history patterns of fish under consideration, as it was shown that salmon precocious parr males produce ejaculates of greater quality (Vladić & Järvi 2001: Vladić et al 2010), in accordance with predicted inverse relationship between fish age and gamete quality (Fleming 1996). Therefore, maximizing number of founders is expected to maximize genetic diversity in the population.

To summarize, genetic effects of captive breeding in hatcheries for supplementation of wild fish after destruction of natural spawning paths due to building of artificial dams for power- plant energy production may create fish that are reproductively inferior in the wild (Fleming et al 2000; Araki et al 2007). Nevertheless, negative effects of fish supplementation might be dependent on species and on the fish strain, since detrimental effects of the seven generation rearing on reproductive performance were not found in the brown trout from the river Dalälven stock (Dannewitz et al 2004).

5. SPERM COMPETITION AND MATERNAL EFFECTS

Rather than paternal, maternal genes were found to influence sperm phenotype (Froman & Kirby 2005), including sperm length in *Callosobrushus maculates* (Gay et al 2009). In addition to effects on sperm size, maternal genes strongly influence sperm motility, notably through predominantly (see Ankel-Simons & Cummins 1996; Rand 2001) maternally inherited haploid mtDNA (Ruiz-Pesini et al 2000; Froman & Kirby 2005). Selective forces are expected to differ between internally and externally fertilizing species due to discrete differences in the physical environment in which sperm compete to fertilize eggs. In externally fertilizing species where fertilization occurs instantaneously and sperm are relatively short-lived, sperm velocity may be naturally selected trait that confers advantage in competition for fertilizations. Nevertheless, in cases when insemination and fertilization are temporally separated as is the case for most internally fertilizing species, different sperm traits might be selected for (Parker 1993). In such cases, sperm longevity should be selected for (Taborsky 1998). Therefore, we expect that selective forces shape different ejaculate traits, which are advantageous depending of the mode of reproduction. In addition, sperm quality and quantity may trade off due to constraints imposed by conflicting life-history demands of simultaneous investment in body growth and gamete quality (Stearns 1992). However, if sperm quality and morphology are polygenic and/or unlinked traits, we should not expect to detect simple correlation between gamete quality and morphology.

Sexual selection selects for the two male traits: intrinsic genetic quality (indirect mechanism) and paternal care (direct mechanism). There is no paternal care in salmonid fishes. Body size is the male trait directly preferred by females (Andersson & Simmons 2006); the mechanism suggested by which a cryptic female choice for male's genetic quality is exerted in externally fertilising fish is through ovarian fluid that facilitates sperm function in the Arctic charr (Turner and Montgomerie 2002) and in the Atlantic cod (Litvak and Trippel 1998). The effect of ovarian fluid on sperm performance depends on the physiological compatibility between male and female partners (Rosengrave et al 2008). Mate choice is believed to optimize variation on Major Histocompatibility Complex (MHC) genes that confer resistance to disease. However, in the brown trout, female eggs were fertilized preferentially by males with intermediate molecular divergence in MHC genes, as males with great amino acid divergence on the MHC loci might have lower adaptation to locally adapted pathogens (Forsberg et al 2007). In the Atlantic salmon, variation in the MHC I class gene was biased toward similar genotypes, suggesting thus suppression of hybridization and outbreeding depression as the possible mechanism in salmonid mate choice (Yeates et al 2009). Thus, local adaptation may be disrupted by interbreeding between wild and escaped farmed fish in small isolated wild populations (reviewed by Hutchings and Fraser 2006).

6. CONCISE SUMMARY

Salmonid reproductive methods are influenced by male intrasexual rivalry for fertilization and the physical condition of females. Human interference in natural spawning, implemented in hatcheries during compensatory breeding programs, may potentially reduce genetic diversity within stocks. Consequently, it is imperative to comprehend the processes that influence the evolutionary mechanisms sustaining various reproductive phenotypes in the extant natural salmonid populations. Several questions remain unanswered:

1. How can we sustain current aquacultural output levels while preventing a decline in the genetic diversity of wild fish stocks caused by interbreeding with escaped farmed fish?

2. Implement a strategy for precociously maturing hatchery parr males that optimizes outcomes or generates ejaculates of superior quality compared to their wild equivalents?
3. What costs, if any, do wild females incur when mating with precociously developed parr males from hatcheries??

7. REFERENCES

Alm, G. (1959) Connection between maturity, size and age in fishes. *Rep. Inst. Freshwater Res. Drottningholm* 40: 5- 145.

Adams, C.E. & Thorpe J.E. (1989) Photoperiod and temperature effects on early development and reproductive investment in Atlantic salmon (Salmo salar L.). *Aquaculture,* 79: 403-409.

Afzelius, B. (1959) Electron microscopy of the sperm tail. Results obtained with a new fixative. *J. Biophys. Biochem. Cytol.,* 5: 269-278.

Andersson, M. & Simmons, L.W. (2006) Sexual selection and mate choice. *Trends Ecol. Evol.,* 21: 296-302.

Ankel-Simons, F. & Cummins, JM (1996) Misconceptions about mitochondria and mammalian fertilization: Implications for theories on human evolution. *Proc. Natl. Acad. Sci.,* 93: 13859- 13863.

Anonymous (2009) *Fisheries, Sustainability and Development.* (eds Wramner, P., Ackefors, H., Cullberg M). Royal Swedish Academy of Agriculture and Forestry (KSLA), Halmstad,

Araki, H., Cooper, B. & M.S. Blouin (2007) Genetic effects of captive breeding cause a rapid, cumulative fitness decline in the wild. *Science,* 318: 100- 103.

Armstrong, J.D. & West, C.L. (1994) Relative ventricular weight of wild Atlantic salmon parr in relation to sex, gonad maturation and migratory activity. *J. Fish Biol.,* 44: 453- 457.

Atkinson, D.E. (1968) The energy charge of the adenylate pool as a regulatory parameter. Interaction with feedback modifiers. *Biochemistry,* 7: 4030- 4034.

Baccetti, B. (1985) plasticity of the sperm cell. in *Biology of Fertilization* (eds. Metz, C. B. & Monroy, A.) pp. 3- 58. Acad. Press, New York.

Bagliniere, J L & Maisse, G (1999) Biology And Ecology Of The Brown Sea Trout. Springer, London Ltd.

Ball, M.A. & Parker, G.A. (1996) Sperm competition games: External fertilization and "adaptive" infertility. *J. theor. Biol.,* 180: 141- 150.

Ball, M.A. & Parker, G.A. (1997) Sperm competition games: inter- and intra-species results of a continuous external fetilization model. *J. theor. Biol.,* 186: 459- 466.

Belding, D.L. (1934) The cause of the high mortality in the Atlantic salmon after spawning. *Trans. Am. Fish. Soc.,* 64: 219- 224.

Billard, R. (1983) Ultrastructure of trout spermatozoa: changes after dilution and deep- freezing. *Cell Tiss. Res.,* 228: 205-218.

Brommer, J.E. (2000) The evolution of fitness in life-history theory. *Biol. Rev.,* 75: 377-404.

Burness, G., Casselman, S.J., Schulte-Hostedde, A.I., Moyes, C.D. & R. Montgomerie (2004) Sperm swimming speed and energetics vary with sperm competition risk in bluegill (*Lepomis macrochirus*). *Behav. Ecol. Sociobiol.,* 56: 65- 70.

Caballero, A., & Toro, M. A. (2000). Systems of mating to reduce inbreeding in selected populations. Interrelations between effective population size and other pedigree tools for the management of conserved populations. *Genet. Res., Camb.* 75: 331-343.

Campton, D.E. (2004) Sperm competition in salmon hatcheries: the need to institutionalize genetically benign protocols. *Trans. Am. Fish. Soc.,* 133: 1277-1289.

Christen, R., Gatti, J-L. & R. Billard (1987) Trout sperm motility. The transient movement of trout sperm is related to changes in the concentration of ATP following the activation of the flagelar movement. *Eur. J. Biochem.,* 160: 667-671.

Cosson, J., Billard, R., Cibert, C., Dréanno, C & Suquet, M. (1999) Ionic fractors regulating the motility of fish sperm. In *The Male Gamete: From Basic Science to ClinicalApplications* (ed. Gagnon, C.) Cache River Press, pp. 161- 186.

Cornwallis, C.K. & Birkhead, T.R. (2007) Changes in sperm quality and numbers in response to experimental manipulation of male social status and female attractiveness. *Amer. Natur.,* 170: 758- 770.

Crow, J. F. & Kimura, M. (1970). *An Introduction to Population Genetics Theory.* New York: Harper & Row.

Dalley, D.L., Andrews, C. W. & Green, R. H. (1983) Precocious male Atlantic salmon parr (*Salmo salar*) in insular Newfoundland. *Can. J. Fish. Aquat. Sci.* 40: 647-652.

Dannewitz, J., Petersson, E., Dahl, J., Prestegaard, T., Löf. A.C. & T. Järvi (2004) Reproductive success of hatchery- produced andc wild-born brown trout in experimental stream. *J Appl Ecol,* 41: 355-364.

Einum, S. & Fleming, I.A. (1999) Maternal effects of egg size in brown trout (*Salmo trutta*): norms of reaction to environmental quality. *Proc. R. Soc Lond.* B, 266: 2095- 2100.

Fjalestad KT, Gjedrem T & B Gjerde (1993) Genetic improvement of disease resistance in fish: an overview. Aquaculture, 111: 65-74.

Fleming, I.A. (1996) Reproductive strategies of Atlantic salmon: ecology and evolution. *Rev. Fish Biol. Fish.*, 6: 379- 416.

Fleming, I.A., Hindar K, Mjølnerød IB, Jonsson, B, Balstad, T & A Lamberg (2000) Lifetime success and interactions of farm salmon invading a native population. *Proc. R. Soc. Lond. B,* 267: 1517-1523.

Froman, D.P. & Kirby, J.D. (2005) Sperm mobility: Phenotype in roosters (*Gallus domesticus*) determined by mitochondrial function. *Biol. Reprod.,* 72: 562-567.

Forsberg, LA, Dannewitz, J, Petersson, E & Grahn, M (2007) Influence of genetic dissimilarity in the reproductive success and mate choice of brown trout-females fishing for optimal MHC dissimilarity. *J evol Biol,* 20: 1859-1869.

Franzén, Å (1956) On Spermiogenesis, Morphology of the Spermatozoon, and Biology of Fertilization Among Invertebrates. *Zoologiska Bidrag från Uppsala,* 31, 355-482.

Gage, M.J.G., Stockley, P. & Parker, G.A. (1995) Effects of alternative male mating strategies on characteristics of sperm production in the Atlantic salmon (*Salmo salar*): theoretical and empirical investigations. *Phil. Trans. R. Soc. Lond. B* 350: 391- 399.

Gage, M.J.G., Macfarlane, C.P., Yeates, S., Ward, R.G., Searle, J.B. & G.A. Parker (2004) Spermatozoal traits and sperm competition in Atlantic salmon: relative sperm velocity is the primary determinant of fertilization success. *Current Biology*, 14: 44- 47.

Garant, D., Dodson, J.J. & L. Bernatchez (2003) Differential reproductve success and heritability of alternative reproductive tactics in wild Atlantic salmon (*Salmo salar* L.). *Evolution,* 57: 1133-1141.

Garcia-Vasquez, E., Moran, P., Perez, J., Martinez, J.L., Izquierdo, J.I., de Gaudemar, B. & E. Beall (2002) Interspecific barriers between salmonids when hybridisation is due to sneak mating. *Heredity,* 89:282-292.

Gay, L., Hosken DJ, Vasudev, R, Tregenza, T & PE Eady (2009) Sperm competition and maternal effects differentially influence testis and sperm size in *Callosobruchus maculatus*. *J evol Biol*, 22: 1143-1150.

Gibbons, I. R. (1981) Cilia and flagella of eukaryotes. Discovery in cell biology. *J. Cell Biol.*91: 107s- 124s.

Gibson, R. J. (1993) The Atlantic salmon in fresh water: spawning, rearing and production. *Rev. Fish Biol. Fish.* 3: 39- 73.

Ginsburg, A. S. (1972) *Fertilization in Fishes and the Problem of Polyspermy.* Akademiya Nauk SSSR, Institut Biologii Razvitiya. (ed. Detlaf, T. A.). Translated from Russian by Israel Program for Scientific Translations, Jerusalem.

Glebe, B.D. & Saunders, R.L. (1986) Genetic factors in sexual maturity of cultured Atlantic salmon (*Salmo salar*) parr and adults reared in sea cages. *Can. Spec. Publ. Fish. Aquat.Sci.,* 89: 24- 29.

Gray, J & Hancock, GJ (1955) The propulsion of sea urchin spermatozoa. *J Exp Biol,* 32: 802- 814.

Gross, M.R. (1996) Alternative reproductive strategies and tactics: diversity within sexes. *Trends Ecol. Evol.,* 11: 92- 98.

Harold, F.M. (1986) *The Vital Force: A Study of Bioenergetics.* W. H. Freeman and Company. Harrison, A & King, SM (2000) The molecular anatomy of dynein. *Essays in Biochemistry,* 35: 75-87.

Heath, D.D., Heath, J.W., Bryden, C.A., Johnson, R.M. & C.W. Fox (2003) Rapid evolution of egg size in captive salmon. *Science,* 299: 1738-1740

Humphries, S, Evans, JP & LW Simmons (2008) Sperm competition: linking form to function. *BMC Evolutionary Biology,* 8: 319.

Hutchings, J.A. (1986) Lakeward migrations by juvenile Atlantic salmon, *Salmo salar. Can. J. Fish. Aquat. Sci.,* 43: 732- 741.

Hutchings, J.A. & Fraser, D.J. (2008) The nature of fisheries- and farming-induced evolution. *Molecular Ecology,* 17: 294-313.

Hutchings, J. A. & Myers, R. A. (1987) Escalation of an asymmetric contest: mortality resulting from mate competition in Atlantic salmon, *Salmo salar. Can. J. Zool.* 65: 766- 768.

Hutchings, J.A. & Myers, R. A. (1988) Mating success of alternative maturation phenotypes in male Atlantic salmon, *Salmo salar. Oecologia,* 75: 169- 174.

Jannsson, H. & Öst, T. (1997) Hybridization between Atlantic salmon(*Salmo salar)* and brown trout (*S. tritta)* in a restored section of the river Dalälven, Sweden. *Can. J. Fish. Aquat. Sci.,* 54: 2033- 2039.

Jamieson B.G.M. (1991) *Fish Evolution and Systematics: Evidence from Spermatozoa.* Cambridge University Press, Cambridge. Jones, J.W. (1959) *The Salmon.* London: Collins.

Jonsson, B., Jonsson, N. & Hansen, L.P. (1990) Does juvenile experience affect migration and spawning of adult Atlantic salmon? *Behav. Ecol. Sociobiol.,* 26: 225- 230.

Jonsson, B. & Jonsson, N. (2005) Lipid energy reserves influence life-history decision of Atlantic salmon (*Salmo salar*) and brown trout (*S. trutta*) in fresh water. *Ecol. Freshw. Fish*, 14: 296- 301.

Jonsson, B. & Jonsson, N. (2006) Cultured Atlantic salmon in nature: a review of their ecology and interaction with wild fish. *ICES J Mar Sci*, 63: 1162-1181.

Kinosita, K. Jr., Yasuda, R. & Noji, H. (2000) F_1-ATPase: a highly efficient rotatory ATP machine. *Essays in Biochemistry*, 35: 3-18.

Levitan, D.R. (2000) Sperm velocity and longevity trade off each other and influence fertilization in the sea urchin *Lytechinus variegatus*. *Proc. R. Soc. Lond.* B, 267: 531- 534.

Litvak, MK & Trippel, EA (1998) Sperm motility pattern of Atlantic cod (*Gadus morrhua*) in relation to salinity: effects of ovarian fluid and egg presence. *Can. J. Fish. Aquat. Sci.*55: 1871-1877.

Lundqvist, H. (1980) Influence of photoperiod on growth in Atlantic salmon parr (*Salmo salar*) with special reference to the effect of precocious sexual maturation. *Can. J. Zool.*, 58: 940- 944.

Lundqvist, H., Borg. B. & Berglund, I. (1989) Androgens impair seawater adaptability in smolting Baltic salmon. *Can. J. Zool.*, 67: 1733- 1736.

Maynard Smith, J. (1982) *Evolution and the Theory of Games*. Cambridge University Press, Cambridge

McGinnity P, Prodöhl P, Ferguson A, Hynes R, O. Maoiléidigh N, Baker N, Cotter D, O'Hea B, Cooke D, Rogan G, Taggart J & T Cross (2003) Fitness reduction and potential extinction of wild populations of Atlantic salmon, Salmo salar, as a result of interactions with escaped farm salmon. *Proc. R. Soc. Lond. B*, 270: 2443–2450

McNamara, J.M. & Houston, A.I. (1996) State-dependent life histories. *Nature*, 380: 215- 221.

McNamara, JM, Trimmer, PC, Eriksson, A, Marshall, JAR & AI Houston (2011) Environmental variability can select for optimism or pessimism. *Ecology Letters*, 14: 58-62.

Mills, D. (1971) *Salmon and trout: A Resource, its Ecology, Conservation and Management.* Oliver & Boyd, Edinburgh.

Mills, D. (1989) *Ecology and Management of Atlantic Salmon.* Chapman and Hall, London.

Mohri H. & Ogawa K. (1975) Tubulin and dynein in spermatozoan motility. in *The Functional Anatomy of Spermatozoon.* (ed. Afzelius, B. A.) pp. 161- 168. Pergamon Press Ltd.

Morisawa, M. (1994) Cell signaling mechanisms for sperm motility. *Zool. Sci.,* 11: 647- 662.

Morisawa, M. & Hayashi, H. (1985) Phosphorilation of a 15 K axonenmal protein is the trigger initiating trout sperm motility. *Biomed. Res.,* 6: 181- 184.

Morisawa, M. & Okuno, M. (1982) Cyclic AMP induces maturation of trout sperm axoneme to initiate motility. *Nature,* 295: 703- 704.

Morisawa, M. & Suzuki, K. (1980) Osmolality and potassium ion: Their roles in initiation of sperm motility in teleosts. *Science,* 210: 1145- 1147.

Morisawa, M., Oda, S., Yoshida, M. & H. Takai (1999) Transmembrane signal transduction for the regulation of sperm motility in fishes and ascidians. In *The male Gamete: FromBasic Science to Clinical Applications* (ed. Gagnon, C.) Cache River Press, pp. 149- 160.

Myers, R.A. (1984) Demographic consequences of precocious maturation of Atlantic salmon *(Salmo salar). Can. J. Fish. Aquat. Sci.,* 41: 1349- 1353.

Ohta, T. (1982) Linkage disequilibrium due to random genetic drift in finite subdivided populations. *Proc. Natl. Acad. Sci.,* 79: 1940- 1944.

Omoto, C.K. (1991) Mechanochemical coupling in cilia. *Int. Rev. Cyt.,* 131: 255- 292.

Omoto, C.K. & Brokaw C.J. (1982) Structure and behaviour of the sperm terminal filament. *J. Cell Sci.,* 58: 385- 409.

Ota, K, Heg, D, Hori, M & M Koda (2010) Sperm phenotypic plasticity in a cichlid: a territorial male´s counterstrategy to spawning takeover. *Behav, Ecol.,* 21: 1293-1300.

Parker, G. A. (1970) Sperm competition and its evolutionary consequences in the insects. *Biol. Rev.* 45: 525- 567.

Parker, G. A. (1990a) Sperm competition games: raffles and roles. *Proc. R.. Soc. Lond. B* 242: 120- 126.

Parker, G. A. (1990b) Sperm competition games: sneaks and extra-pair copulations. *Proc. R. Soc. Lond. B* 242: 127- 133.

Parker, G. A. (1993) Sperm competition games: sperm size under adult control. *Proc. Roy. Soc. Lond. B.* 253: 245- 254.

Parker, G.A. (1998) Sperm competition and the evolution of ejaculates: towards a theory base. in *Sperm Competition and Sexual Selection* (eds. T.R. Birkhead and A.P. Møller), pp. 3- 54.

Petersson, E. & Järvi, T. (1997) Reproductive behaviour of sea trout (*Salmo trutta*)- the consequences of sea- ranching. *Behaviour,*134: 1- 22.

Pizzari, T., Cornwallis, C.K. & D.P. Froman (2007) Social competitiveness associated with rapid fluctuations in sperm quality in male fowl. *Proc.Roy.Soc.Lond.B,* 274: 853- 860.

Rand, DM (2001) The units of selection on mitochondrial DNA. *Annu. Rev. Ecol. Syst.,* 32: 415-448.

Reusch, T.B.H., Häberli, M.A., Aeschlimann, P.B. & Milinski, M. (2001) Female sticklebacks count alleles in a strategy of sexual selection explaining MHC polymorphism. *Nature,* 414: 300- 302.

Retzius, G (1904) Zur Kenntnis der Spermien der Evertebraten. Biologische Untersuchungen. N.F., 11: 1-32.

Ridley, M (1993) Evolution, Blackwell.

Rosengrave, P., Gemmell, N.J., Metcalf, V., McBride, K. & Montgomerie, R. (2008) A mechanism for cryptic female choice in chinook salmon. *Behav. Ecol.,* 19, 6, 1179- 1185

Rosengrave, P., Montgomerie, R., Metcalf, V. & Gemmell N.J. (2009) Sperm traits in Chinook salmon depend upon activation medium: implications for studies of sperm competition in fishes. *Can J Zool,* 87: 920-927.

Rudolfsen, G., Figenschou, L., Folstad, I., Tveuten, H. & M. Figenschou (2006) Rapid adjustments of sperm characteristics in relation to social status. *Proc.Roy.Soc. Lond.B,* 273: 325- 332.

Rueffler, C., Van Dooren, T.J.M., Leimar, O. & PA Abrams (2006) Disruptive selection and than what? *Trends Ecol. Evol.* 21: 238- 245.

Ruiz-Pesini, E., Lapeña, A.C., Díez-Sánchez, C., Pérez-Martos, A., Montoya, J., Alvarez, E., Díaz, E., Urriés, A., Montoro, L., López-Pérez, M.J. & J.A. Enríquez (2000) Human mtDNA haplotypes associated with high or reduced spermatozoa motility. *Am. J. Hum.Genet.,* 67: 682-696.

Stearns, S.C. (1976) Life-history tactics: A review of the ideas. *Quart. Rev. Biol.,* 51: 3-47. Stearns, S.C. (1992) *The Evolution of Life Histories.* Oxford University press.

Stockley, P., Gage, M.J.G., Parker, G.A. & A.P. Møller (1997) Sperm competition and the evolution of testis size and ejaculate characteristics. *Am. Nat.,* 149: 933-954.

Stoss, J. (1983) Fish gamete preservation and spermatozoan physiology. in *Fish Physiology. Vol. IX. Reproduction Part B. Behaviour and Fertility Control* (eds. Hoar, W. S., Randall, D. J. and Donaldson, E. M.) pp. 305- 350. Academic Press, Inc. New York, London.

Taborsky, M. (1994) Sneakers, sattelites, and helpers: parasitic and cooperative behavior in fish reproduction. *Adv. Study Behav.* 23: 1- 100.

Taborsky, M. (1998) Sperm competition in fish: 'bourgois' males and parasitic spawning. *Trends Ecol. Evol.* 13: 222- 227.

Taylor, GI (1951) Analysis of the swimming of microscopic organisms. *Proc R Soc Lond A* 209: 447-471.

Thomaz, D., Beall, E & Burke T. (1997) Alternative reproductive tactics in Atlantic salmon: factors affecting mature parr success. *Proc. R. Soc. Lond. B,* 264: 219-226.

Thorpe, J.E. (1986) Age at first maturity in Atlantic salmon, *Salmo salar*: Freshwater period influences and conflicts with smolting. *Can. Spec. Publ. Fish. Aquat. Sci.,* 89: 7-14.

Thorpe, J.E., Mangel, M., Metcalfe, N.B. & Huntingford, F.A. (1998) Modelling the proximate basis of salmonid life-history variation, with application to Atlantic salmon, *Salmo salar* L. *Evol. Ecol.,* 12: 581- 599.

Tomkins, J. L. & Hazel, W. (2008) The status of conditional evolutionary stable strategy. *Trends Ecol. Evol.* 22: 522-528.

Turner, E. & Montgomerie, R. (2002) Ovarian fluid enhances sperm movement in Arctic charr. *J. Fish Biol.,* 60:1570-1579.

Vladić, T. (2000) The effect of water temperature on sperm motility of adult male and precocious male parr of Atlantic salmon and brown trout. *Verh. Int. Verein. Limnol.* 27: 1070- 1074.

Vladić, T. (2001) Gonad and Ejaculate Allocation in Alternative Reproductive Tactics of Salmon and Trout with Reference to Sperm Competition (Thesis, Stockholm University, 68 pp.).

Vladić, T. (2006) Sperm quality and egg size in the brown trout: implications for sperm competition and cryptic male choice. *Verh. Int. Verein. Limnol.* 29: 1331-1340.

Vladić, T. & Järvi, T. (1997) Sperm motility and fertilization time span in Atlantic salmon and brown trout- the effect of water temperature. *J. Fish Biol.,* 50: 1088- 1093.

Vladić, T.V. & Järvi, T. (2001) Sperm quality in alternative reproductive tactics of Atlantic salmon: the importance of the loaded raffle. *Proc. Roy. Soc. Lond., B,* 268, 2375- 2381.

Vladić, T.V., Afzelius, B.A. & Bronnikov, G.E. (2002) Sperm quality as reflected through morphology in salmon alternative life histories. *Biol. Reprod., 66,* 98-105.

Vladić, T., Forsberg, LA & Järvi, T (2010) Sperm competition between alternative reproductive tactics of the Atlantic salmon in vitro. *Aquaculture,* 302: 265-269.

Wedenkind, C., Rudolfsen, G., Jacob, A., Urbach, D. & R. Müller (2007) The genetic consequences of hatchery-induced sperm competition in a salmonid. *Biol Conserv*, 137: 180-188.

Withler, R. E. (1988) Genetic consequences of fertilizing Chinook salmon (*Oncorhynchus tschawyutscha*) eggs with pooled milt. *Aquaculture,* 68: 15- 25.

Withler, R.E. & Beacham, T.D. (1994) Genetic consequences of the simultaneous or sequential addition of semen from multiple males during hatchery spawning of chinook salmon (*Onchorhynchus tschawytscha*). *Aquaculture,* 126: 11- 23.

Witman, G.B. (1990) Introduction to cilia and flagella. in *Ciliary and Flagellar Membranes* (ed.R.A. Bloodgood), Plenum Publishing Corporation, pp. 1- 30.

Woolley, D. (2000) The molecular motors of cilia and eukaryotic flagella. *Essays in Biochemistry,* 35: 103-115.

Woltereck, R (1909) Weitere experimentelle Untersuchungen über Artverenderung, speziell über das Wessen quantitative Artunterschiede bei Daphniden. *Verhandlungen der Deutchen Zoologischen Gesellschaft,* 110-172.

Yeates, S, Searle, J, Ward, RG & MJG Gage (2007) A two-second delay confers first male fertilization precedence within in vitro sperm competition experiments in Atlantic salmon. *J Fish Biol*, 70: 318-322.

Yeates, SE, Einum, S, Fleming, IA, Megens, H-J, Stet, RJM, Hindar, K, Holt, WV, Van Look, KJW, & MJG Gage (2009) Atlantic salmon eggs favour sperm in competition that have similar major histocompatibility alleles. *Proc. R. Soc. Lond. B* 276, 559-566

5 Chapter

EMBRYONIC AND LARVAL DEVELOPMENT OF FRESHWATER FISH

1. INTRODUCTION: AN OVERVIEW

In recent years, the natural population of freshwater fish has been swiftly diminishing due to multiple anthropogenic and natural factors. As per IUCN (2000), of the 266 species, 14 are poised for extinction, the condition of 12 has significantly deteriorated, and 28 are classified as critically endangered. Furthermore, the fish face threats from the desiccation of low-lying regions and the indiscriminate application of fertilizers and pesticides. Insufficient knowledge exists regarding the early development of freshwater fishes. It is essential to conduct a thorough study to delineate the distinct stages of embryonic and larval development in order to comprehend the biological clock and cultivation techniques of these species (Rahman et al, 2009). Embryonic studies substantiate phylogenetic evolution by providing evidence to ascertain an organism's ancestral forms. For instance, it elucidates evolutionary development by addressing various phenomena, such as the gill cleft in lower vertebrates (fish), which is observed in nearly all mammalian embryos at early embryonic stages. This phase of fish life is utilized in several experimental studies, particularly in aquaculture and toxicological research (Rahman et al, 2009).

Life commences with the fusion of male and female gametes. Upon fertilization of the egg by a sperm, a zygote is created, initiating embryonic development, which culminates in hatching. The hatchlings subsequently undergo organogenesis and resemble their parents, so concluding the larval phases. Ovarian egg development is maternally sourced and predetermined within the ovary, although its genetic complexity is established at the moment of fertilization (Rahman et al, 2009).

This section analyzed the embryonic and larval development of freshwater fish. The text describes female fish utilized for egg production, the hormones involved in this process, the fertilization phase, which constitutes the first foundation of the organism, and many environmental elements influencing this process.

In this context, following the fertilization of freshwater fish, the embryonic developmental stages examined included the fertilized egg, cleavage, morula, blastula, gastrula, formation of the embryonic body, optic vesicle and auditory vesicle development, closure of the blastopore, tail formation, and hatching stages. During the larval development phase following hatching, until the conclusion of yolk sac absorption (pre-larvae), and subsequently until the completion of metamorphosis (postlarval), formations are exhibited.

2. DETERMINING THE METHODS OF LARVAL DEVELOPMENT

I wonder if there are different stages in fish lives? To understand this it is needed to observe the life stages of fish. The life stages of fish is formed of 5 stages (Demir, 2006).

1. Embryonic Phase
2. Larval Phase
3. Fry Phase
4. Ripe Phase
5. Senescent Phase

1. Embryonic Phase: It is the stage which start from insemination of egg to the end of vitellus' absorbtion. The period between fertilization of egg and outlet of organism is called incubation period. This is called aclation. Incubation ends with hatching. For that purpose egg capsule is softened with some kind of chemicals and enzymes and these are generallly excreted by ectodermal glands which are situated on anterior face of embryo or some glands on pharynx. Factors like increasing of heat and light increases the activities of embryo and embryo comes out from softened capsule with these chemicals. Embryo always comes out with head and front part of the body. Tail comes out at last and pushes to make embryo come out easier (Langeland and Kimmel., 1997).

It is divided into two:

a. Embryo stage in egg
b. Embryo stage outside the egg

The general characteristic of embryonic stage is being fed by vitellus which is an alternate nutrient.

The embryonic stage, may refer to different stages in eggs. Stages of embryonic development in eggs varies according to different species can be summarized as follows: fertilized egg, cleavage, morula, blastula, gastrula, embryonic body formation, optic vesicle and auditory vesicle formation, blastopore closing, tail formation and hatching stages. In the period of larval development after hatching, until the end of the yolk sac absorption period (pre-larvae) and subsequently until the end of metamorphosis (postlarval) formations are shown (Langeland and Kimmel., 1997).

The stages of embryonic development of Şabut (*Barbus grypus* H, 1843) and Masopotamian spiny eel (*Mastacembelus mastacembelus* B&S, 1794) were given as following (Şahinöz et al., 2006; Şahinöz et al., 2007).

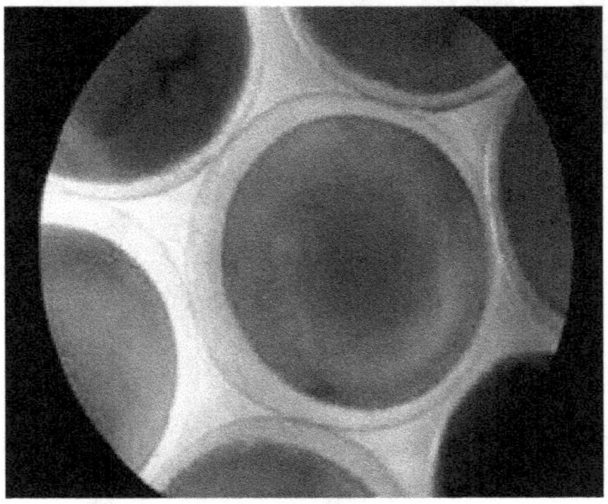

Fertilized eggs and perivitellin space formation after the fertilization. (*Barbus grypus* H, 1843)

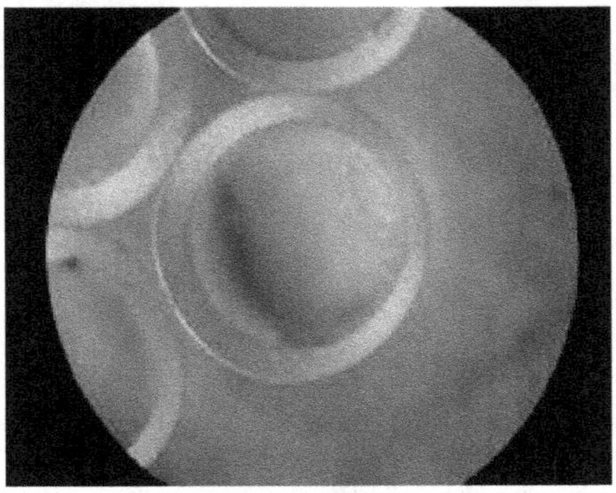

Germinal disc formation (x10). (*Barbus grypus* H, 1843)

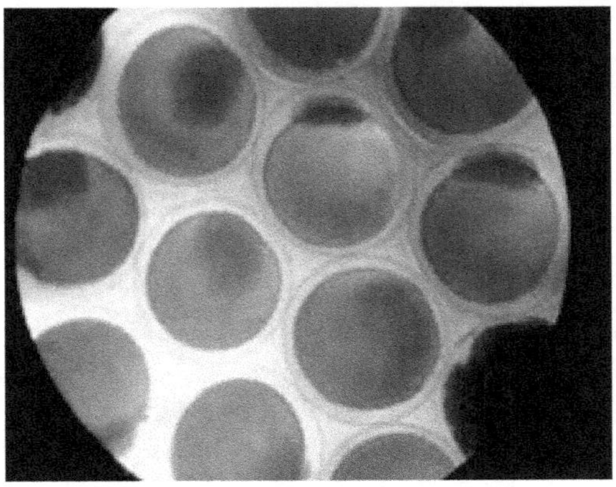

Morula phase and blastomer formation (x10). (*Barbus grypus* H, 1843)

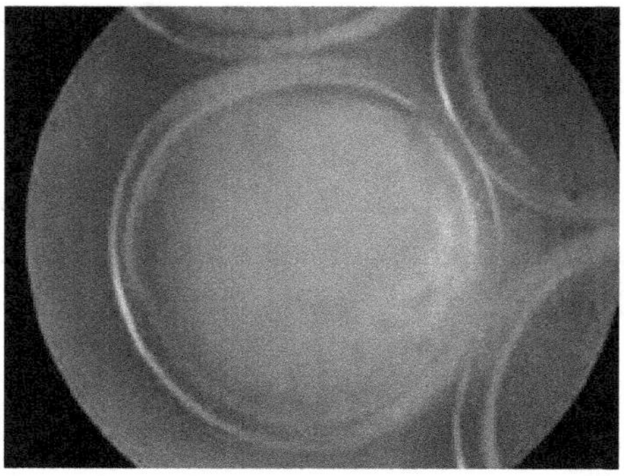

Gastrula phase (x10). (*Barbus grypus* H, 1843)

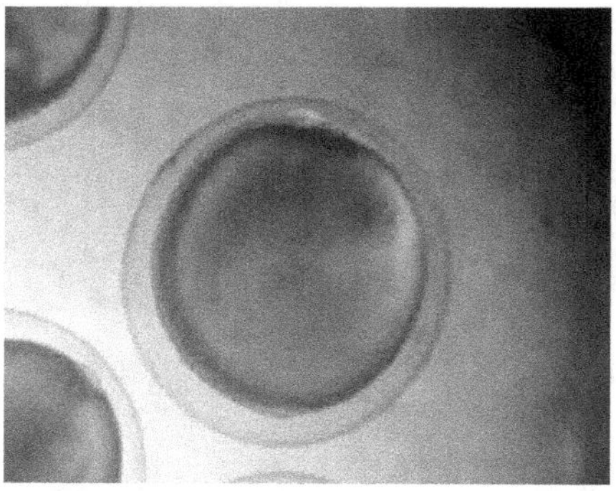

Embryonic body formation (x10). (Barbus grypus H, 1843)

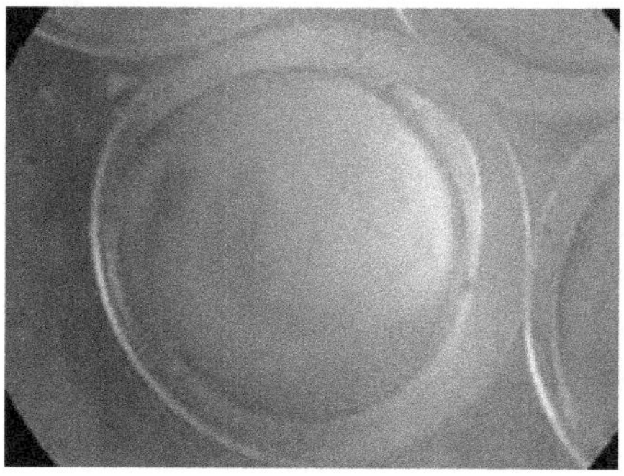

Head formation (x10). (*Barbus grypus* H, 1843)

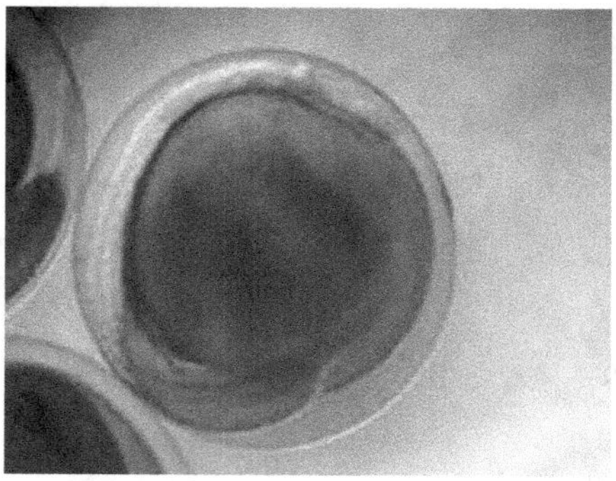

Eyes and otic vesicle formation. (*Barbus grypus* H, 1843)

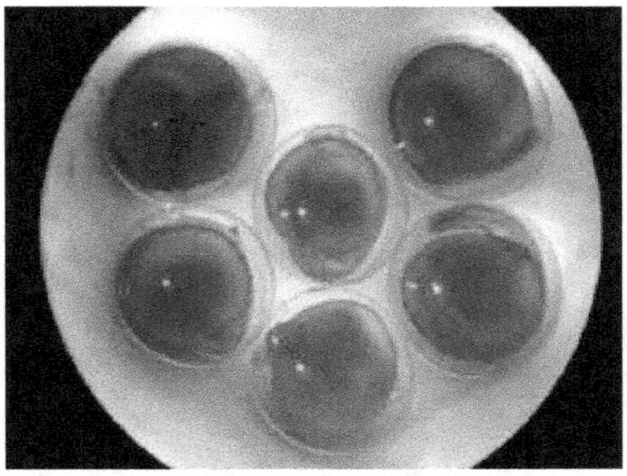

First heart beat and movement (x10). (Barbus grypus H, 1843)

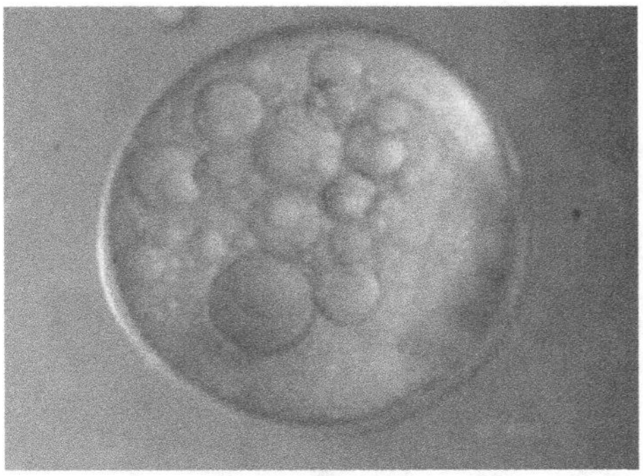

Perivitelline space formation of *M. mastacembelus* (Bank&Solender, 1794) (x10).

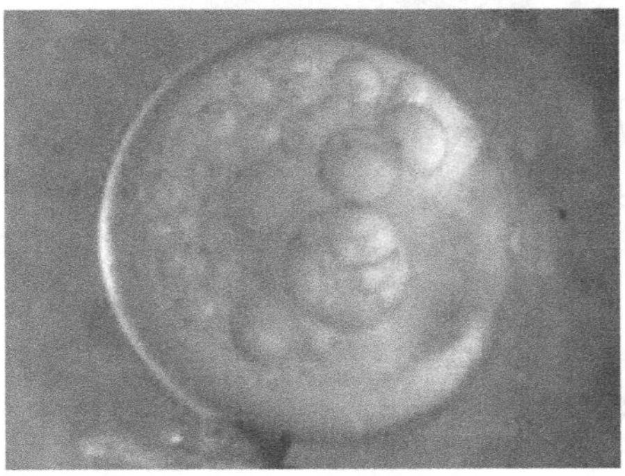

First division in egg cell (x10).

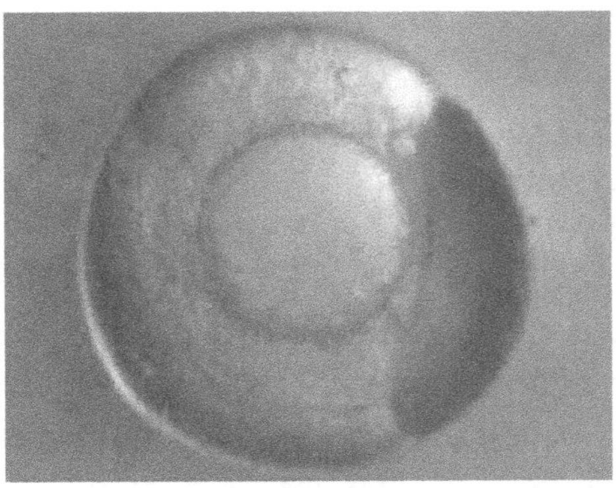

1 oil droplet formation (x10).

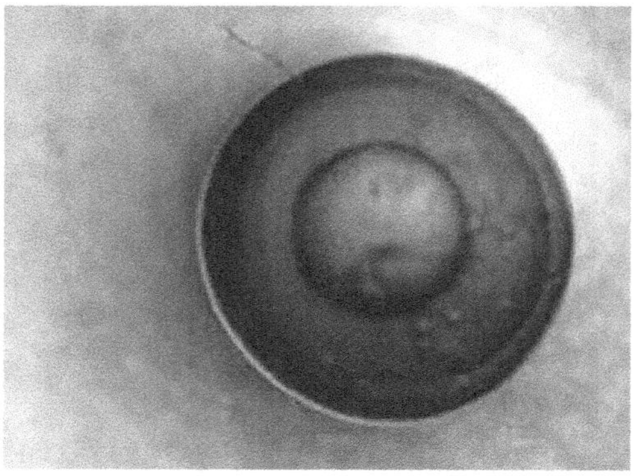

Embryonic shield formation (x10).

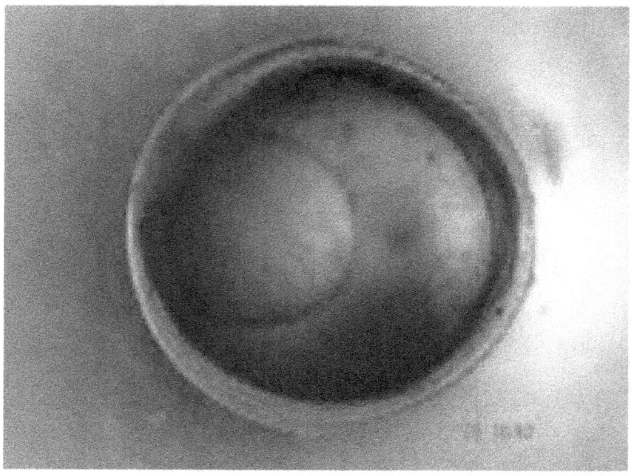

Embryonic body formation and notocorda (x10).

2. Larval Phase: During the larval stage which follows embryonic stage, nutrition occurs outside. Larvae as the general appereance may differ from mature in many ways in both outer and iner structure. After coming out from the egg, especially teleosts gothrough some stages. These stages are named differently by different scientists. The most common among them is one which Hubbs (1934) proposed. There are 3 phases of fish from coming out egg to fry stage.

- 2.1.Prelarval Phase
- 2.2.Postlarval Phase

2.1 Prelarval Phase: It is the period which starts from coming out from egg to the end of absorbtion of yolk sac. The most important charecteristic of prelarval stage is the existence of yolk sac. This yolk sac is located in anterior and ventral of the body. At the beginnig of tje prelarval phase mouth, anus and digestive tube is like straight pipe. Head is smaller then body, eyes are big and nonpimented. The double-walled sacs in the form of external sensory organ of balance otoliths are on both sides of the head. Nostrils are not developed under eye. Towards the end of this phase mouth and anus opens. Eyes are pigmented and nutrition mouth starts at the outside part. During prelarval phase only the pectoral fins which appears in adults. This draft is positioned horizontally first, then becomes vertically by the progress. Promordial fin remains the surface pelajik by increasing the surface and consists of two layers of skin bend. This double layer of skin is called the space between the subdermal space. In some species after completin prelarval phase and taking the characteristic shape of adult called alevin in Salmonidae fishes. Postlarval stage followed by the majority of the fish prelarval phase.

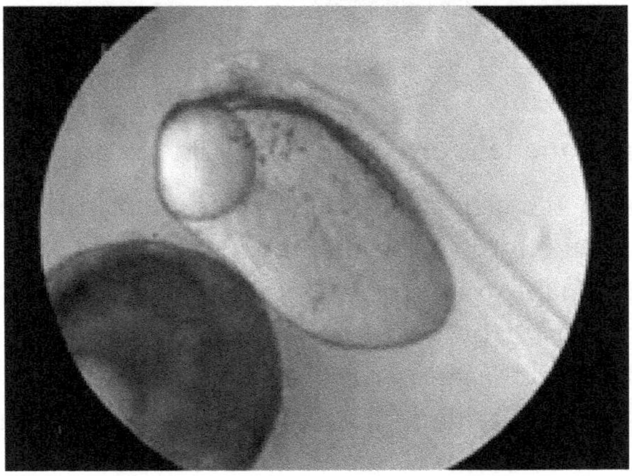

A newly hatched larva. (*M. mastacembelus* B&S, 1794)

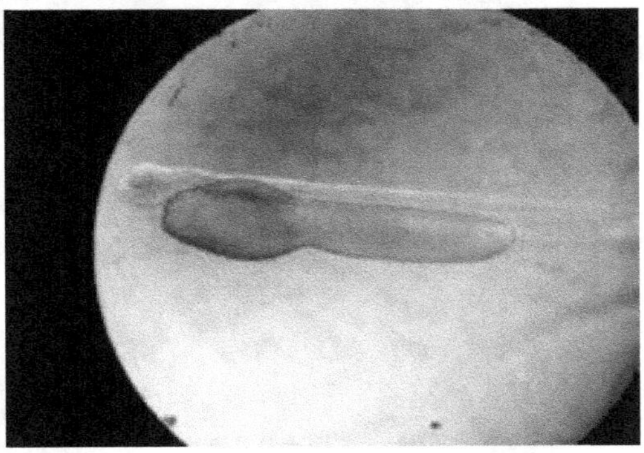

A newly hatched larva. (*Barbus grypus* H, 1843)

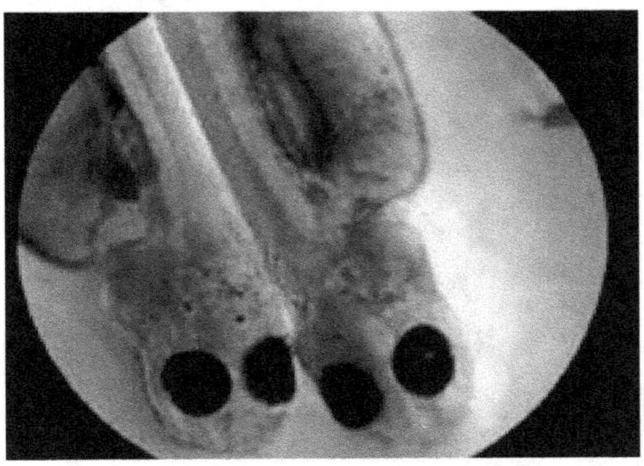

Air bladder formation (x10). (*Barbus grypus* H, 1843)

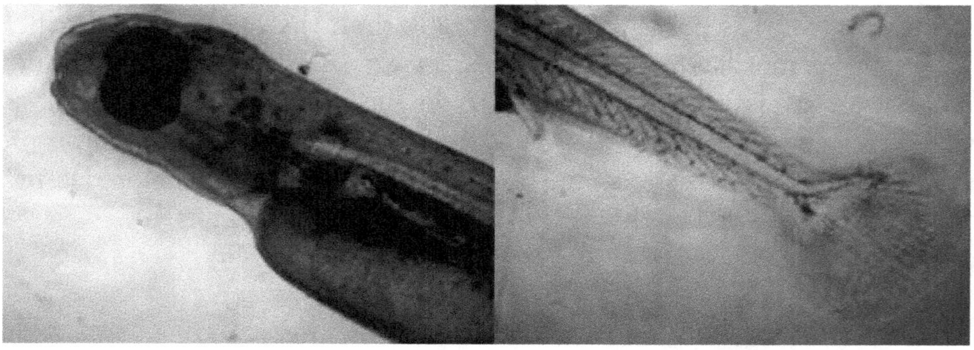

Mouth openning, anus and tail formation (x10). (*Barbus grypus* H, 1843)

2.2 Postlarval Phase: It is the time starts after absorbtion finished to the end of metamorphosis. The length of the time changes species to species. It varies to species according to shape, size, body ratio, fin size, pigmentation in different sizes and order, shape and time of organ formation in postlarval phase. Some organs form in postlarval phase in order to make easier to stay pelagic. Nutrition takes place entirely from outside in postlarval phase. Nutritions are fitoplancton, zooplancton or mixture of both.

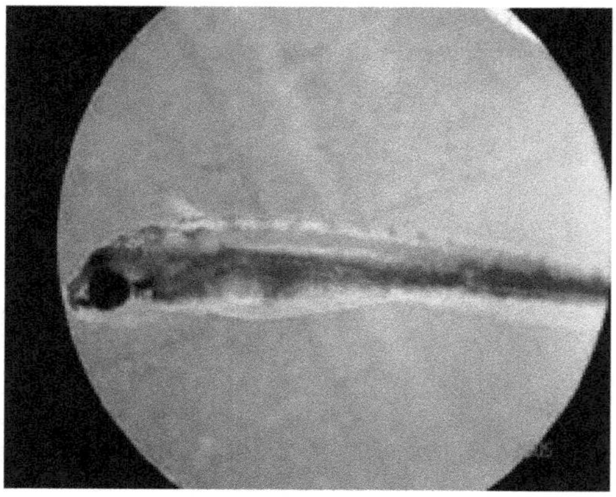

Blood vessel, digestion channels, anus. (*M. mastacembelus* B&S, 1794)

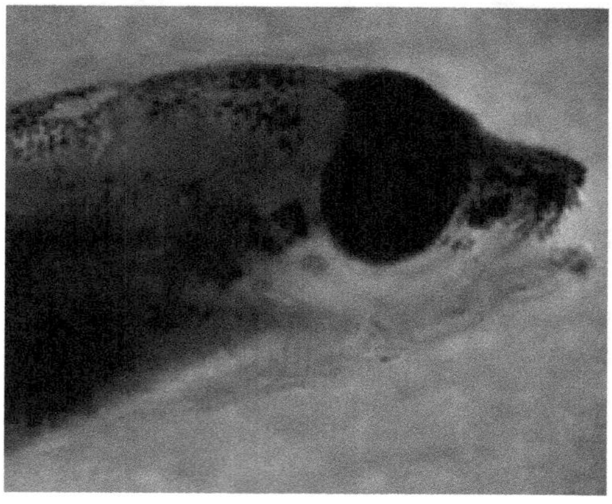

Mouth openning. (*M. mastacembelus* B&S, 1794)

Pigmentation. (*M. mastacembelus* B&S, 1794)

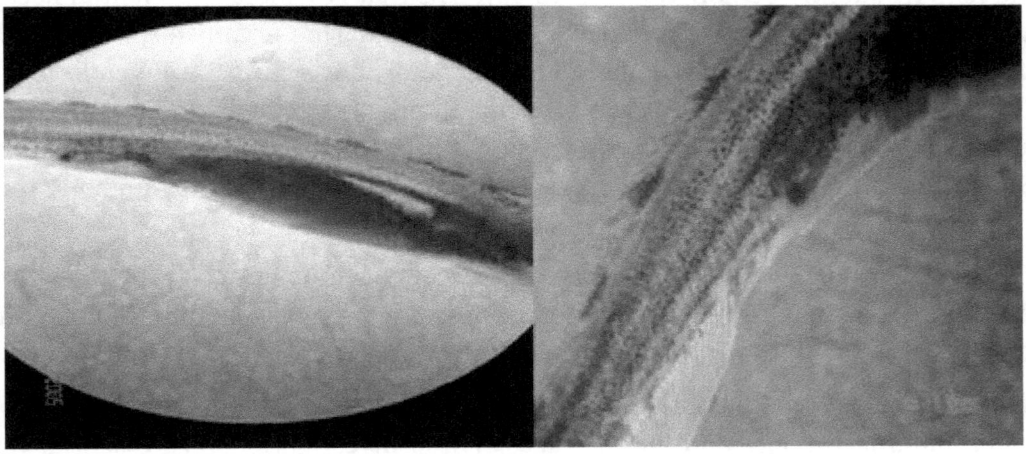

Development of visceral organs. (*M. mastacembelus* B&S, 1794)

3. Fry Phase: At the end of post larval phase youth phase starts. Except mugillidae family at other teleosts formation of scales determine the result of postlarval phase. Formation of scales shows the first phase of youth phase. Scales start to develop early. The other event is developing of lateral line. This can be seen at the end of postlarval phase and during metagenezis. Circulation system is formed in many species. The first view of hemoglobine developes during metamorphosis again. Organism is a litlle copy of the mature in youth stage. However some differences are observed in body ratios and colour. In gonads secondary characters does not exist yet if there are. Like in salmonidae some species do not show remarkable metamorphosis during postlarval phase and the individuals belong to this family pass to youth phase directly. At the end of post larval phase with the completion of metamorphosis fish has the ability to move freely and passed to nekton from plancton now on.

4. Ripe Phase: Fish gonads are fully developed and have capability to reproduce capability at intervals. Secondary characters are developed in some fish species (if any).

5. Senescent Phase: Firstly sexual activities decreases then completely disappears. Similarly the growth of height slows down or stops.

3. CONCISE SUMMARY

A primary objective in population dynamics is to comprehend the mechanisms that determine which individuals reach reproductive maturity. Mortality schedules are intricately associated with the life-history traits of the organism and its reproductive strategy (Stearns 1992).

Organisms exhibiting intricate life cycles, including insects, marine invertebrates, and fish, frequently demonstrate death patterns that resemble exponential decay functions (Hunt and Scheibling 1997). For several marine organisms, nearly 99% of individuals perish before undergoing metamorphosis into the juvenile stage (Gosselin and Qian 1997). A crucial element affecting the dynamics of these populations is the fluctuation in the amount of juveniles entering the population (Stearns and Koella 1986). The processes that dictate success throughout this transition are intricate and not entirely comprehended.

From conception, people exhibit varying rates of development and growth, which predispose some to a diminished likelihood of surviving subsequent developmental stages.

Fisheries biology examines a fish population, emphasizing the significance of understanding the egg and larval stages. These investigations are essential for identifying spawning seasons and locations, assessing temporal variations in spawning periods, estimating the quantity of ovulating mature stock, predicting mortality rates at the conclusion of the spawning period, and analyzing environmental correlations. (Yüksek ve Gücü, 1994).

4. REFERENCES

Erdinç Şahinöz, Zafer Doğu, Faruk Aral, 2006. Development of Embryos in *Mastacembelus mastacembelus* (Bank&Solender, 1794) (Mesopotamian Spiny Eel) (Mastacembelidae). Aquaculture Research. Blackwell Science, 37 (16): 1611-1616.

Erdinç Şahinöz, Zafer Doğu, Faruk Aral, 2007. Embryonic and Pre-larval Development of Shabbout (*Barbus grypus* H.). The Israeli Journal of Aquacultre. Bamidgeh. 59 (4): 236–239.

Gosselin, L. A. and Qian, P. 1997. Juvenile mortality in benthic marine invertebrates. Mar. Ecol. Prog. Ser., 146: 265-282.

Hubbs, C. L. 1934. Terminology of early stages of fishes. Copeia, 260.

Hunt, H. L. and Scheibling, R. E. 1997. Role of early postsettlement mortality in recruitment of benthic marine invertebrates. Mar. Ecol. Prog. Ser., 155: 269.

IUCN, 2000. Red Book of Threatened Fishes of Bangladesh. 116.

Langeland, J. and C. B. Kimmel. 1997. The embryology of fish. *In* S. F. Gilbert and A. M. Raunio (eds.), Embryology: Constructing the Organism. Sinauer Associates, Sunderland, MA, 383–407.

Necla Demir, 2006. İhtiyoloji, Nobel Yayn, Ankara, 252.

Rahman, M.M., Miah, M. I., Taher, M. A., Hasan, M. M. , 2009. Embryonic and larval development of guchibaim, *Mastacembelus pancalus* (Hamilton) J. Bangladesh Agril. Univ. 7 (1): 193–204.

Stearns, S. C. and Koella, J. C. 1986. The evolution of phenotypic plasticity in life history traits: predictions of reaction norms for age and size at maturity. Evolution. 40: 893- 913.

Yüksek, A. Ve Gücü, A.C., 1994. Balk Yumurtalar Tayini İçin Bir Bilgisayar Yazlm (Karadeniz Pelajik Yumurtalar), Karadeniz Eğitim-Kültür ve Çevre Koruma Vakf, İstanbul, 51.

6 Chapter

HISTOPATHOLOGY AND ALTERNATIVE TECHNIQUES FOR IDENTIFYING VIRAL HEMORRHAGIC SEPTICEMIA (VHS) IN CERTAIN IRANIAN RAINBOW TROUT FARMS

1. INTRODUCTION: AN OVERVIEW

Viral haemorrhagic septicaemia (VHS) represents a significant economic challenge in European salmonid aquaculture (Jorgensen, 1974). VHS illness is known to induce significant mortality in farmed rainbow trout (Oncorhynchus mykiss) across all age groups. This illness has recently been identified in rainbow trout cultivated in the Pacific Ocean and in farms raising marine species. Numerous instances of VHS illness have been documented in rainbow trout aquaculture facilities in Iran. Water temperature significantly influences the progression of a VHS outbreak (Ahne & Thomsen, 1985), with the disease posing a greater threat at temperatures below 15-16°C (McAllister, 1979). Disease outbreaks typically transpire at water temperatures ranging from 4 to 14°C. The initial outbreaks of VHS illness transpired in Iranian aquaculture in November 2005 at Rodsar, Gilan province, in northern Iran, where water temperatures were recorded at 10°C.

Viral haemorrhagic septicaemia (VHS) is regarded as the most severe disease affecting rainbow trout in aquaculture. It may result in mortality rates of up to 80% in impacted populations. Historically, Iran was devoid of this ailment. In November 2005, a VHS outbreak resulted in significant mortality at the Rodsar trout farm in Gilan province. This outbreak caused significant economic losses for numerous farmers. Viral hemorrhagic septicemia virus (VHSV) belonging to the family Rhabdoviridae. VHS was initially documented at a rainbow trout farm in Denmark and has subsequently resulted in considerable losses in rainbow trout farms across numerous nations in continental Europe. It has also been documented in Japan and Russia. A very virulent novel strain of VHS has been documented in many freshwater fish species in the USA. Viral hemorrhagic septicemia (VHS) is an acute to chronic viral affliction of salmonids that results in significant economic challenges for rainbow trout farming in many European nations. This disease, albeit undetected in North America, is listed due to its considerable concern. The ailment was initially identified by Schaeperclaus in Germany in 1938. In 1949, the ailment was designated as Egtved illness following an outbreak in Denmark in the village of the same name. In 1966, the Office International des Epizooties proposed changing the name to viral hemorrhagic septicemia to mitigate confusion arising from the nomenclature employed in several European nations where it is present. (Roberts 1978).

Viral hemorrhagic septicemia is caused by a bullet-shaped virus similar in size and shape to the virus that causes infectous hematopoietic necrosis (IHN) in North American salmonids. The VHS virus can be distinguished from IHN by specific serum neutralization tests (McAllister et al. 1974). The VHS rhabdovirion, genus *Novirhabdovirus*, is an enveloped, bullet-shaped particle about 180 nm long and 60 nm in diameter (Haghighi et al., 2008c).

The complete virion consists of five structural proteins and contains a single segment of single-stranded RNA (de Kinkelin & Scherrer, 1970; McAllister, 1974; de Kinkelin, 1983). Infected rainbow trout fingerlings frequently experience losses exceeding 90% due to VHS. When fish are exposed for the first time at advanced ages, the resulting disease is more chronic and has a lengthy progression. In cold water under crowded and stressful conditions, losses can be significant, even among mature fish. The disease is transferred via touch and from fish to fish through water (Rasmussen 1965). Transmission from parent to offspring is believed to occur as a contaminant during spawning processes, rather than through the virus residing within the egg (Roberts 1978). VHS outbreaks are predominantly prevalent and severe throughout the winter season. Losses diminish in spring as water temperatures increase, halt during summer, and then reemerge intermittently in fall. Similar to other viral infections, there is no treatment available for VHS. Avoidance is the sole effective control strategy. (Ghittino 1965).

2. ETIOLOGY

A variety of clinical signs and histopathological changes may be apparent in fish infected with VHS virus. Some fish show frank clinical manifestations of disease, whereas others appear healthy. Historically, clinical and pathologic signs of VHS have been catalogued into acute, chronic, and latent forms. Acute signs are typically accompanied by a rapid onset of heavy mortality, whereas the later or chronic stage is associated with lower mortalities, which occur over a protracted time period. It has also been reported that apparent virus carriers can demonstrate no clinical signs of VHS (Roberts, 2001). Disease signs of VHS include; fish that appear lethargic, dark in colour, exophthalmic, and anaemic. Haemorrhages are often evident in the eyes, skin, gills and at the bases of the fins. Internally, profuse haemorrhaging can be found in periocular tissues, skeletal muscle, and viscera, while congestion of the liver tissue and necrosis of the haematopoietic tissues can also be found (Bruno & Poppe, 1996).

Horizontal transmission of VHSV through contact with infected stocks or VHSV contaminated water supplies has been associated with several VHS outbreaks in European salmonid aquaculture. Other potential transmission routes include; viruscontaminated nets, boots, egg crates or other equipment. Feeding cultured fish with wet fish food prepared from contaminated fish stocks can also transmit the virus. Such practices are considered high risk and should not be encouraged (Roberts & Shepherd, 1974). At present there is no evidence to suggest that VHSV can be vertically transmitted from parent to progeny.

There is no cure for VHS. Therefore, the best methods for the prevention of disease are based on prohibiting the use of the wet fish foods, minimising fish farm escapes, and ensuring that appropriate biosecurity practices are applied on all aquaculture facilities (Haghighi et al., 2007). In the case of a VHS outbreak on a farm, the site should be fallowed immediately and all holding ponds/tanks should be dried and disinfected.

Disinfection should be performed at least three months before restocking of the aquaculture facility (Ahne & Thomsen, 1985).

3. THE SPECIES AFFECTED FROM VHS

VHS is principally a disease of farmed rainbow trout, but most salmonid fish are considered susceptible, as are whitefish (Coregonus spp), grayling and pike. The disease has also been reported in farmed turbot, wild Pacific herring and numerous other marine fish species (Jensen, M.H. 1965).

4. GEOGRAPHIC DISTRIBUTION

Research samples were collected from farmed rainbow trout populations that had symptoms of VHS disease. Fish were sampled between October 2004 and October 2006, from 100 rainbow trout sites in 10 provinces in Iran. Samples for diagnostic screening were aseptically collected from liver, kidney, spleen, bronchia, heart, intestine, and pancreas of rainbow trout with symptoms of VHS disease and all were screened by pathological Technique. The first confirmed VHS disease outbreaks occurred in Iranian aquaculture in November 2005 at Rodsar of Gilan province in the north of the country. In this study of 100 rainbow trout fish farms, 15 positive sites were identified by Histopathology. The results of this study reveal a potential high frequency of VHS virus in some centers of rainbow trout aquaculture in Iran and therefore the control and diagnosis of VHS disease is vitally important to the development of the Iranian rainbow trout aquaculture industry. This is the Important study of VHS infection in rainbow trout aquaculture sites in Iran and highlights the need for routine diagnostic screening of aquaculture facilities to control the spread of VHS disease.

5. TRANSMISSION

The disease is transmitted horizontally through contact with infected fish or water. Large numbers of virus are shed in the faeces, urine and sexual fluids. There is no vertical transmission of the virus (Amend, D.E 1975).

There is no treatment for VHS. As a notifiable disease there is a legal obligation to report any suspected outbreaks of VHS to the Fish Health Inspectorate.

If the disease were found on a farm, movement restrictions would be applied to all farms on the same river catchment and all contact sites. Attempts would be made to eradicate the disease and all contact sites would be investigated to look for evidence of the source and spread of the virus (Bellet, R. 1965).

The approved status of any infected area would be lost until a testing programme had been undertaken to confirm the eradication of the disease.

6. CLINICAL SIGNS

The Office International des Epizooties (OIE), Manual of Diagnostic Tests for Aquatic Animals recommend that the following methods are used for VHS diagnosis:

1. Clinical signs of VHS disease.
2. Clinical methods (grosspathology - microscopicpathology - electron microscopy
3. /cytopathology).
4. Agent detection and identification methods include; microscopic examination of histological sections, isolation of VHSV in cell culture then identification by one of the following confirmatory tests:- neutralisation assay, indirect fluorescent antibody test (IFAT), enzyme-linked antibody test (ELISA) and reverse-transcription polymerase chain reaction (OIE 2006).

Disease signs of VHS include; fish that appear lethargic, dark in colour, exophthalmic, and anaemic. Haemorrhages are often evident in the eyes, skin, gills and at the bases of the fins. Internally, profuse haemorrhaging can be found in periocular tissues, skeletal muscle, and viscera, while congestion of the liver tissue and necrosis of the haematopoietic tissues can also be found (Bruno & Poppe, 1996). Horizontal transmission of VHSV through contact with infected stocks or VHSV contaminated water supplies has been associated with several VHS outbreaks in European salmonid aquaculture. Other potential transmission routes include; viruscontaminated nets, boots, egg crates or other equipment. Feeding cultured fish with wet fish food prepared from contaminated fish stocks can also transmit the virus. Such practices are considered high risk and should not be encouraged. At present there is no evidence to suggest that VHSV can be vertically transmitted from parent to progeny. There is no cure for VHS. Therefore, the best methods for the prevention of disease are based on prohibiting the use of the wet fish foods, minimising fish farm escapes, and ensuring that appropriate biosecurity practices are applied on all aquaculture facilities (Haghighi et al., 2007). In the case of a VHS outbreak on a farm, the site should be fallowed immediately and all holding ponds/tanks should be dried and disinfected. Disinfection should be performed at least three months before restocking of the aquaculture facility (Ahne & Thomsen, 1985).

The Office International des Epizooties (OIE), Manual of Diagnostic Tests for Aquatic Animals recommend that the following methods are used for VHS diagnosis: 1. Clinical signs of VHS disease. 2. Clinical methods rosspathology – microscopicpathology – electron microscopy/cytopathology). 3. Agent detection and identification methods include; microscopic examination of histological sections, isolation of VHSV in cell culture then identification by one of the following confirmatory tests:- neutralisation assay, indirect fluorescent antibody test (IFAT), enzyme-linked antibody test (ELISA) and reverse- transcription polymerase chain reaction (OIE 2006). This manuscript describes the one of the detection of VHSV from farmed rainbow trout in Iran by Histopathological identification method.

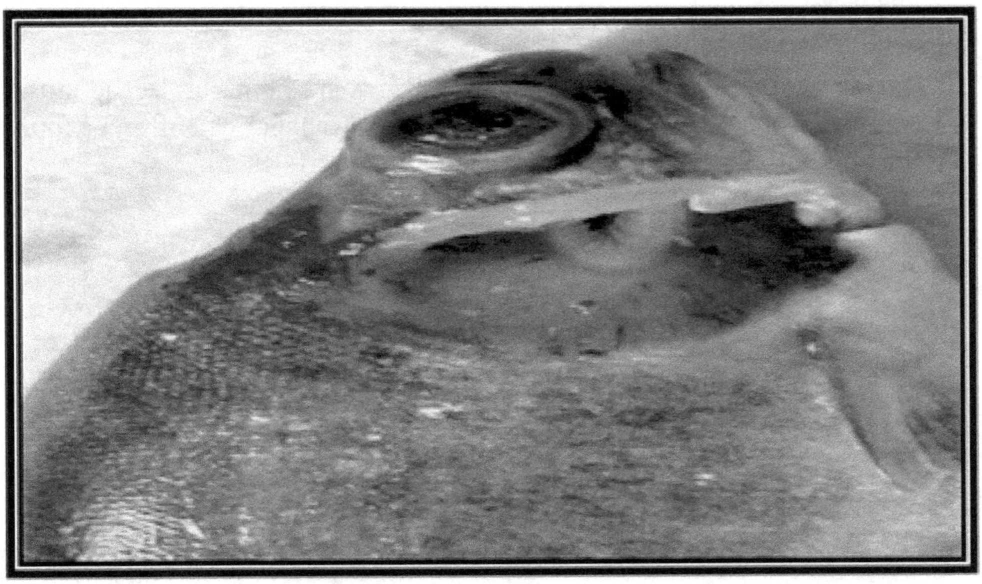

Fig. 1. An acute anemia in the gills with pale in color.

The VHS virus has also been isolated from the marine environment in the Baltic and North seas, the Atlantic Ocean and off the Pacific coast of North America.

Outbreaks of VHS in rainbow trout typically occur between temperatures of 7°C and 14°C. The disease progresses in three stages. The acute stage sees a rapid onset of high mortalities often with severe clinical signs such as darkening of body colour, exophthalmia (popeye), bleeding around eyes and fin bases, pale gills and petechial (pin point) haemorrhaging on the surfaces of the gills, viscera and in the muscle. During the second sub acute, or chronic stage, the body continues to darken, exophthalmia may become more pronounced but haemorrhaging around the eyes and fin bases is often reduced. Fish are severely anaemic and paleness is particularly evident in the abdomen. Fish may develop a spiralling swimming motion, corkscrewing around the body axis.

The final, nervous stage sees reduced mortality and clinical signs are usually absent. The corkscrew swimming motion becomes more pronounced.

External Signs:

1. Rainbow trout involved in acute outbreaks of VHS are dark in color, lethargic, and exhibit hemorrhages in the fin sockets. Exophthalmia (popeye) is common and persists throughout the course of the disease.
2. As the disease progresses, affected tish become nearly black. An acute anemia develops, and the gills are pale in color.
3. After several months, the mortality finally may cease and some of the remaining fish often display whirling behavior, erratic swimming and nervousness.

Internal Signs:

1. During acute outbreaks, small hemorrhages are common in the musculature, gills, and visceral organs. Massive hemorrhages can often be found in the abdominal cavity of freshly dead fish (Roberts 1978).

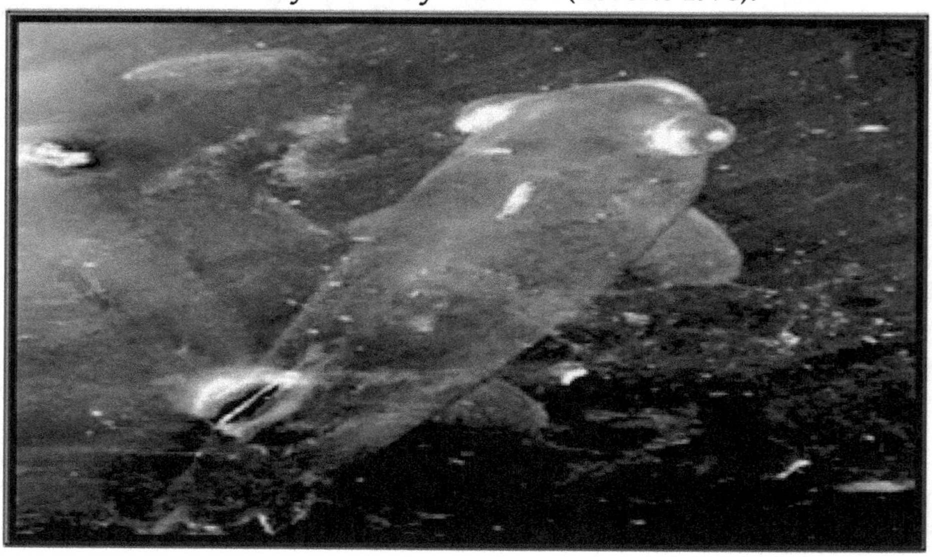

Fig. 2. Exophthalmia (popeye) with bleeding around eyes and dark in color are common signs of the disease.

2. During mid-stages of a disease outbreak, internal organs become very pale. Visceral, intramuscular, and gill hemorrhages develop as distinctive signs of the disease.
3. In late stages of the disease, kidneys become swollen and discolored.

Results

Province (state) Mazandaran Sample size Fish farms 25/Positive cases 2/Percent positive cases 8%

Province (state) Ardebil Sample size Fish farms 6/Positive cases 3/Percent positive cases 50%

Province (state) Lorestan Sample size Fish farms 8/Positive cases (-)/Percent positive cases 0%

Province (state) Markazi Sample size Fish farms 4/Positive cases 2/Percent positive cases 50%

Province (state) North Khorasan Sample size Fish farms 6/Positive cases 2/Percent positive cases 33.3%

Province (state) Kordestan Sample size Fish farms 10/Positive cases 3/Percent positive cases 30%

Province (state) Kohkiloeh Sample size Fish farms 12/Positive cases 1/Percent positive cases 8.3%

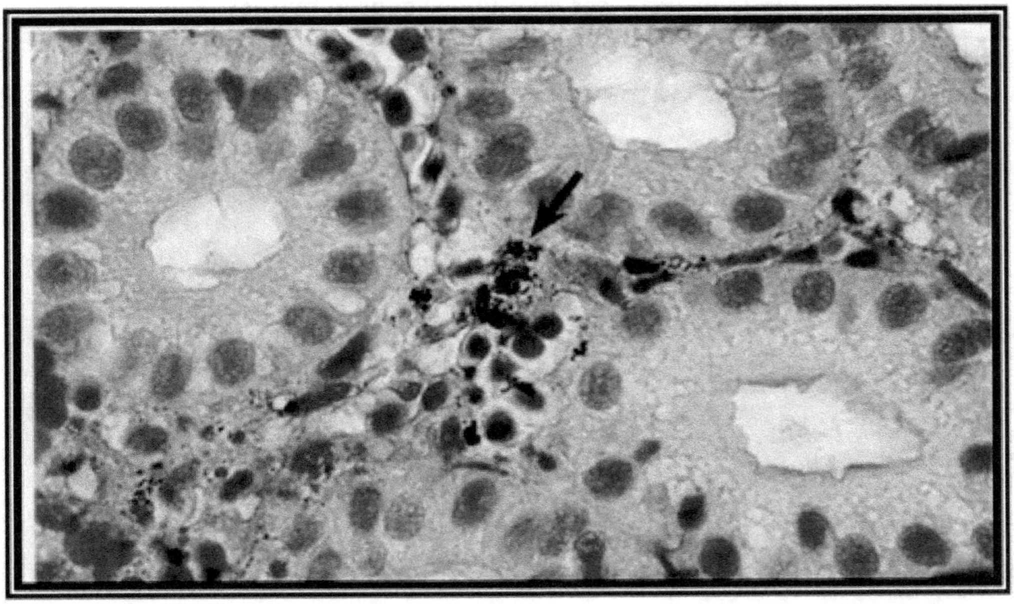

Fig. 3. The histopathological section of kidney, staining with immunoperoxidase assay technique indicates the localized virus in kidney haematopoietic tissue (arrow).

Province (state) West Azarbayjan Sample size Fish farms 8/Positive cases (-)/ Percent positive cases 0%

Province (state) Esfahan Sample size Fish farms 12/Positive cases (-)/Percent positive cases 0%

Total Sample size Fish farms 100/Positive cases 15/Percent positive cases 15%

Fig. 4. Dissection of infected tissue with the typical signs as haemorrhage in swim bladder and visceral adipose caused by VHS.

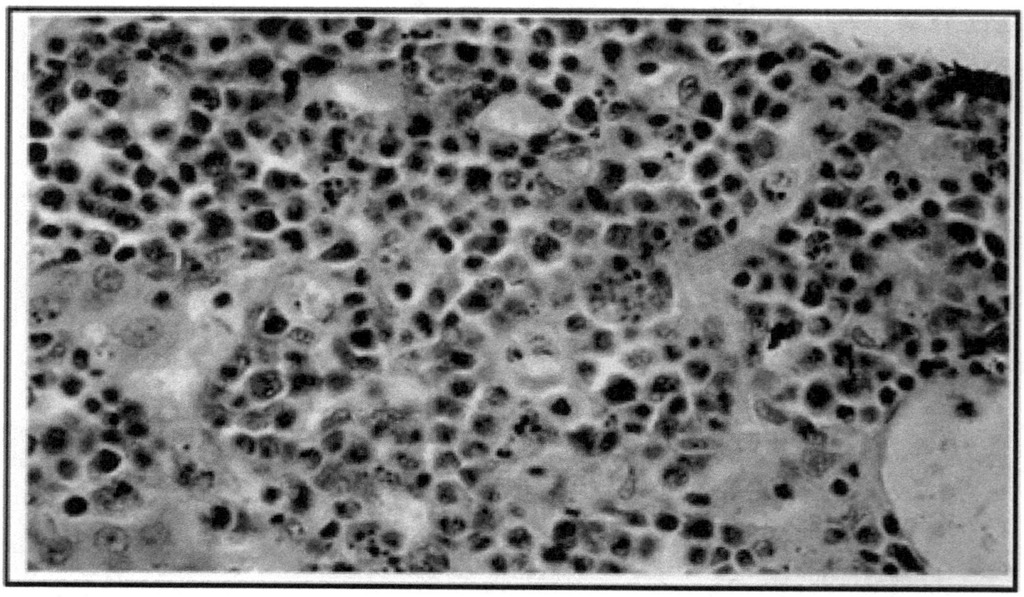

Fig. 5. The histopathological section of kidney, severe cellular necrosis with disintegrated nuclei (H&E staining).

7. LABORATORY TESTS

7.1 Sample collection

All research samples were collected from farmed rainbow trout populations that had symptoms of VHS disease. Sixty fish were sampled from each farm, producing a 95% confidence that the disease would be detected at a 5% prevalence level, between October 2004 and October 2006. In total 100 rainbow trout farms in 10 provinces in Iran were sampled. Most of the fish sampled were rainbow trout fry, weighting between 3-5 grams, however from some sites in the Gilan, Mazandaran and Markazi (Arak) provinces rainbow trout broodfish were tested for VHSV. Disease outbreaks usually occurred between the months of November and March, at lower water temperatures between 4-14°C. Tissue samples for the Pathological method was aseptically dissected from anterior kidney, liver, spleen, heart, bronchi, intestine, pancreas, and muscle. Samples for Histopathological examination were stored in saline formalin 10% (Merck, Iran). All tissues were processed within 24h of sampling.

Table 1. Sampling of Iranian rainbow trout fish farms for VHS virus by Histopathological method.

Results	Positive Cases	Negative Cases	Percent positive cases
Histopathology	15	85	15
Clinical Signs	18	82	18

7.2 Serology

One of the important Serological techniques such as the ELISA method (Haghighi et al., 2007) is also useful diagnostic tools for screening.

Serological techniques use to confirmatory tests for identification such, neutralisation assay (SNT), enzyme-linked antibody test (ELISA).

7.3 Virus isolation in cell culture

One of the important and careful methods for detection and isolation of the virus pathogen agent is the cell culture. To study the pathogenicity and to obtain isolates of VHS virus the establishment of a laboratory with virus culture equipment is necessary (de Kinkelin, 1983).

7.4 Immunohistochemistry

This method was done as standard method, for eliminating cellular peroxide, slides were kept in a room temperature in 10%H_2O_2 for 10 min, were washed three times with TBS (0.05 M Tris base, 0.15 M NaCl, pH 7.6). In order to block unspecific sites, incubation done with normal serum for 10 min (Diluted in 1:10 TBS). About 50μL of VHS monoclonal antibody (diluted by TBS 1:800) was added to slides and was kept at 37 °C for one hour. Slides were washed by TBS for 3 other times and then the conjugate (F(ab) 2 Rabbit Anti Mouse IgG-HRP) was added and kept for 30 min in 37°C. Slides were washed again for 3 times by TBS buffer and the substrate containing chromosole (DAB) Diaminobenzidine tetra hydro chloride 3) was added after 10 min slides were put in water in order to stop the reaction. About 3 to 5 min they were stained by hematoxiline (0.4 g sodium iodate, 100 g potassium alum, 100 g chloral hydrate, 2 g citric acid, 2000 ml ddH_2O) and were washed for 10 min by the water. To dehydration of tissue, slides were put in 70 % ethanol for 3 min, in 95% ethanol for 5 min, and again twice for 5 min in 100% ethanol. In order to make a transparent tissue it was put in xylol for 5 min. For microscopic study, slides were mounted and brown and golden structures in infected tissues were studied by magnifying lenses of 20, 40 and 100 (Haghighi et al., 2007).

7.5 Immunofluorescence

Indirect fluorescence Antibody test (IFAT) and Immunoperoxidase assay techniques (OIE, 2006) are important methods for screening test. In this methods by preparing of histopathological section from kidney or other tissues (spleen, liver), and staining with specialized Immunofluorescence and immunoperoxidase assay techniques indicates the localized virus in kidney haematopoietic tissue (arrow in figure 3).

7.6 Histopathology

Histopathology signs for diagnostic studies of the disease is one of the golden test. A standard tissue preparation method was applied, whereby all tissues from live or moribund fish were fixed in 10 % formalin for 24 hours. Samples were then embedded in a paraffin block and sectioned with digital microtome at 5-7 micrometer thickness. Slides were stained by haematoxilin and eosin, then treated with mounting media (Roberts, 2001). A small number of samples were confirmed VHSV positive by immunoperoxidase assay as previously described (Haghighi et al., 2007). Samples were screened by pathological technique. The results of this study are summarized in table 1. Out of 100 fish farms, 15 sites tested VHSV positive by pathology method and 18 positive cases with clinical signs that, refered to VHS for further research electron microscopy and cytopathology tests (TEM & SEM) are proposed.

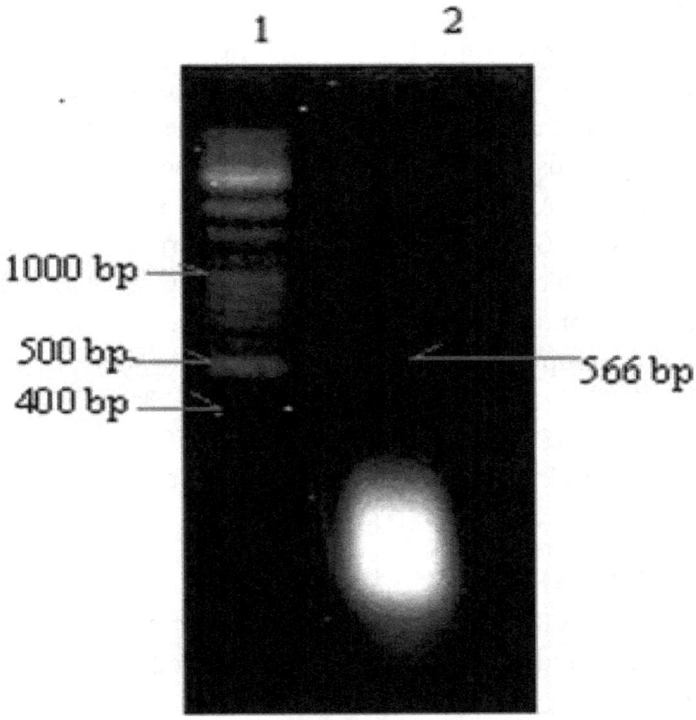

Fig. 6. Agarose gel electrophoresis. Lane 1: 100 bp ladder DNA marker. Lane 2: 566 bp PCR product related to VHS virusglycoprotein gene (agarose gel prepared on 2 %).

7.7 PCR method

The other way for diagnosis of VHS in Iran is PCR method.for this technique need **RNA extraction**, that Viral RNA extraction was done by RNXplus buffer following the manufacturers protocol (CinnaGen, Iran).

Briefly, about 1 cubic mm of fish tissue was transferred to 1.5 ml micro tube, then 200 µl RNX plus buffer was added. The mixture was incubated for 5 min at room temperature, and then 50 ml of chloroform was added and centrifuged at 12000 rpm for 15 min at 4°C.

Total tissue RNA (including viral RNA) was precipitated by ethanol, and then dissolved in 10 ml of diethyl pyrocarbonate treated water (Kazemi et al., 2004).

7.7.1 Reverse transcription reaction

Reverse transcription (RT) was performed as previously described (Pfeffer, 1988). Briefly, template RNA (1 ml) was incubated in a 20 ml reaction mixture containing: 40 pico mol of specific antisense external primer (VHS R 5‘- TTT TGG AGT CAG TTT CCT CGC G - 3‘), 100 unit of reverse transcriptase enzyme (RT) (Fermentas, Lithuania), 20 unit RNasine (Fermentas, Lithuania), 1x RT buffer, 0.2 mM dNTP, for 1 h at 42°C.

A nested-PCR was used to amplify a fragment of the viral glycoprotein gene. The first PCR reaction mixture contained 10 µl of synthesized cDNA, 1.5 mM MgCl2, 0.1 mM dNTP, 1X PCR buffer, 40 pico mol each forward and reverse primers and 1.25 unit of Taq DNA polymerase (CinnaGen, Iran). Primers for nested PCR were designed based on virus glycoprotein gene, Accession number Z93412 (Nest I primers: VHS F 5'- GTC CCA ACT CAG ATC ATC CAT C - 3' and VHS R 5'- TTT TGG AGT CAG TTT CCT CGC G - 3', amplified 617 bp of viral glycoprotein gene). The PCR was performed by 30 cycles of: denaturation at 94°C for 30 seconds, annealing at 58°C for 30 seconds and extension at 72°C for 40 seconds. 1 micro liter of PCR product was used as template DNA for the second PCR. The second PCR reaction was also performed within 30 cycles (Nest II primers: VHS2 F 5' - GCT ATC AGT CAC CAG CGT CTC – 3'and VHS2 R 5' - GGT CCT GTA ACC TGG ATC AGG – 3', amplified 566 bp of viral glycoprotein gene). Neither of the primer sets have been referenced previously (Haghighi et al., 2008b).

7.7.2 Detection of PCR product

Electrophoresis of the PCR product was performed on a 2% agarose gel, stained by ethidium bromide and DNA banding was observed by UV light under UV Transilluminator.

8. CONCISE SUMMARY

This study's results demonstrate the presence of VHS infection in rainbow trout hatcheries and broodstock locations in Iran. The severity of VHS disease is contingent upon the virulence of the viral strain, the immunological response of individual fish, stressors, and environmental circumstances associated with season, temperature, and pH fluctuations (Ahne & Thomsen, 1985). An enhanced understanding of on-site cleanliness and biosecurity protocols, along with the screening of broodstock populations for specific infections and the isolation and quarantine of diseased fish or those exhibiting abnormal behavior, has significantly contributed to curbing the development of VHS in Iran. Detecting asymptomatic carriers of VHSV is often challenging, which may result in the proliferation of infection. Consequently, the development of sensitive and precise methodologies for the detection of VHSV, such as PCR, is essential (Haghighi et al., 2008a).

Fish populations affected by VHSV. Disease occurs in certain farms of rainbow trout breeding and propagation in Iran, and histopathological techniques are effective for disease diagnosis.

Prevention necessitates the avoidance of VHS, which is the most effective control strategy (Jorgensen, P.E.V. 1977). Fish or eggs must not be sourced from regions where VHS has been identified. There is no signal for VHS. Prior to the identification of the viral etiology of VHS, numerous European veterinarians argued that the disease stemmed from dietary inadequacies. The identification of a viral agent elucidated the ineffectiveness of nutritional supplements and antibacterial treatments un mitigating losses. An eradication plan is necessary for every country experiencing an outbreak of VHS.

9. REFERENCES

Amend, D.E 1975. Detection and transmission of infectious hematopoietic necrosis virus in rainbow trout. J. Wildl. Dis. 11: 471-478.

Ahne W & Thomsen I (1985). Occurrence of viral hemorrhagic septicemia virus in wild whitefish *Coregonus* sp. Zentralbl. Veterinaermed. Reihe B 32, 7375.

Bruno DW & Poppe TT (1996). A colour atlas of salmonid diseases. Academic Press, London. ISBN 0-12-137810-1.

de Kinkelin P & Scherrer R (1970). Le Virus d'Egtved I. Stabilite, developpement et structure du virus de la souche danoise F1. *Annales de Recherches Veterinaires (Paris)* 1, 1730 (in French).

de Kinkelin P (1983). Viral haemorrhagic septicaemia. *In* "Antigens of fish pathogens" (D.P. Anderson, M. Dorson and Ph. Dubourget, Eds.), pp. 5162. Fondation Marcel Merieux, Lyon, France.

Bellet, R. 1965. Viral hemorrhagic septicemia (VHS) of rainbow trout in France. Am. N.Y. Acad. Sci. 126(l): 461-467.

Ghittino, P 1965. Viral hemorrhagic septicemia (VHS) of rainbow trout in Italy. Ann. N.Y. Acad. Sci. 126(l): 468-478.

Jensen, M.H. 1965. Research on the virus of Egtved disease. Ann. N.Y. Acad.Sci. 126(l): 422.426.

Jorgensen, P.E.V. 1977. Surveillance and eradication of diseases from hatcheries, p. 72-73. In Proc. Int. Symp. Dis. Cult. Salm., Tavolek, Inc., Seattle. WA.

Jorgensen, P.E.V. 1974. A study of viral diseases in Danish rainbow trout, their diagnosis and control. Ph.D. thesis. Danish Royal Vet. and Agri. Univ Copenhagen. 101 p.

McAllister, F.E., J.L. Fryer, and K.S. Pilcher. 1974. An antigenic comparison between infectious hematopoietic necrosis virus (OSV strain) and the virus of hemorrhagic septicemia of rainbow trout, Salmo gaivdneri (Denmark strain) by cross neutralization. J. Wild. Dis. 10: 101-103.

Rasmussen, C.J. 1965. A biological study of the Egtved disease (INUL). Ann. N.Y. Acad. Sci. 126(l): 427.460.

Roberts, R.J. 1978. The virology of teleosts, p. 128.130. In R.J. Roberts (ed.) Fish pathology, R.J. Roberts (Ed.), Balliere Tindall, London.179

Haghighi A, Soltani M, Kazemi B, Sohrabi I & Sharifpour I (2007). Use of Immunohistochemical and PCR Methods in Diagnosis of Infection Haematopoietic Necrosis Disease in some Rainbow Trout Hatcheries in Iran. *Pakistan Journal of Biological Sciences* 10(2), 230- 234.

Haghighi, A., Z. Sharifnia, M. Bandehpour and B. Kazemi, (2008a). The first report of spring viraemia of carp in some rainbow trout propagation and breeding by pathology and molecular techniques in Iran. AJAVA., 3: 263-268.

Haghighi, A., M. Bandehpour, Z. Sharifnia and B. Kazemi, (2008b). Diagnosis of viral haemorrhagic septicaemia (VHS) in Iranian rainbow trout aquaculture by pathology and molecular techniques. Bull. Eur. Assoc. Fish Pathol., 28: 171-175.

Haghighi - Khiabanian Asl, M. Azizzadeh, M. Bandehpour, Z. Sharinia and B. Kazemi (2008c). The First Report of SVC from Indian Carp Species by PCR and Histopathologic Methods in Iran, *Pakistan Journal of Biological* (PJBS) 2008 | Volume: 11 | Issue: 24 | Page No.: 2675-2678 DOI: 10.3923/pjbs.2008.2675.2678

Kazemi B, Bandehpour M, Yahyazadeh H, Roozbehi M, Seyed N, Ghotasloo A & Taherpour A (2004). Comparative study on HCV detection in Iranian patients by RT- PCR and ELISA techniques during 2001-2003. *Journal of Medical Science* 4(2), 132- 135.

The Office International des Epizooties (OIE) (2006). Manual of Diagnostic Tests for Aquatic Animals. Fifth edition, Notifiable disease chapter 2.1.5. pp. 141-158.

Pfeffer U (1988). One Tube RT PCR with sequence Specific Primers. *In* "RNA isolation and characterization protocols", (R. Rapley and D.L. Manning, Eds.), pp. 143-151. Humana Press.

Roberts RJ (2001). Fish pathology. Bailliere Tindall, London. Virology pp. 238-243 and pp. 386-387.

7 Chapter

ENHANCEMENT OF STURGEON FISH STOCKS IN IRAN

1. INTRODUCTION: AN OVERVIEW

Understanding production costs and their fluctuations is crucial for the creation and administration of a sturgeon farm, highlighting the primary areas where cost reduction is worthwhile. An analysis of production factor costs in a sturgeon hatchery may assist the manager in decision-making and adapting to changes. The production cost encompasses all expenditures incurred throughout the production process. A fundamental limitation in the advancement of sturgeon aquaculture in Iran is the absence of dependable economic data regarding inputs and outputs at the farm level, both in quantitative and monetary terms. This chapter addresses the determinants of the microeconomic structure in various sturgeon hatcheries over different years, given that sturgeon farming is presently the most significant sub-sector of aquaculture in the region and its rapid development has garnered substantial attention for stock enhancement over the past two decades.

Consequently, a thorough analysis of the economics of sturgeon aquaculture would be advantageous for both farm managers and policymakers. The aquaculture production process is influenced by biological, technological, economic, and environmental factors, and can be analyzed through the interactions of technological and biological elements within the culture environment (Bjorndal, 1990). According to Shang (1990), issues including biology, technology, feed and nutrition, engineering, fish pathology, and institutional features all influence production economics. From a microeconomic perspective, the principal motivation of a fish farm may be profit generation, while other factors, such as stock enhancement, may also be considered. Investigating the economics of sturgeon aquaculture is crucial for its future advancement. The economic assessment serves as a foundation for decision-making by farm management and for the development of governmental aquaculture and improvement programs. Economic analysis is crucial for assessing investment viability, determining resource allocation efficiency, enhancing management practices, evaluating new cultural technologies, assessing market potential, and identifying research areas with significant potential returns (Shang, 1990).

Various factors, including illegal fishing, habitat degradation, dam construction, sand extraction from riverbeds, petroleum contamination, and pollution from industrial, agricultural, and domestic sources, contribute to the deterioration of aquatic habitat quality and adversely impact fish stocks, including the sturgeons of the Caspian Sea. (Khodorevskaya *et al.*,1996, Lukyanenko *et al.*,1999, Ivanov *et al.*,1999; Salehi, 2006; Pourkazemi, 2006; and Moghim *et al.*, 2006). As a result, the harmful human impact has grown at an especially rapid rate. To overcome such devastating effects, one of the options is to construct hatcheries to produce large quantities of fingerlings for stock enhancement. Recently 10% of fingerlings were also marked. Many countries with different methods and various objectives are involved in the stock enhancement activities, or reconstruction of economically valued species. All Caspian countries are involved to protect the Sea and manage the fishermen. To assess the success of stocking, now, in Iran, the best are usually landings result. Since 90% of released fingerlings were Persian sturgeon the contribution of this fish was increased over the recent years. The fish sampling method is trawler research ship. Iran contributes to these efforts through the reproduction and enhancement of thirteen native species, releasing more than 500 million fingerlings into the Caspian Sea and the Persian Gulf annually (Bartley, 1995 & 1999; Shehadeh, 1996; Bartley and Rana, 1998; Abdolhay, 1998 & 2006; Tahori, 1998; PDD, 2006 and Salehi, 2002, 2004, 2005 & 2006). As Figure 1 shows total fingerling production of sturgeon increased from 1.1 million in 1982 to more than 24 million in 1998, declined to 21 million in 2004. The production declined to 10 million in 2005 and 13.9 million in 2006 (Salehi, 2001 & 2005 and PDD, 2006 & 2007). Over the 2000-2004, the contribution of *A. persicus* was 79% of total sturgeon fingerlings production in Iran, followed by *A. nudiventris* 7.5%, *Huso huso* 6.6%, *A. gueldenstaedti* 4.2% and *A. stellatus 3%* (Table 1). On average yearly production of *A. persicus* was more than 15 million fingerlings over the 2000-04 periods, with the maximum production of 18.4 million in 2003 from the lowest level of 12.3 million in 2002. Yearly production of *Huso huso* was 1.57 million fingerlings over the same period, however, the highest level of *H. huso* production was 2.4 million in 2002 and the lowest level was 42075 in 2003, the production of species such as *Huso huso* was mainly depend on availability of brood-stock.

Considering the background of sturgeon species stock enhancement and fishing data, it appears that the rise in the contribution of Persian sturgeon to the total catch in Iran may have been influenced by stock enhancement efforts. Keyvan (2002) observed that the existing sturgeon enhancement strategy in Iran plays a crucial role in safeguarding the Caspian Sea sturgeon. This study, from an economic perspective, could propose a developing policy to enhance the productivity and breeding processes of sturgeon hatchery production in the southern regions of the Caspian Sea, specifically in the Guilan, Mazandaran, and Golestan provinces of northern Iran, where the Iranian Fisheries Organization has invested significantly over the past two decades and requires further investment for future advancement. The disparity between the released fingerlings and the capacity of the Caspian Sea is substantial, as estimated by research groups.

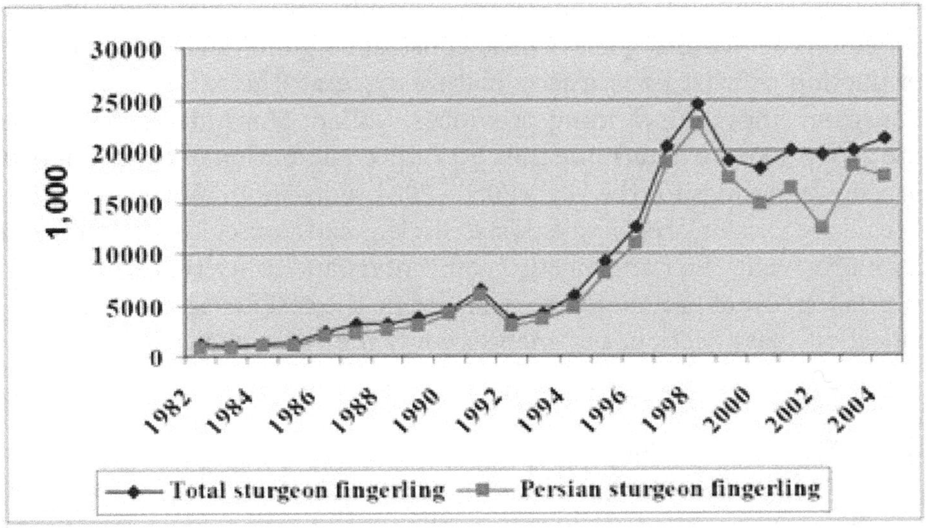

Sources: Abdolhay, 1998 & 2006, Salehi, 2005 and PDD, 2005 & 2006.

Fig. 1. Number of total fingerling production and Persian sturgeon over the 1982-2004 in Iran.

Table 1. Number of sturgeon fingerlings production by species over the 2000-04 in Iran.

Species/year	2000	2001	2002	2003	2004	Average yearly	%
Huso huso	1900919	640963	2,372,794	42,075	1570000	1,246,938	6.6
A. Persicus	13711199	16278595	12,331,354	18,420,205	17398000	15,177,803	78.7
A. stellatus	226373	820136	1,182,902	196,082	322000	635,893	3
A. nudiventris	1113826	1782914	1,178,582	1,414247	1300000	1,532,668	7.5
A. gueldenstaedti	1327480	447855	1,564,273	-	610000	1,197,307	4.2

Sources: Abdolhay, 1998 & 2006, Salehi, 2005 and PDD, 2006.

2. STUDY STRUCTURE AND METHODS

A study of fingerling production of sturgeons, input costs and the contribution of cost factors was carried out to help clarify sturgeon fingerling production costs and their difference with location and year. Most of the fe-male fish were killed to separate the eggs and released the sperms on the eggs in tanks. Data were collected for more than two decades, fingerlings were 2-3 months old and 3-5 grammas. Since the relations between the Caspian Sea countries are not developed, unfortunately, the optimal stocking protocol is not developed but totally every countries were involved to protect the Sea and manage the fishermen. Attention is also directed to addressing questions such as: which input is significant in explaining outputs from various regions or year categories? What constraints inhibit increased productivity and production of existing sturgeon culture system? The study cover the three main sturgeon fingerling farming provinces, Guilan, Mazandaran, and Golestan. Over the 2001 - 2005, to determine the costs and contribution of production factors for sturgeon fingerlings for the years 2000 - 2004, a questionnaire was prepared. An expert team comprising of economists, statisticians and aqua culturists completed the questionnaire, while referring all sturgeon centers and other related departments. The criteria for cost of production are calculating the cost for brood stocks. Though, data collection, classification, and analysis cover the years 2000 2004.

3. RESULTS AND DISCUSSIONS

Over the 2000-04, yearly production of fingerlings of sturgeon were more than 19.7 million fingerlings. On average 41% of sturgeon fingerlings production belongs to the province of Golestan, followed by 39% in the province of Guilan, and the balance were produced by Mazandaran province. In 2004, two sturgeon centers in Golestan province with the total production of 8.8 million fingerlings were the largest producer, followed by centers in Guilan province with the production of 7.8 million fingerlings, and the sturgeon center in Mazandaran province with the production of 4.7 million fingerlings. Results show, on average, yearly production of sturgeons were 8.1, 7.6 and 4 million fingerlings over the period respectively (Table 2).

Table 2. Number of sturgeon fingerlings production in hatcheries over the 2000-04 in Iran.

Province/ Year	2000	2001	2002	2003	2004	Average yearly	%	SD
Guilan	8006482	6996793	5,218,715	9,932,584	7781357	7587186	39	1708261
Mazandaran	4022885	3076093	3,457,803	4,850,203	4666528	4014702	20	760613
Golestan	6250430	9897577	10,559,387	5,258,579	8756815	8144558	41	2301952
All	18279797	19970463	19,235,905	20,041,366	21200000	19745506	100	1079547

SD: Standard deviation.

As Table 3 shows, over the 2000-04, total costs per sturgeon fingerling production was averaged Rials 1970 in the province of Guilan, varying from Rials 1224 to Rials 2655. Though, on average, total costs per fingerling production increased 27% per year over the period. Due to expansion of fingerlings production from 5.2 million in 2002 to 9.9 million by 2003, total costs per fingerling production was drastically declined by 35%. Average cost of fertilized eggs were 15% of total costs, varying from 4% in 2000 to 24% in 2004. Average cost of Labor and salary are 52% of total costs, varying from 60 % to 42% over the same period. The other main costs are the cost of 'maintenance' and 'depreciation' averaging 8% and 10% of total costs respectively. There is a little difference in the cost of feed and fertilizer, which averaged only more than 5% of total costs. As Table 4 shows, total costs per fingerling production of sturgeon was averaged Rials 1245 in the Mazandaran province, varying from Rials 477 to Rials 1958. Over the 2000-04, on average, sturgeon center in Guilan paid 59% more than the center of Mazandaran per fingerling production. Though, on average, total costs per fingerling production per year increased dramatically by 73% over the same period. Average cost of fertilized eggs were 24% of total costs, varying from 6% in 2000 to 40% in 2004. Average cost of Labor and salary were 45% of total costs, varying from 48 % to 37% over the same period. The other main costs were the cost of 'feed and fertilizer' and 'depreciation' averaging 10% of total costs. Due to production of other produts such as kutum and carp fingerlings in Mazandaran hatchery, total cost per fingerling production of sturgeon were lower than their counterparts in Guilan and Golestan provinces. As Table 5 shows, over the 2000-04, total costs per fingerling production was accounted Rials 1837 in Golestan province, varying from Rials 1027 to Rials 3358. Though, on average, total costs per fingerling production increased 46% per year over the period. Average cost of fertilized eggs were 28% of total costs, varying from 8% in 2000 to 43% in 2002 and 40% in 2004. Due to production of *Huso huso*, the hatcheries in Golestan province paid more than the average for fertilized eggs. Average cost of Labor and salary were 36% of total costs, varying from 46 % to 30% over the 2000-2004 period. The other main costs were the cost of 'feed and fertilizer' and 'depreciation' averaging 10% and 9% of total costs respectively. Overall, total costs per fingerling production of sturgeon was averaged Rials 1667 in Iran, varying from Rials 992 to Rials 2623 over the years 2000-2004. Though, on average yearly growth for the cost of a sturgeon fingerling production was accounted 50% in Iran, varying from 27% in Guilan to 46% in Golestan and 73% in the province of Mazandaran.

Table 3. Average costs (Rials[1] per fingerling) of sturgeon fingerling production over the 2000-04 in Guilan province.

Year	2000		2001		2002		2003		2004		Average	
CF	Rials	%	Rials	%	Rials	%	Rials	%	Rials	%	Rials	%
E	49	4	57	4	611	23	394	20	615	24	345	15
F&F	98	8	101	7	106	4	39	2	51	2	79	5
L&S	734	60	877	61	1301	49	946	48	1076	42	987	52
H&Ph	12	1	14	1	27	1	59	3	51	2	33	2
W&E	37	3	43	3	53	2	39	2	77	3	50	3
Main	110	9	129	9	133	5	177	9	256	10	161	8
Misc	37	3	57	4	159	6	138	7	231	9	124	5
D	147	12	158	11	266	10	177	9	205	8	191	10
TC	**1224**	**100**	**1437**	**100**	**2655**	**100**	**1971**	**100**	**2562**	**100**	**1970**	**100**
TC-D	**1077**		**1279**		**2389**		**1794**		**2357**		**1779**	

CF: Cost Factors, E: Fertilization eggs, F&F: Feed &Fertilizer, L&S: Labor & Salary, H&Ph: Harvesting & post harvest, W&E: Water and energy, Main: Maintenance, Misc: Miscellaneous, D: Depreciation and TC= Total Costs.

Table 4. Average costs (Rials per fingerling) of sturgeon fingerling production over the 2000-04 in Mazandaran province.

Year	2000		2001		2002		2003		2004		Average	
CF	Rials	%	Rials	%	Rials	%	Rials	%	Rials	%	Rials	%
E	29	6	38	8	518	29	551	36	783	40	384	24
F&F	105	22	85	18	18	1	61	4	98	5	73	10
L&S	229	48	250	53	840	47	658	43	724	37	540	45
H&Ph	14	3	9	2	54	3	15	1	39	2	26	2
W&E	14	3	14	3	18	1	15	1	39	2	20	2
Main	19	4	19	4	71	4	61	4	78	4	50	4
Misc	10	2	9	2	89	5	31	2	59	3	40	3
D	57	12	47	10	179	10	138	9	137	7	112	10
TC	**477**	**100**	**471**	**100**	**1787**	**100**	**1530**	**100**	**1958**	**100**	**1245**	**100**
TC-D	**420**		**424**		**1608**		**1392**		**1821**		**1133**	

CF: Cost Factors, E: Fertilization eggs, F&F: Feed &Fertilizer, L&S: Labor & Salary, H&Ph: Harvesting & post harvest, W&E: Water and energy, Main: Maintenance, Misc: Miscellaneous, D: Depreciation and TC= Total Costs.

Average cost of Labor and salary were accounted 44% of total costs, varying from more than 50% in 2000 to 36% in 2004 (Table 6). Average cost of fertilized eggs were accounted 22% of total costs, varying from 6% in 2000 to 35% in 2004. The other major inputs costs were the cost of 'feed and fertilizer', depreciation and maintenance averaging 8%, 9% and 6% of total costs respectively. Over the 2000-2004, among the operation costs factors the highest variability belongs to the fertilized eggs followed by labor and salary. As Tables 3, 4 and 5 Show, the cost per sturgeon fingerling production in Guilan is higher than the other provinces, at Rials 1,970 followed by Golestan with Rials 1, 837, and only R 1,242 in Mazandaran.

1 On average, 1US$= 8,000 Rial over the 2000-04.

Table 5. Average costs (Rials per fingerling) of sturgeon fingerling production over the 2000- 04 in Golestan province.

Year	2000		2001		2002		2003		2004		Average	
CF	Rials	%	Rials	%	Rials	%	Rials	%	Rials	%	Rials	%
E	82	8	73	10	1055	43	557	38	1343	40	622	28
F&F	175	17	73	10	250	5	65	9	269	8	166	10
L&S	472	46	342	47	805	30	389	29	1007	30	603	36
H&Ph	51	5	36	5	56	3	39	2	101	3	57	4
W&E	31	3	22	3	56	3	39	2	101	3	50	3
Main	62	6	51	7	167	4	52	6	134	4	93	5
Misc	41	4	44	6	167	3	39	6	168	5	92	5
D	113	11	87	12	222	9	117	8	235	7	155	9
TC	1027	100	728	100	2775	100	1296	100	3358	100	1837	100
TC-D	914		641		2553		1179		3123		1682	

CF: Cost Factors, E: Fertilization eggs, F&F: Feed &Fertilizer, L&S: Labor & Salary, H&Ph: Harvesting & post harvest, W&E: Water and energy, Main: Maintenance, Misc: Miscellaneous, D: Depreciation and TC= Total Costs.

Table 6. Average costs (Rials per fingerling) of sturgeon fingerling production over the 2000- 04 in Iran.

Year	2000		2001		2002		2003		2004		Average	
CF	Rials	%	Rials	%	Rials	%	Rials	%	Rials	%	Rials	%
E	60	6	66	7	561	32	629	31	918	35	447	22
F&F	159	16	112	12	53	3	101	5	131	5	111	8
L&S	506	51	506	54	736	42	811	40	944	36	701	44
H&Ph	30	3	28	3	35	2	41	2	52	2	37	2
W&E	30	3	28	3	35	2	41	2	79	3	43	3
Main	60	6	66	7	70	4	122	6	157	6	95	6
Misc	30	3	37	4	88	5	101	5	157	6	83	5
D	119	12	103	11	175	10	183	9	184	7	153	10
TC	992	100	937	100	1753	100	2028	100	2623	100	1667	100
TC-D	873		834		1578		1845		2439		1514	

CF: Cost Factors, E: Fertilization eggs, F&F: Feed &Fertilizer, L&S: Labor & Salary, H&Ph: Harvesting & post harvest, W&E: Water and energy, Main: Maintenance, Misc: Miscellaneous, D: Depreciation and TC= Total Costs.

Average cost of Labor and salary were accounted 44% of total costs, varying from more than 50% in 2000 to 36% in 2004. Average cost for fertilized eggs were accounted 22% of total costs, varying from 6% in 2000 to 35% in 2004. The other major inputs costs were the cost of 'feed and fertilizer', depreciation and maintenance averaging 8%, 9% and 6% of total costs respectively. Over the 2000-2004, among the operation costs factors the highest variability belongs to the fertilized eggs followed by labor and salary.

As Figure 2 shows, total costs per fingerling production of sturgeon was averaged $US 0.21 in Iran, varying from $US 0.12 to $US 0.33 over the years 2000-2004. Though, on average, total costs per fingerling production of sturgeon was increased %175 over the period. The main reason for the growth of a sturgeon fingerling cost might be the effects of inflation rate in Iran, which affected all operation costs factors and the growth of the price of fertilized eggs, which affected by the global growth price of caviar in recent years.

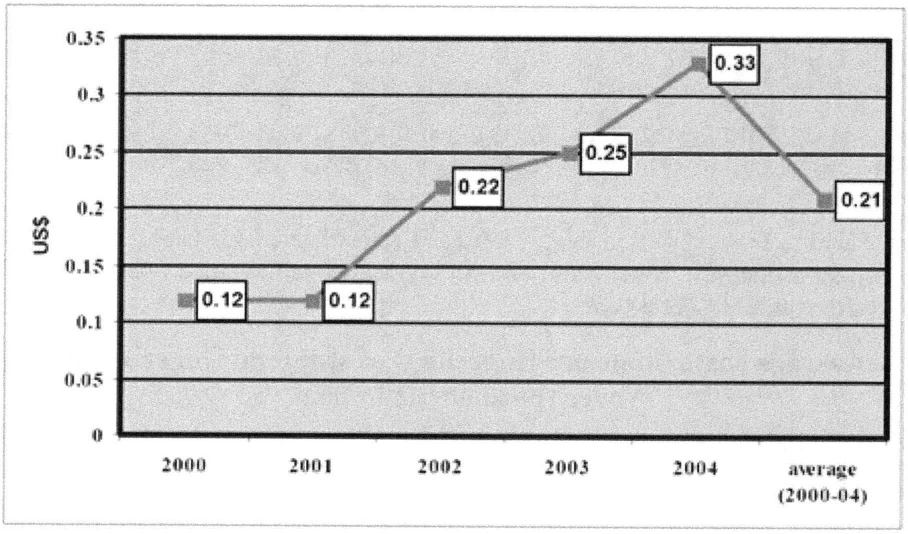

Fig. 2. Average costs ($US) of a sturgeon fingerling production for the years 2000-04 in Iran.

Comparing with other aquaculture activities, the share of labor and salary in hatcheries are very high, which noted by Salehi (Salehi, 1999, 2002 & 2005) for carp farming 12%, trout farming 13%, shrimp farming 17% and shrimp hatcheries due to using foreign experts 26% and sturgeon farming in USA were accounted 12% (Katherine *et al.*, 1985). It seems, the main reason for this higher labor and salary cost, could be justify by inactivity of hatcheries during almost 6 months off season, which could be reduced by adopting extra activities in such hatcheries. The costs sensitivity of hatcheries production of sturgeon shows labor and salary is the most sensitive, followed by fertilized eggs (Figure 3).

Although the total sturgeon catch in the South Caspian Sea has diminished, the proportion of Persian sturgeon has increased, reflecting the beneficial impact of fingerling releases by the Fisheries of Iran over the past two decades. Furthermore, the rise in the contribution of sturgeon fish recorded by the beach seine net cooperatives indicates a good impact of stock enhancement (Abdolhay 1998, 2006; Tahori 1998; Fadaee 2002; Pourkazemi 1999, 2006; Moghim et al. 2006). This study addresses the significance of stock restoration broadly, with a specific focus on sturgeon augmentation as a method for biodiversity preservation and a catalyst for economic activity. The ongoing production and improvement of sturgeon fingerlings, along with substantial investments made by the Fisheries Organization of Iran, indicate that this industry is likely to gain significance in the forthcoming years. The future production of sturgeon fingerlings will significantly depend on the capacity to acquire broodstock from the Caspian Sea and the operational cost potential of Iranian Fisheries. The sturgeon rehabilitation sector could gain from research focused on creating technically feasible production and enhancement systems, as previously, alongside advancements in nutrition, genetic enhancement, illness avoidance, water quality, and industry management. The collaboration among beach seine net cooperatives and relevant organizations in Iran, along with sustained cooperation with other coastal nations in the Caspian Sea, is anticipated to significantly impact stock enhancement and biodiversity conservation of sturgeon fish in the forthcoming years.

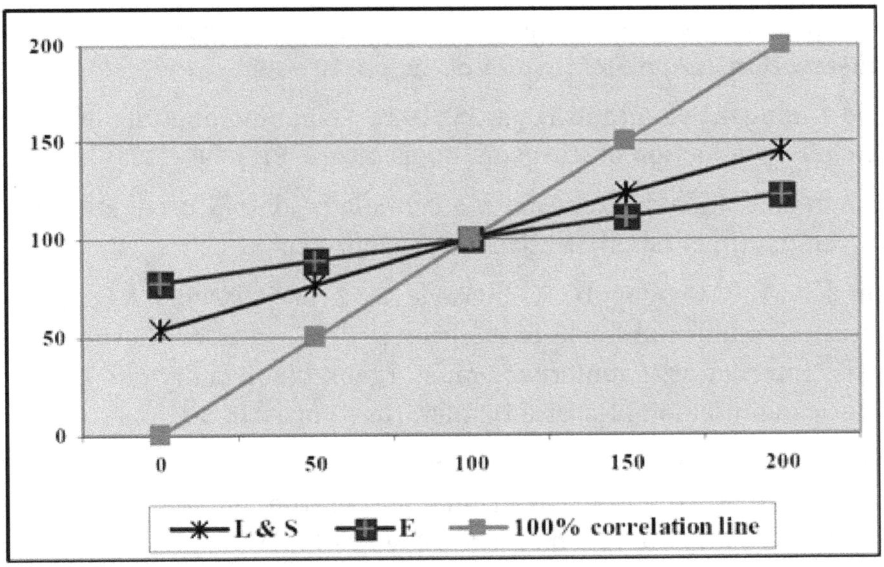

Fig. 3. The cost sensitivity of fingerling production of sturgeons for the years 2000-04 in Iran.

4. REFERENCES

Abdolhay H. 1998, Artifical hatching for stock rehabilitation, 7th National Fisheries conferences, Responsible Fisheries, (in Persian), Shilat, Tehran, Iran, p187-205.

Abdolhay H. 2006, Fingerling production and release for stock enhancement in the Southern Caspian Sea: an overview, Journal of Applied Ichthyology, Dec. 2006- Vol. 22, P 125-131.

Bartley M.D. 1995, Marine and coastal area hatchery enhancement programmes: Food security and Conservation of Biological Diversity, Paper prepared for Japanese/FAO Conference on Fisheries and Food security Kyoto, Japan, Dec. 1995, p15.

Bartley M.D. and K. Rana, 1998, Stocking inland waters of the Islamic Republic of Iran, FAO, 5p.

Bartley, M.D. 1999, Marine ranching: current issues, constraints and opportunities. In Marine ranching; FAO Fish. Circ., No. 943, pp28-43.

Bjorndal T. 1990. The economics of salmon aquaculture. Blackwell scientific. publications. London. UK. 118p.

Fadaee B. 2002, The survey of probability role beach siene net on sturgeon resources, ISRI., (in Persian),Rasht, Iran, 25p.

Ivanov, V.P.; Vlasenko, A.D.; Khodorevskaya, R.P. and Raspopov, V.M., 1999. Contemporary status of caspian sturgeon (Acipenseridae) stock and its conservation, J. Appl. Ichthyol. Vol. 15, pp. 103-105.

Katherine J. Shigekawa S. and Logan S. 1985, Economic analysis of commercial hatchery production of sturgeon. Aquaculture, 51, p299-312.

Keyvan A. 2002, Introduction on biotechnology of cultured sturgeon, Azad University, (in Persian), Lahijan, Iran, 270p.

Khodorevskaya R.P., Dovgopol G.F., Zhuravleva O.L. and Vlasenko A.D. 1996, Present status of commertial stocks of sturgeon in the Caspian Sea basin, Proceedings of the international conference on sturgeon biodiversity and conservation, American Museum of Natural History, New York p28-30.

Lukyanenko, V.I.; Vasilev, A.S.; Lukyanenko, V.V. and Khabarov, M.V., 1999. On the increasing threat of extermination of the unique caspian sturgeon population and the urgent measures Moghim, M., D. Kor, M., Tavakolieshkalak, and M. B. Khoshghalb, 2006: Stock status of Persian Sturgeon along the Iranian coast of the caspian Sea, Journal of Applied Ichthyology, Dec. 2006- Vol. 22, P 99-107.

PDD, 2005, Annual report of Shilat, Shiat, (in Persian),Tehran, Iran, 40p.

PDD, 2006, Fishing data information, Shiat, (in Persian, unpublished),Tehran, Iran, 35p.

PDD, 2007, Annual report of Shilat, Shiat, (in Persian, unpublished),Tehran, Iran, 65p.

Pourkazemi, M. 1999, Management and enhancement of sustainable resource, Aquaculture Department, (in Persian), Shilat, Tehran, Iran, No. 18. p17-30.

Pourkazemi, M. 2006. Caspian Sea sturgeon observation and fisheries: past, present and future, Journal of Applied Ichthyology, Dec. 2006. Vol. 22, P 12-16.

Salehi H. 1999, A strategic analysis of carp culture development in Iran, PhD.Thesis, University of Stirling. Stirling, UK, 328p.

Salehi H. 2001, Economics analysis of sturgeon fingerlings production in Iran, Paper presented in Second national conferences, Zibakenar, Rasht, Iran (unpublished in Persian), 5p.

Salehi H. 2002, Economics Analysis of Kutum (*Rutilus frisii kutum*) Fingerling Production and Releasing In Iran, (unpublished in Persian), 10p.

Salehi H. 2004, An economic analysis of carp culture production costs in Iran, Iranian Journal of Fisheries Sciences, IFRO., Tehran, Iran, p1-24.

Salehi H. 2005, An economic analysis of fingerling production of sturgeon in the south Caspian Sea over the years 2002-03, 5th International Symposium on sturgeon, Ramsar, 9-13 May., Iran. p.319-324.

Salehi H. 2006, Comparative Economics of Kutum (*Rutilus frisii kutum*) Fingerling Production and Releasing over the 2001-2003 in North of Iran (submitted for Pajohesh and sazandagi magazin), 13p.

Shehadeh, Z. H. 1996, Major trends in global aquaculture production and summary overview of the Gulfs (Persian Gulf and Gulf of Oman) area (1984 to 1994), TOFC Committee for development and management of the fishery resources of the Gulfs, Cairo, Egypt, 1-3 October, 8 p.

Tahori B. H. 1998, Sturgeon hatching in South Caspian Sea, 7th National Fisheries conferences, Responsible Fisheries, (in Persian), Shilat, Tehran, Iran, pp221-244.

Chapter 8

ENVIRONMENTAL IMPACT OF UTILIZING CONTAMINATED WATER IN AQUACULTURE

1. INTRODUCTION: AN OVERVIEW

Water and soil contamination is one of the most perilous threats to the environmental conditions in arid and semi-arid nations such as Egypt. Owing to the scarcity of freshwater for irrigation, nations in arid and semi-arid regions are compelled to utilize marginal waters for agricultural and aquacultural purposes. The utilization of low-quality water poses significant risks to the environmental condition. Moreover, nations such as Egypt are encountering significant challenges in eliminating untreated wastewater and industrial effluents, as well as drainage water. The discharge of such waters into drains and lakes will significantly disrupt the ecosystem and the environment overall. The utilization of these waters for irrigation or aquaculture will exacerbate the problem due to the scarcity of freshwater resources. The utilization of contaminated water in aquaculture poses a significant environmental threat to land and groundwater. The water level in aquaculture facilities exceeds the original land elevation. As a result, water from fish farms flows to the neighboring land, potentially causing issues if the water is contaminated.

The Bahr El-Baqar drain is regarded as one of the most contaminated drains in Egypt. (Abdel-Shafy & Aly 2002). It receives and carries the greatest part of wastewater (about 3 BCM/year) into Lake Manzala through a very densely populated area of the Eastern Delta passing through four highly populated Governorates. Unfortunately, at the last decades, great areas on both sides of the drain were using its polluted water for irrigation and raising fish. As a polluted drain with high risk to the surrounding environment, Bahr El Baqar has received considerable concern by many scientists. Ali et al. (1993), Abdel-Azeem et al. (2007) studied the effect of prolonged use of drain water for irrigation on the total heavy metals content of south Port-Said city soils. They found that using such kind of water will cause high concentration of heavy metals in soil and plants roots and shots. Water quality, chemical composition, and hazardous effects on Lake Manzala water and living organisms caused by Bahr El-Baqar drain water has also been studied by several investigators like: Rashed & Holmes (1984), Khalil (1985) and Ezzat (1989). Special attention has been paid to the effect of environmental pollution from microbiological and toxicological points of view (Zaki 1994). Fish farms located on both sides of the Bahr El-Baqar drain are using the polluted water for rising fish since long time ago. Furthermore, many agricultural lands located in these areas are also using such kind of water for irrigation. The hazardous of such kind of water on environment is enormous. Not to mention to the fish itself coming from these farms. The fish production from these farms goes directly to the market. Consequently, the risk to the human health is very high.

Photo 1. Fish farms adjacent to Bahr El-Baqar drain in Egypt (From Hamed et al. 2011).

Hamed 2008, studied the effect of fish farms on both of soil structure and soil salinity in area near the study area of the current research and with the same type of soil (heavy clay soil). He used blue dye (food-grade Vitasyn-Blau AE 90 from Swedish Hoechst Ltd) mixed with water in order to test the soil structure properties. He found that fish farming does not contribute to decrease in the soil salinity. He concluded that increasing fish farming activities may lead to increasing soil salinity problems in agricultural lands. The results showed also that there is no evidence that soil properties are enhanced by fish farming. On the contrary, the soil nutrient state appears to be decreasing. A layer of 10 cm thickness of mud layer with cracks is formed at the surface when the farm dried (Photo 2.).

Photo 2. High amount of macropores and cracks at the mud layer accumulated at the surface In fish farm site (From Hamed 2008).

Shang et al. (1998) concluded that extensive fish farming (shrimp) rapidly depleted the soil organic matter content. Other studies (Beverage & Phillips, 1993; Deb, 1998; Flaherty et al., 1999) have reported that intensive and semi-intensive fish farming result in high volume of organic and inorganic effluents and toxic chemicals to the ecosystem that gives hyper- nutrification, eutrophication, and high soil toxicity. In Bangladesh, studies have shown that fish farming destroys the mangrove forest, increases soil acidity, salinity (farmers use sea water), and water pollution (Deb, 1998; Guimaraes, 1989; Hossain et al., 2004 Rahman, 1994). Ali (2006) examined the impact of shrimp farming on rice ecosystem in a village in southwestern Bangladesh. He reported that prolonged shrimp farming using sea water for a 5-, 10-, and 15-year period increased soil salinity, acidity, and significantly degraded the area's soil quality. It also drastically reduced the rice production and destroyed the aquatic and non-aquatic habitat inherent in the rice ecosystem.

A national project called El Salam Canal has recently finished. It relies on mixing water from the Damietta Branch of the Nile River with water from two major agricultural drains to be used for irrigation of 600,000 Feddan in the western side of Suez Canal and North Sinai. A total annual water requirement of 4.45 billion m^3 of mixed water is required to irrigate 600,000 Feddan as follows:

- 2.11 billionm3 fresh water from the Damietta branch,
- 0.435 billionm3 drainage water from the Elserw drain, and
- 1.905 billionm3 drainage water from the Bahr Hadous drain.

The 200000 Feddan located within the service area of El Salam Canal in western part of Suez Canal spill their drainage water into Bahr El Baqar drain. Some agricultural areas located near the drain use El-Salam Canal water for irrigation. Due to the lack of the canal water, fish farms in the area forced to use other source of water. Unfortunately, the most easier source is the polluted Bahr El-Baqar drain.

The objective of this chapter is to conduct an integrated environmental assessment for fish farms located adjacent to the most polluted drain in northeastern Egypt and use its water for raising fish. A comparison between the pollution level in new/old fish farms using polluted water and agricultural lands use the same kind of waters will be conducted. The comparison will include also agricultural lands using fresh water from El-Salam Canal, lands subjected to fill from the drain and moor lands which play a reference role. An investigation of the available clues of the problem include using another drain to decrease pollution will be conducted. The level of pollution in soil and water will be recorded as a result of using the polluted water from the drain for irrigation and raising fish. The problem of seepage from the drain to the adjacent fish farms or from fish farms to the adjacent lands will be investigated. In order to achieve that water and soil samples have been collected and analyzed in order to calculate the concentrations of five main heavy metals (Pb, Zn, Cd, Cu and Mn). Samples were obtained from various depths between 1m and 4m at 24 distinct locations within the study area. To do a comprehensive integrated environmental assessment of the issue, various sites have been selected in both new and old fish farms, moorlands, landfill areas, and agricultural fields utilizing contaminated drainage water and fresh water from the canal.

2. LOCATIONS OF SAMPLES COLLECTED

The field experiments were conducted by Salem et al 2011 in year 2009 in area located at the last 20 km of the Bahr El Baqar drain before it spills its water in Manzala Lake. The study area consists of fish farms, agricultural lands, lands subjected to fill from the drain and moor lands. The area is located within the service area of the national project El Salam Canal south of Port Said city.

Total of 24 boreholes were dug in 8 horizontal sections for different locations at the study area. Every horizontal section has length of 120-160 m from the Bahr El Baqar drain side and contains three boreholes. Fig 1, shows the location of the horizontal sections while Fig 2. shows the locations of the boreholes.

Four sections were taken at each side of the drain. The depth of each borehole is 2-4 m. Five of boreholes were conducted in new/old fish farms. Three boreholes were conducted near the old fish farms (near the border of the farms). Six boreholes were conducted in moor lands and seven boreholes in fill land adjacent to the drain. Three boreholes were conducted in agricultural lands; two of them in agricultural land using the polluted water and one in agricultural land using fresh canal water.

The agricultural land using fresh canal water has used polluted water from the drain for irrigation for 20 years and changed to use fresh water from the canal 5 years ago. The soil type in the study area is heavy clayey soil with more than 60% clay. Boreholes taken in moor lands will be used for comparison.

3. SOIL/WATER SAMPLES

Two soil samples were collected from each borehole. Chemical analyses were carried out within 24 hr of sampling at the laboratory. The concentrations of six heavy metals (Pb, Zn, Cu,Mg, and Cd) were analyzed for each soil sample. Heavy metals were analyzed by the total adsorbed metals method according to USEPA (1986) using atomic spectrophotometer (model PYE UNICAM SP9, England).

The fish farms and the agricultural lands in the area are supposed to use fresh water from El-Salam canal or its branches but due to water shortage especially in summer most of the lands use the drain water for irrigation or for raising fish. There is a branch drain called Sarhan drain parallel to Bahr El Baqar drain at the eastern side located 100 m far from the drain. It collects the drainage water from the area nearby and spills it again to Bahr El-Baqar drain.

Boreholes were taken by rotary boring method at 8 sections, 4 sections at each side of Bahr El Baqar drain. Every section contains three bore holes. Two soil samples were taken from every borehole. One soil samples was taken above water table and the other taken under water table. The total depth of each borehole is 2-4 m.

Table 1. shows locations and depths of boreholes. It shows also the type of land use of each borehole.

3.1 Results of water samples from the drain

Table (2) shows the concentrations of heavy metals for water samples taken from Bahr El Baqar drain at the same study area one year before the study was conducted (Hamed et al 2011). The results reflect the size of pollution of the drain water along the whole year.

3.2 Comparison due to land use

The objective of this section is to compare between new/old fish farms using polluted water and different land uses with different water quality used. The comparison will be between old fish farms, new fish farms, lands adjacent to old fish farms, agricultural lands use polluted water from the drain and agricultural lands use fresh water from Elsalam canal and fill resulted from the excavation of the drain sides and bottom and finally natural lands (moor).

Table 1. Locations and depths of boreholes and type of land uses (U: upper layer. L: lower layer) (From Salem et al. 2011).

Sector	Side	Pore hole	Distance From Bahr El-Baqar drain(m)	Symbol / no.	Depth (m)	using	Water table depth
S1	right	S1b1	60	S1b1U	1.5	Old Fish farm(10 years old)	1.65
				S1b1L	4		
		S1b2	142	S1b2U	1	land adjacent to old fish farm	1.2
				S1b2L	3		
		S1b3	162	S1b3U	1	New Fish farm(two years old)	1.3
				S1b3dL	4		
S2	right	S2b1	32.5	S2b1U	0.8	Old Fish farm(10 years old)	2.2
				S2b2L	2.5		
		S2b2	65	S2b2U	0.7	Old Fish farm(10 years old)	2.5
				S2b2L	2.75		
		S2b3	131	S2b3U	0.7	Fill from Sarhan drain(5 years old)	1.5
				S2b3L	2		
S3	left	S3b1	10	S3b1U	0.5	Fill from Bahr El-Baqar drain (15 years old)	0.7
				S3b2L	2		
		S3b2	27	S3b2U	0.5	moor	0.85
				S32b2L	1.5		
		S3b3	44	S3b3U	0.5	moor	1
				S3b3L	1.5		
S4	right	S4b1	35	S4b1U	0.5	Fill from Sarhan drain (5 years ago)	1
				S4b1L	3		
		S4b2	70	S4b2U	1	moor	1
				S4b2L	3.5		
		S4b3	106.5	S4b3U	1.5	Cultivated land using El-salam canal water	2.3
				S4b3L	3.5		
S5	left	S5b1	35	S5b1U	1	Cultivated lands using Bahr El- Baqar drain water	3.5
				S5b2L	4		
		S5b2	72	S5b2U	1	land adjacent to old fish farm	2.3
				S5b2L	3		
		S5b3	120	S5b3U	0.5	New Fish farm(two years old)	0.7
				S5b3L	3		

S6	right	S6b1	35	S6b1U	0.5	Fill from Bahr El-Baqar drain (20 years old)	0.7
				S6b2L	2		
		S6b2	67	S6b2U	0.5	moor	3.6
				S6b2L	3.75		
		S6b3	114	S6b3U	0.5	Cultivated land using Bahr El-Baqar drain	1.75
				S6b3L	2.5		
S7	left	S7b1	35	S7b1U	1	moor	1
				S7b1L	3		
		S7b2	63	S7b2U	0.8	moor	1
				S7b2L	2		
		S7b3	73	S7b3U	0.5	land adjacent to old fish farm	0.7
				S7b3L	2		
S8	left	S8b1	40	S8b1U	1	Fill from Bahr El-Baqar drain (15 years old)	2.25
				S8b1L	2.75		
		S8b2	80	S8b2U	1	Fill from Bahr El-Baqar drain (15 years old)	2.75
				S8b2L	3		
		S8b3	115	S8b3U	1	Fill from Bahr El-Baqar drain (20 years old)	1.9
				S8b3L	2.2		

Table 2. The values of heavy metals in drain water for one year (From Hamed et al. 2011).

Sites	Concentration in mg/liter												
	Year 2008				Year 2009								
	Sep.	Oct.	Nov.	Dec.	Jan.	Feb.	Mar.	Apr.	Jul.	Aug.	Sep.	Oct.	Nov.
Cu	0.151	0.022	0.204	0.245	0.062	0.005	0.045	0.106	0.083	0.014	0.012	0.008	0.011
Pb	0.749	0.235	0.273	0.195	0.088	0.030	0.287	0.041	0.033	0.053	0.009	0.031	0.066
Zn	0.139	0.549	2.066	1.438	0.431	0.333	0.688	0.031	0.095	0.199	0.024	0.028	0.051
Cd	0.069	0.061	0.017	0.062	0.101	0.199	0.032	0.128	0.001	0.025	0.021	0.016	0.016
Mn	2.573	3.935	0.084	0.065	0.281	0.56	0.278	0.519	0.064	0.010	0.011	0.125	0.816

Fig. 1. Locations of the study area (From Salem et al. 2011).

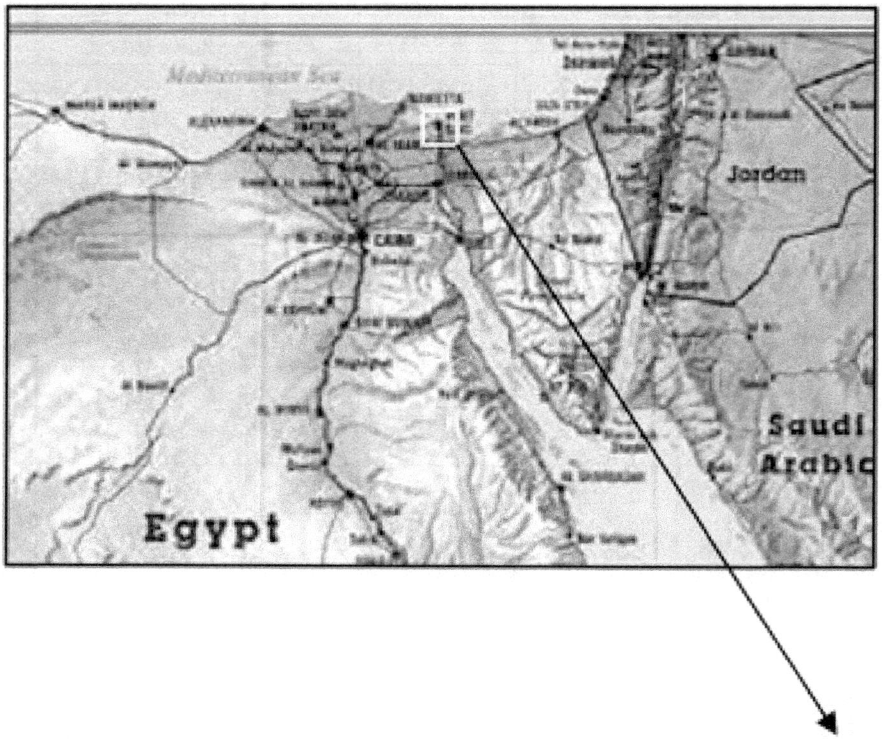

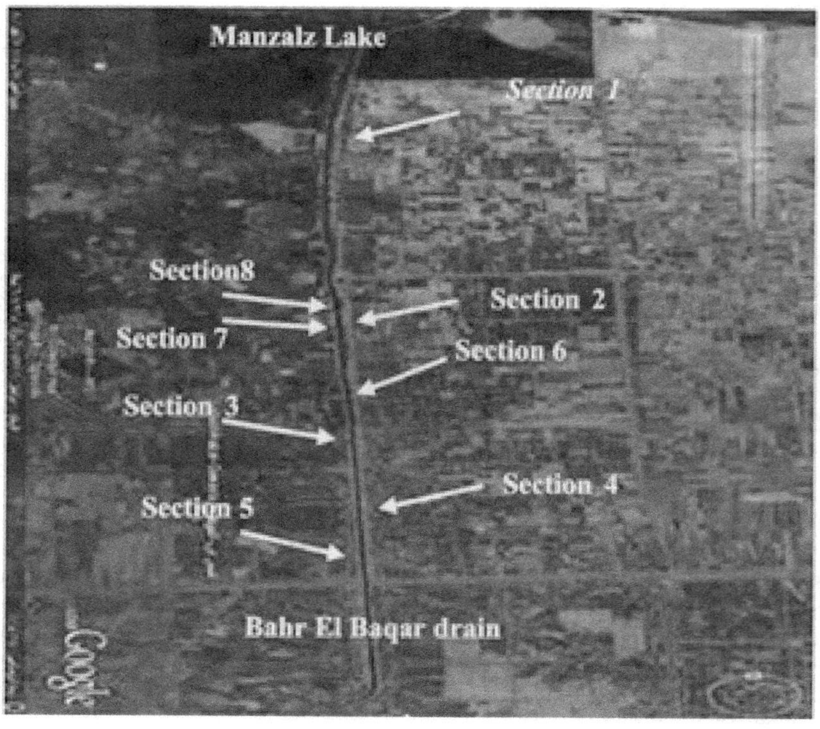

Fig. 2. Layout of the boreholes (From Salem et al. 2011).

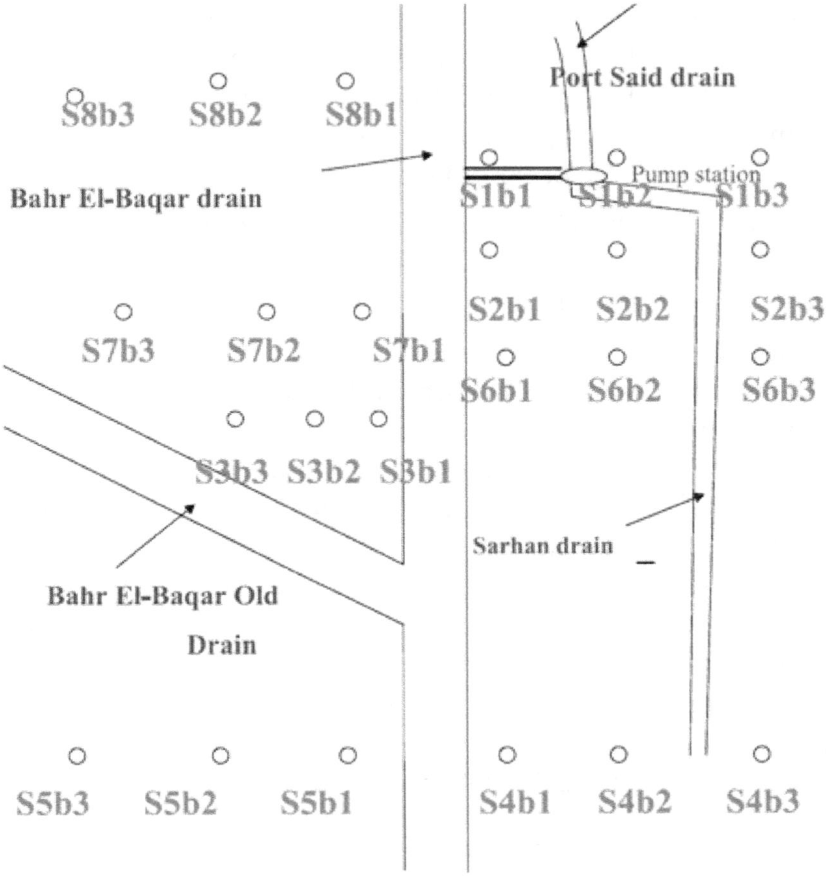

Fig (3) shows the difference between the concentrations of heavy metals in samples of top soil as a result of different land uses. The results showed that the old fish farms that used Bahr El-Baqar drain water have the highest percentage of heavy metals. And the second highest rate exists in the agricultural land irrigated with polluted water from Bahr El-Baqar drain for along time. And the third highest concentration of heavy metals is in the land adjacent to the fish farm that uses Bahr El-Baqar drain water. It indicates that the fish farms using polluted water significantly affect the neighboring land.

Next is land exposed to fill which exists only in top soil. In this case the concentration of heavy metals depends mainly on the age of the fill and its original location. There are two types of fill, fill comes from polluted Bahr-ElBaqar drain and fill comes from the agricultural minor drain (Sarhan drain). For fill from Bahr El-Baqar drain, the concentration is decreased with the increase of the fill age. The lowest concentration is for the fill has age of 20-25 years. It could be attributed to the increase in pollution of the drain bottom soil by time. The reason could be the increase in population and the increase in industry spread in the recent decades which cause increase in untreated waste water and industrial disposal spilled into the drain. In spit of the new fill comes from the relatively less polluted drain, Sarhan drain, it has a high concentration ratio of heavy metals. It is known that Sarhan drain receives drainage water from fish farms and agricultural lands using polluted water from Bahr El Baqar drain for irrigation and raising fish.

Next are the new fish farms that used Bahr El-Baqar drain water, and finally, the lowest heavy metal concentrations were found in both of the agricultural land uses fresh water from ElSalam Canal and the natural lands (moor lands).

Fig (4) shows the difference between the concentrations of heavy metals in samples of lower soil as a result of different land uses. Nearly the same results as in top soil were obtained except for that the land adjacent to the fish farms which used Bahr El-Baqar drain water has the highest percentage of heavy metals. This is a good proof that the fish farms using polluted water significantly affect the neighboring land especially in lower soil layers. The old fish farms that used Bahr El-Baqar drain water receive the second highest rate in the concentrations of heavy metal. And the third highest concentration of heavy metals is the agricultural land irrigated with water from Bahr El-Baqar drain for along time, Next is the new fish farms that used Bahr El-Baqar drain water, and the finally is the moor lands.

Fig. 3. Mean heavy metals concentrations in upper soil samples for different land uses, mg/liter (From Salem et al. 2011).

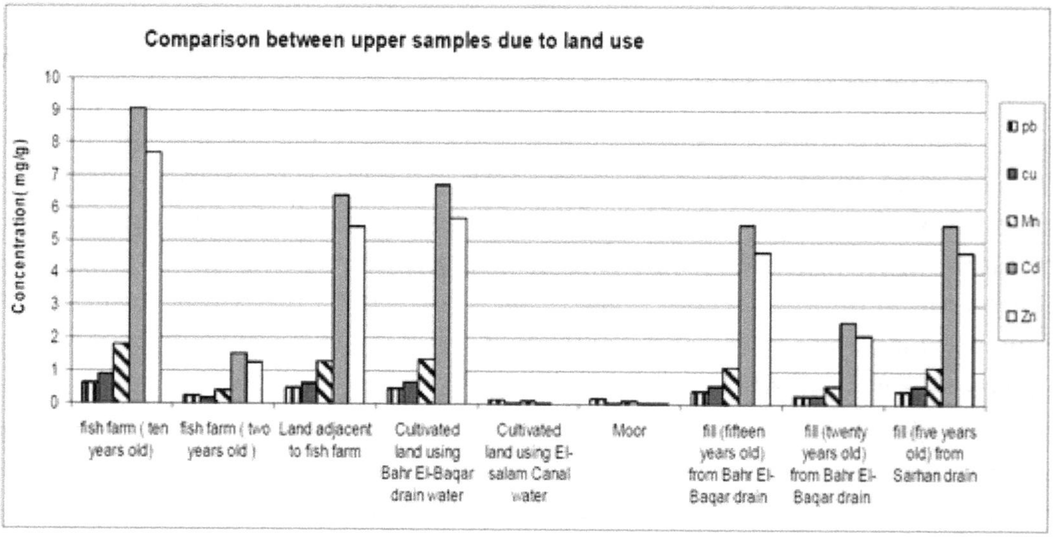

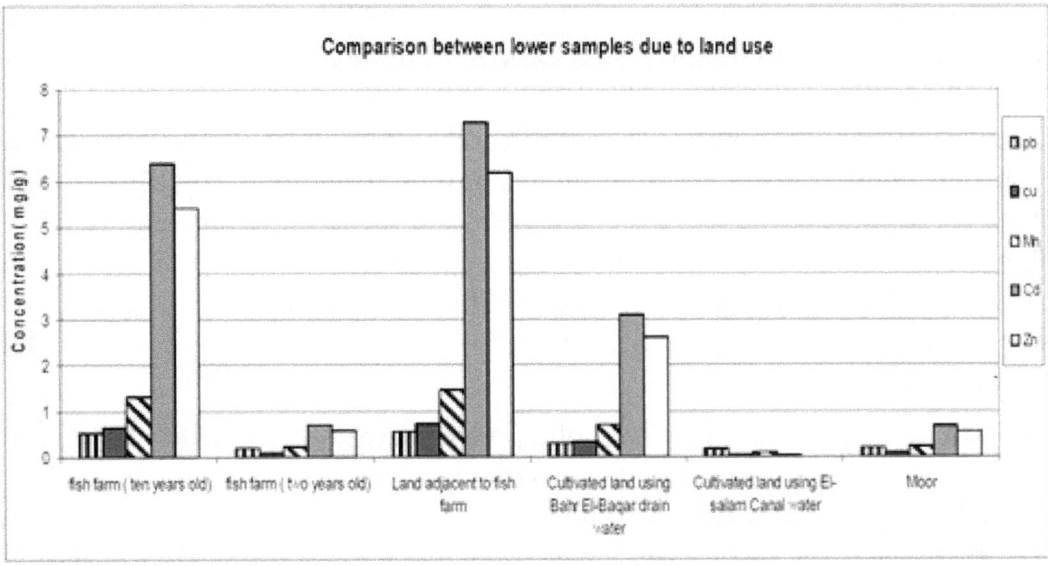

Fig. 4. Mean heavy metals concentrations in lower soil samples for different land uses, mg/liter (From Salem et al. 2011).

3.3 Comparison due to location from Bahr El-Baqar drain

In the current section, comparison of new/old fish farms and other land uses will be investigated in details for each section and each borehole. Moreover, effect of seepage from the polluted drain or from fish farms to the adjacent lands will be studied. Boreholes were dug in different distances and locations in lands with different land uses. The distances from the drain were kept nearly constants for different sections. Here, the effect of minor drain (Sarhan drain) parallel to Bahr El-Baqar drain and at 100 m far will be investigated. There is a question needs to be answered, will the minor drain work as a defend barrier for seepage from the polluted drain to the lands located on the minor drain side? Will it affect the fish farms using polluted water? For the coming section, only the data for the lower soil layer will be used in order to investigate the seepage from the drain and fish farms.

Fig (5) shows heavy metals concentrations of boreholes in section (1). It shows concentrations in boreholes S1b1, S1b2 and S1b3 located in old fish farms, land adjacent to old fish farms and new fish farms respectively. It is clear from the figure that the seepage factor from the main drain is not dominated here. The highest heavy metal concentrations are found in bore hole S1b2 (land adjacent to old fish farm) not in S1b1 the closest borehole to the main drain. The effect of the minor drain (Sarhan drain) is not clear here since borehole S1b2 has a high ratio of heavy metals concentration. It is probably due to the influence of pump station near the section. It seems that the type of land use is the dominated factor for the pollution concentration in this section.

Fig (6) shows mean heavy metals concentrations of boreholes in section (2). It shows concentrations in boreholes S2b1, S2b2 and S2b3 located in old fish farms for the first two boreholes and fill from Sarhan drain for the third one. The highest heavy metal concentrations are found in borehole S2b2. Also, in this section it is clear that the seepage factor from the main drain is not dominated. The values of concentrations in old fish farm (10 years old) 65 m far from the drain are higher than that the corresponding values in old fish farms 32.5 m away. The pollution in fill from Sarhan drain (5 years old) is lesser than that in old fish farms. It is probably due to the high pollution accumulation in old fish farms and the fill location near Sarhan drain.

Fig (7) shows heavy metals concentrations in section (3). It shows concentrations in boreholes S3b1, S3b2 and S3b3 located in fill from Bahr El Baqar drain for the first borehole and moor land for the second and the third ones. The dominated factor here is the land use factor. The higher value of concentration exists in second borehole. The ratio of concentration in this section is small compared with other sections. It is probably due to moor lands (natural land) which contain less amount of pollution and the low effect of drain seepage. In fill borehole, the effect of fill is small since the soil sample was taken in 2 m depth.

Fig (8) shows heavy metals concentrations in section (4). It shows concentrations in boreholes S4b1, S4b2 and S4b3 located in fill from Sarhan drain (5 years ago), moor land and agricultural lands using water from ElSalam canal respectively. The ratio of concentration in this section is small compared with other sections. The seepage factor here also is not dominated. The less concentration is found in agricultural lands using water from ElSalam canal. The location of Sarhan drain adjacent to the agricultural lands could be another factor contributing for lower pollution concentration in its soil.

Fig (9) shows heavy metals concentrations in section (5). It shows concentrations in boreholes S5b1, S5b2 and S5b3 located in cultivated lands using Bahr El-baqar drain water, land adjacent to old fish farm and New Fish farm (two years old) respectively. The higher concentrations of heavy metals are located in both of cultivated land with Bahr El baqar water and land adjacent to fish farms. These results reflect the bad effect of using polluted water for irrigation and the effect of fish farms on the adjacent lands. The relatively less concentration is located in new fish farms. However, only two years of fish farming using polluted water has raised the pollution concentration many times (see Fig (3) moor land and new fish farms).

Fig (10) shows heavy metals concentrations in section (6). It shows concentrations in boreholes S6b1, S6b2 and S6b3 located in Fill from Bahr El-Baqar drain (20 years old), moor lands and cultivated land using Bahr El-Baqar drain respectively. Soil sample was taken at 2 m depth in fill borehole. Consequently, the effect of fill is small. The ratios of concentration in moor land are nearly the same as that for old fill from Bahr El Baqar drain. The unexpected results here are the lower ratio of concentration for cultivated land using Bahr El-Baqar drain although the soil sample was taken from the root zone area. It is probably due to the effect of Sarhan drain adjacent to the lands. However, the ratio of concentrations still relatively high compared with land using fresh water for irrigation. Again, seepage is not a dominated factor here.

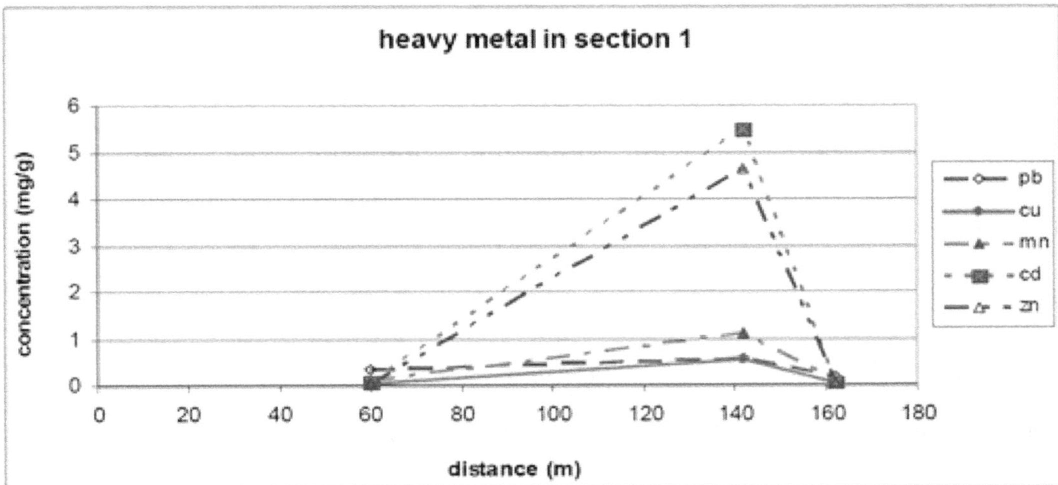

Fig. 5. Heavy metals concentrations in section (1), mg/liter (From Salem et al. 2011).

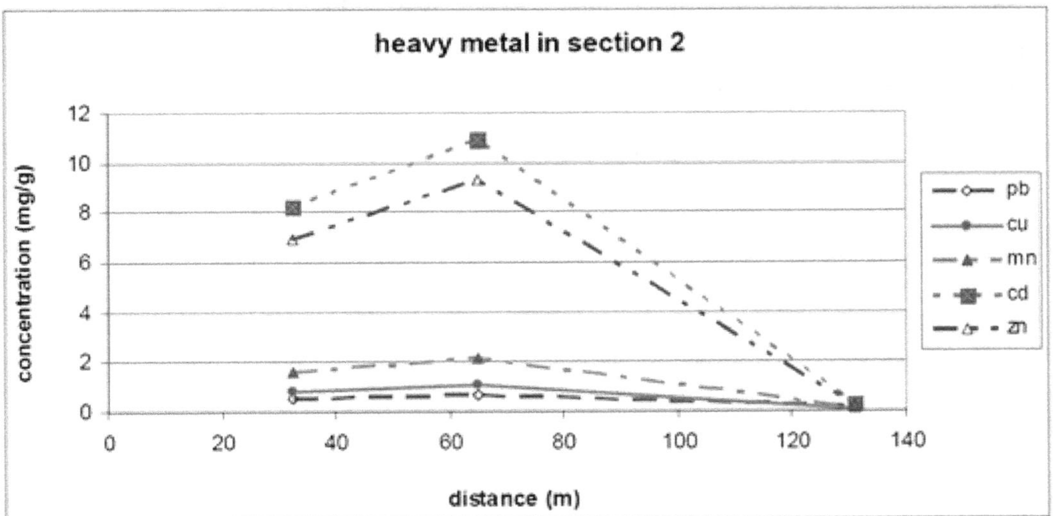

Fig. 6. Heavy metals concentrations in section (2), mg/liter (From Salem et al. 2011).

Fig (11) shows heavy metals concentrations in section (7). It shows concentrations in boreholes S7b1, S7b2 and S7b3 located in moor lands for the first two boreholes and in land adjacent to old fish farms for the third one. The effect of old fish farm on the lands nearby is quit clear here. The difference in heavy metals concentrations between land adjacent to fish farms and moor land is quite high. It reflects the damage effect of fish farms not only on its own soil but also on the soil nearby. It is probably due to the high polluted water level in fish farms which infiltrate to the adjacent lower level lands.

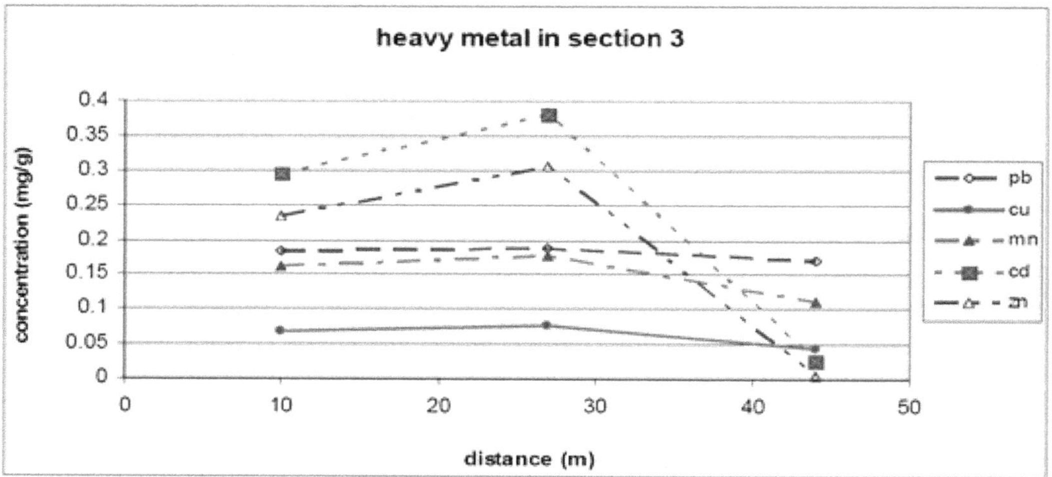

Fig. 7. Heavy metals concentrations in section (3), mg/liter (From Salem et al. 2011).

Fig (12) shows heavy metals concentrations in section (8). It shows concentrations in boreholes S8b1, S8b2 and S8b3 located in fill from Bahr El-Baqar drain for 15, 15 and 20 years old respectively. Unlike other sections, seepage factor could be effective here. Another possible reason is the existence of some fill traces in deep layers.

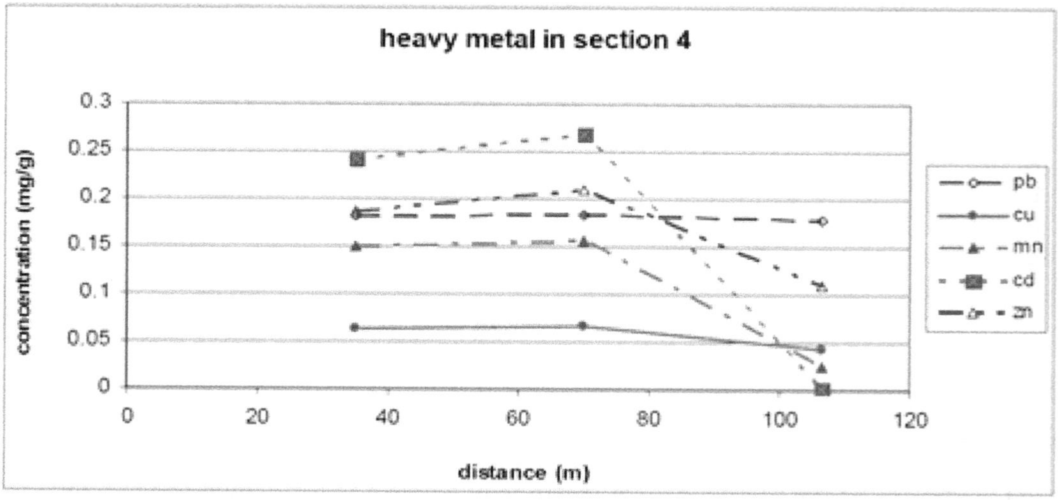

Fig. 8. Heavy metals concentrations in section (4), mg/liter (From Salem et al. 2011).

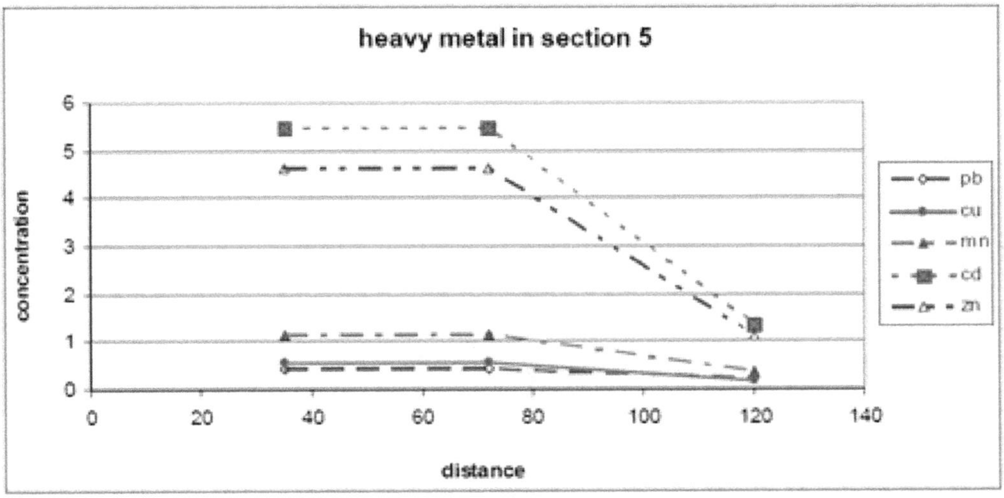

Fig. 9. Heavy metals concentrations in section (5), mg/liter (From Salem et al. 2011).

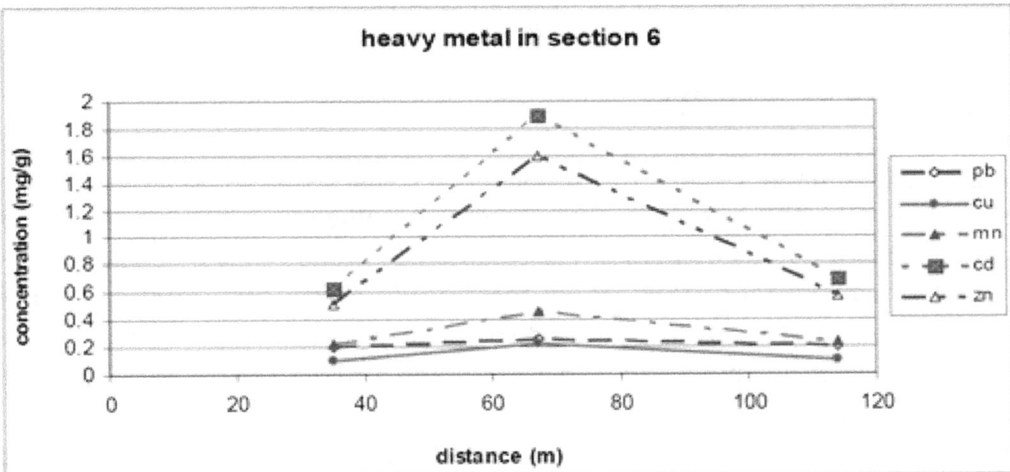

Fig. 10. Heavy metals concentrations in section (6), mg/liter (From Salem et al. 2011).

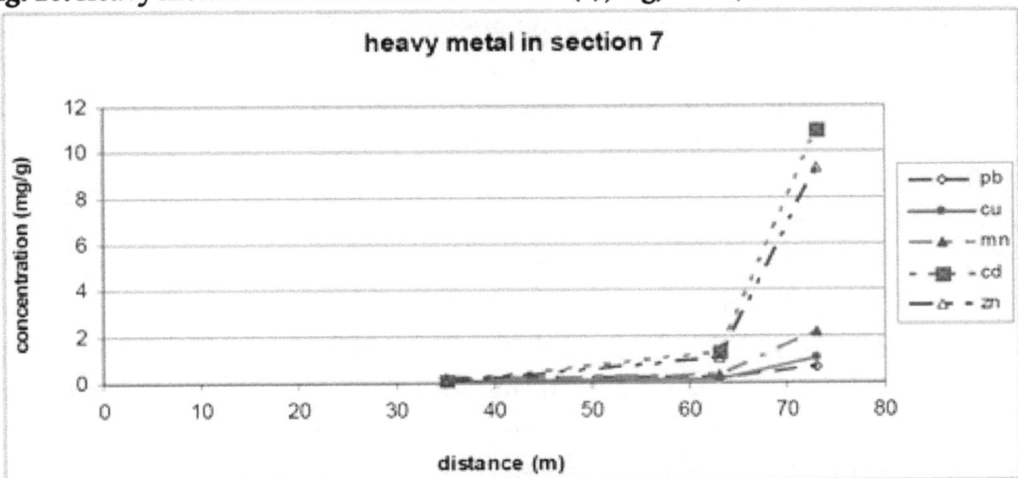

Fig. 11. Heavy metals concentrations in section (7), mg/liter (From Salem et al. 2011).

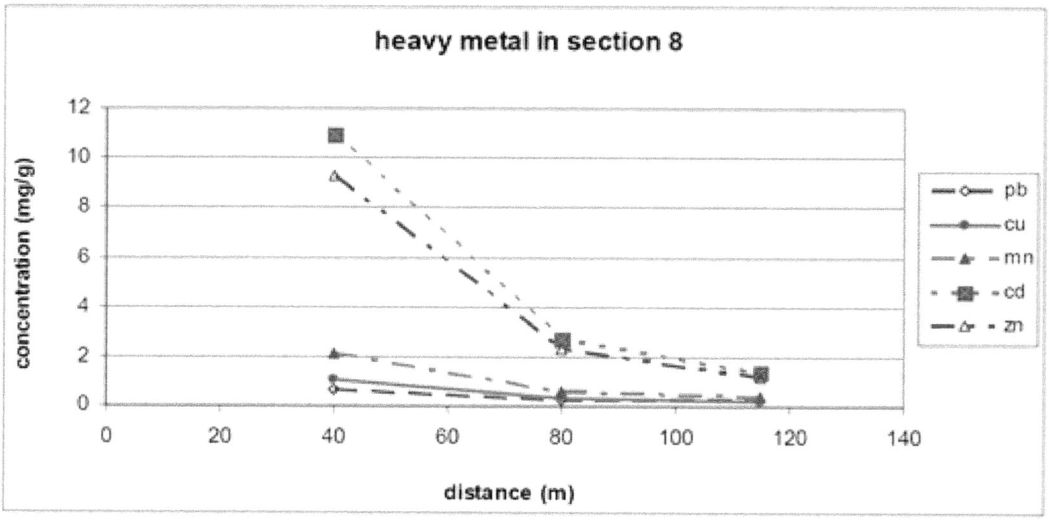

Fig. 12. Heavy metals concentrations in section (8), mg/liter (From Salem et al. 2011).

4. CONCISE SUMMARY

This chapter presents an integrated environmental assessment of fish farms utilizing contaminated water in the vicinity of Egypt's most polluted drainage system, the Bahr el Baqar drain. The chapter examines two prior studies: Salem et al. (2011) and Hamed et al. (2011). A total of 24 boreholes have been excavated in 8 distinct sections on either side of the drain to obtain soil samples at depths between 1 and 4 meters. Samples were dispatched to the laboratory to quantify the amounts of five heavy metals (Pb, Zn, Cd, Cu, and Mn). Boreholes were excavated across various land uses and distances from the contaminated drain to facilitate a comprehensive comparison between fish farms utilizing polluted water and other land uses.

The findings indicated that the most contaminated regions are the decade-old fish farms utilizing dirty drainage water for aquaculture, together with the surrounding land. It illustrates the detrimental impact of prolonged use of contaminated water for aquaculture, affecting both the farm soil and the surrounding soil close to the fish farms. Consequently, the pollution will contaminate fish sourced from these farms, thereby impacting human health. Furthermore, even in newly established fish farms less than two years old, the rise in heavy metal concentrations in the soil is significantly pronounced over a brief period.

The pollution levels in fish farms utilizing contaminated water far exceed those in agricultural fields irrigated with filthy drainage water. Both fish farms and agricultural lands have utilized dirty water from the drain for the same duration. The chapter cautions against utilizing contaminated water for irrigation in agricultural fields. It demonstrated the detrimental impact of this type of water on soil and, consequently, on plants, as the excavation depth falls inside the root zone (1-3).

This conclusion will oppose individuals advocating for the utilization of contaminated drainage water for cultivation or aquaculture.

The inflow from both Bahr ElBaqar drain and the agricultural drain (Sarhan drain) has the third highest level of pollution. The grade of the older fill from the contaminated drain surpasses that of the newer fill. The likely cause is the escalating concentration of pollutants in the sediment of the drainage system over time. The concentration differential is excessively elevated within a relatively brief timeframe. This indicates the swift decline of the environmental conditions of the Bahr El Baqar drain over time. Fish farms and agricultural fields utilizing contaminated water are discharging into the minor drain (Sarhan drain), resulting in the inflow from this drain containing elevated concentrations of heavy metals. Regrettably, numerous aquaculture proprietors utilize this type of material for constructing embankments around their facilities. They utilize the fill to elevate the level of their farm's bed soil.

For agricultural fields that have utilized contaminated water for irrigation over an extended period (20 years) and subsequently transitioned to fresh water for a relatively brief duration (5 years), the enhancement of soil quality is evident. The reduction of heavy metals resulting from the utilization of high-quality water is quite substantial. It will provide a hopeful perspective for acquiring insights into pollution in the region. Moreover, the presence of a minor drain parallel to the major polluted drain, situated at a relatively short distance (90-100m), generally mitigates pollution for the lands located downstream of the minor drain. Therefore, the local government should mandate that fish farm proprietors utilize freshwater for an extended duration to cleanse their properties prior to reusing them for aquaculture.

The Chapter disclosed that the total environmental condition in the vicinity of the drain is significantly hazardous. Five hazardous heavy metals with varying amounts have been detected in every soil sample, whether at the surface or at greater depths. This high amount of pollution adversely affects both the animals and vegetation in the area.

The Chapter ultimately disclosed that aquaculture utilizing contaminated water had detrimental effects on the surrounding ecosystem, as well as on the fish themselves. Furthermore, prior research has demonstrated that prolonged fish farming in high-quality water adversely affects soil structure and salinity. Consequently, we may envision the extent of environmental degradation resulting from the utilization of contaminated water in aquaculture.

5. REFERENCES

Abdel-Azeem, A. M; Abdel-Moneim, T. S.; Ibrahim, M. E.; Hassan, M. A. A. & Saleh, M. Y. (2007). Effects of Long-Term Heavy Metal Contamination on Diversity of Terricolous Fungi and Nematodes in Egypt - A Case Study. Water Air Soil Pollut. Journal, No. 186, pp.:233–254.

Abdel-Shafy, H. I., & Aly, R. O. (2002). Water issue in Egypt: Resources, pollution and protection endeavors. CEJOEM, 8(1), 3–21.

Ali, A.M.S., 2006. Rice to shrimp: land use/land cover changes and soil degradation in Southwestern Bangladesh. Journal of Land Use Policy 23, 421–435.

Ali, O. M.; El-Sikhry, E. M., & El-Farghal, W. M. (1993). Effect of prolonged use of Bahr El Baqar drain water for irrigation on the total heavy metals content of South Port Said soils. In: Proc. 1st Conf. Egypt. Hung. Env. Egypt, pp. 53–57.

Beverage, M. & Phillips, M., (1993). Environmental impact of tropical inland aquaculture. In: Pullin, R., Rosenthal, H., Maclean, J. (Eds.), Environment and Aquaculture. Center for Tropical Aquaculture Research, Manila, pp. 213–236.

Deb, A.K., (1998). Fake blue revolution: environmental and socioeconomic impacts of shrimp culture in coastal Bangladesh. Ocean and Coastal Management 41, 63–88.

Ezzat, A. I. (1989). Studies on phytoplankton in some polluted areas of Lake Manzala. Bulletin of the National Institute of Oceanography and Fisheries, ARE, 15(1), 1–19.

Flaherty, M., Vandergeest, P. & Miller, P., (1999). Rice paddy or shrimp pond: tough decisions in rural Thailand.World Development 27 (12), 2045–2060.

Guimaraes, J.P. de Compos, (1989). Shrimp culture and market incorporations: a study of shrimp culture in paddy fields in Southwest Bangladesh. Development and Change 20 (4), 333.

Hamed, Y. (2008). Soil structure and salinity effects of fish farming as compared to traditional farming in northeastern Egypt. Land Use Policy Journal 25(3) pp 301- 308, July 2008.

Hamed, Y., Shawky, T., Abd-Elrehim, M., ElKiki, M., Berndtsson, R. & Persson, K., (2011) Case Study: Investigation of different potential causes of pollution in Lake Manzala northeastern of Egypt. Article in Press.

Hossain, S., Alam, S.M.N., Lin, C.K., Demaine, H., Khan, Y.S.A., Das, N.G. & Roup, M.A., (2004). Integrated management approach of shrimp culture development in the coastal environment of Bangladesh. World aquaculture development in the coastal environment of Bangladesh. World Aquaculture 35 (4), 35–44.

Khalil, M. T. (1985). The effect of sewage and pollutional wastes upon Bahr El-Baqar Drain and the southern area of Lake Manzala, Egypt. Egyptian Journal of Wildlife and Natural Resources, 6, 162–171.

Rahman, A., 1994. The impact of chrimp culture on the coastal environment. In: Rahman, A.A., Haider, R., Huq, S., Jansen, E.G. (Eds.), Environment and Development in Bangladesh. University Press Ltd., Dhaka, pp. 490–524.

Rashed, I. G., & Holmes, P. G. (1984). Chemical survey of Bahr El Bakar Drain system and its effects on Manzala Lake. In: Proceedings of the 2nd Egyptian Congress of Chemical Engineering, (pp. 1–10), Cairo, Egypt, March 18–20, 1984.

Salem, Sh, Hamed,Y., Sheshtawy, A, and Ali, A(2011). Environmental assessments for areas located both sides of Bar El-Baqar polluted drain northeastern Egypt. Article in Press

Shang, Y.C., Leung, P. & Ling, B.H., (1998). Comparative economics of shrimp farming in Asia. Aquaculture 164 (1–4), 183–200.

U. S. Environmental Protection Agency, USEPA (1986). Test methods for evaluating solid waste: physical/chemical methods. SW-846. Washington, D. C.: USEPA, Office of Solid Waste and Emergency Response.

Zaki, M. M. M. (1994). Microbiological and toxicological study of the environmental pollution of Lake Manzala (108 pp). MSc Thesis, Faculty of Science, Suez Canal University, Ismailia, Egyptd

Chapter 9

PHOTOBACTERIUM DAMSELAE SUBSP. DAMSELAE, AN EMERGING PATHOGEN IMPACTING NEWLY CULTURED MARINE FISH SPECIES IN SOUTHERN SPAIN

1. INTRODUCTION: AN OVERVIEW

Aquaculture is the most rapidly expanding food production sector, fulfilling about 50% of global demand for food fish. Given the anticipated population increase over the next twenty years, it is projected that an extra 40 million tons of aquatic food will be necessary by 2030 to sustain current per capita consumption (NACA/FAO, 2001). In 2004, marine aquaculture production reached 30.2 million tons, accounting for 50.9% of worldwide aquaculture production (FAO, 2004). Fish constitutes the predominant category, accounting for 47.4% by quantity and 53.9% by value. According to the World Aquaculture Society (WAS, 2006), the future of this sector must rely on advancements in scientific and technical improvements, sustainable practices, and primarily, the diversification of cultured fish species. Consequently, the European Union has formulated an innovative strategy to enhance the cultivation of new marine fish and shellfish species while sustaining the output of established species (UE, 2010).

Marine fish farming is a very important activity of Spanish aquaculture industry. The main marine fish species intensively cultured are gilt-head seabream (*Sparus aurata*), European seabass (*Dicentrarchus labrax*), and turbot (*Scophthalmus maximus*), achieving production percentages of 47.91, 12.5 and 18.62%, respectively (MAPA, 2008). In last 7 years, several new marine fish species are being evaluated as potential candidates for aquaculture production. In Southern Spain, studies on the reproductive cycles, nutrition, growth, histology and immune system of species such as Senegelese sole (*Solea senegalensis*), redbanded seabream (*Pagrus auriga*), common seabream (*Pagrus pagrus*), white seabream (*Diplodus sargus*), and meagre (*Argyrosomus regius*) are ongoing (Cardenas & Calvo, 2003; Prieto et al., 2003; Ponce et al., 2004; Manchado et al., 2005; Fernandez-Trujillo et al., 2006; 2008; Martin-

Antonio et al., 2007; Cardenas & Manchado, 2008). However, the intensive culture of these new fish species has favoured the appearance of several outbreaks with varied mortality rates.

The development of a fish disease is the result of the interaction among pathogen, host and environment. Therefore, only multidisciplinary studies involving the virulence factors of the pathogenic microorganisms, aspects of the biology and immunology of the fish, as well as a better understanding of the environmental conditions affecting fish cultures, will allow the application of adequate measures to control and prevent the microbial diseases limiting the production of marine fish. According to Toranzo et al. (2005), several aspects would be raised regarding the infectious diseases caused by bacteria in marine fish: (i) only a relatively small number of pathogenic bacteria are responsible of important and significant economic losses in cultured fish; (ii) several classical diseases considered as typical of fresh water aquaculture are today important problems in marine culture; (iii) clinical signs (external and internal) provoked by each pathogen depend on the host species, fish age and stage of the disease; (iv) there is no correlation between external and internal signs of the disease; and (v) the severity of the disease and the mortality are higher in cultured fish that in wild fish populations, because to the lack of the stressful conditions that usually occur in the culture facilities.

This paper provides a detailed account of the epidemics and the characterization of the implicated etiological agents. The data indicated that Photobacterium damselae subsp. damselae was the most prevalent pathogenic bacterium involved in these outbreaks. This microbe is identified as a pathogen for numerous aquatic creatures, including crustaceans, mollusks, fish, and cetaceans. This bacterial pathogen has been documented to induce illnesses in humans and, hence, may be regarded as a zoonotic agent. We have reassessed the taxonomic classification, as well as the morphological and molecular attributes of this microbe. Furthermore, we elucidate the virulence characteristics and pathogenic pathways of P. damselae subsp. damselae.

2. MICROBIOLOGICAL STUDY OF NEWLY CULTURED MARINE FISH SPECIES IN SOUTHERN SPAIN

2.1 Macroscopic signs of disease in finding fish

Labella (2010), studying the microbial origin of diseases affecting new cultured marine fish species in Southern Spain, reported the occurrence of 9 epizootic outbreaks (from 2003 to 2006) affecting cultures of redbanded seabream (7 outbreaks), common seabream (3 outbreaks), white seabream (2 outbreaks), and meagre (1 outbreak). The mortality of these outbreaks varying between 5 and 94%, depending on season, affected fish species, and fish age (Table 1). Gross external and internal signs varied depending on the outbreak, being similar to those previously described for vibriosis in several fish species (Fouz et al., 1992; Balebona et al., 1998b). The main external signs were exophthalmia, dark skin pigmentation, and pale gills and eroded fins (Fig. 1), whilst in some specimens from outbreaks 3, 5 and 8 haemorrhagic areas and epidermic ulcers were observed. Internally, the predominant infection signs were the presence of a fatty liver, with or without petechiae, and abdominal swelling with ascistic liquid. Splecnomegaly and visceral fat accumulation were also recorded (Fig. 1).

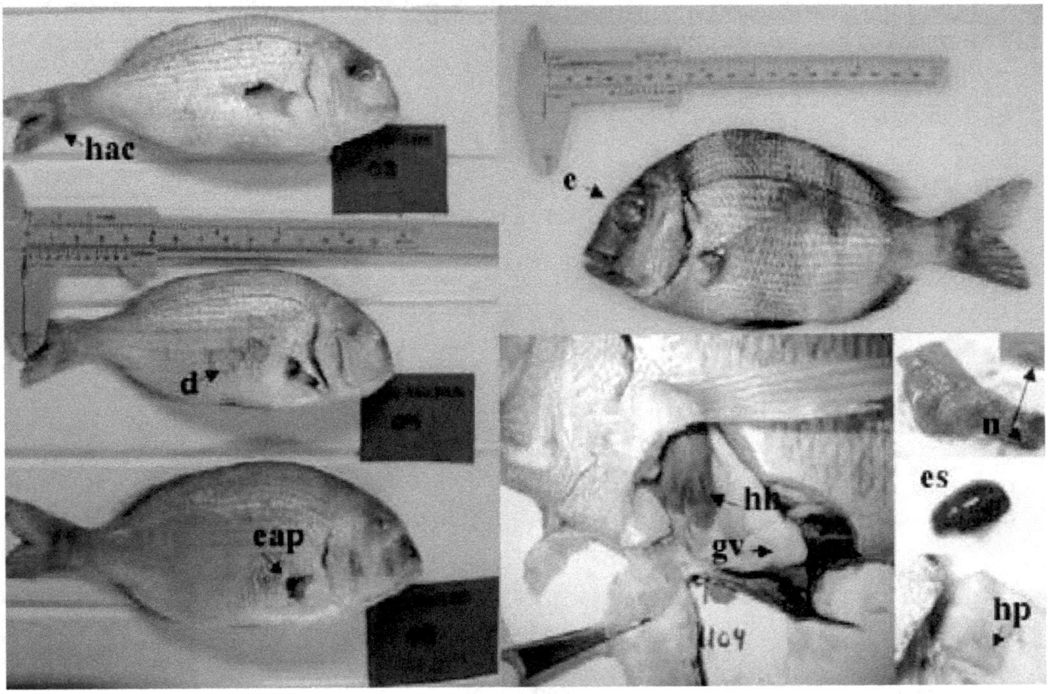

Fig. 1. Main external symptoms and pathological signs in the internal organs observed in affected fish. (hac) haemorrhages in caudal fin; (d) skin desquamation; (eap) fin erosion; (e) exophthalmia; (hh) haemorrhagic liver; (gv) visceral fat accumulation; (n) necrosis; (es) splecnomegaly; (hp) petechiae.

2.2 Organs used for bacterial culture and its isolation

Affected or moribund specimens were killed with an overdose of MS-222 in seawater, and immediately processed for bacteriological analyses. Samples collected from the skin, eyes, brain, liver, spleen and kidney were seeded onto several routine bacteriological culture media for isolation of bacterial fish pathogens. The bacterial isolates were subjected to phenotypic characterization by using the tests specified in Table 2, according to Bergey's Manual of Systematic Bacteriology (Thyssen & Ollevier, 2005). All isolated bacteria were Gram-negative short rods or cocobacilli, motile, oxidase and catalase positives, glucose fermenters, and sensitive to vibriostatic agent pteridine (O/129, 150 μg). Bacterial identification was confirmed by the analysis of 16S rRNA gene sequences, amplified and sequenced as previously described by Labella et al. (2006). Table 1 shows the bacterial species identified and confirmed by 16S rDNA sequences.

Table 1. Epizootic outbreaks affecting newly cultured marine fish species in Southern Spain and identification of the isolated bacterial strains.

Outbreak no. (date)	Affected species	Weight (g)±SD	Cumulative mortality (%)	Bacterial species
1 (December 2003)	*Pagrus auriga*	210 ± 80	22.0	*P. damselae* subsp. damselae
2 (April-June 2003)	*P. auriga*	277 ± 93	26.0	*P. damselae* subsp. *damselae*
2 (May-June 2003)	*Pagrus pagrus*	213 ± 60	*26.5*	*V. ichthyoenteri* *V. harveyi* *V. alginolyticus* *V. fischeri* *V. splendidus*
3 (July 2004)	*P. auriga*	35 ± 7	20.0	*V. harveyi*
3 (August 2004)	*P. pagrus*	300 ± 55	15.0	*V. harveyi* *V. alginolyticus* *V. splendidus*
4 (October December 2004)	*P. auriga*	555 ± 137	28.0	*P. damselae subsp. damselae*
5 (February 2005)	*P. auriga*	627 ± 110	13.0	*V. splendidus*
5 (February 2005)	*P. pagrus*	311 ± 62	25.0	*V. splendidus* *V. ichthyoenteri*
6 (May 2005)	*Diplodus sargus*	2 ± 0.33	80.0	*P. damselae subsp. damselae* *V. fischeri*
7 (August 2005)	*D. sargus*	600 ± 70	94.0	*P. damselae subsp. damselae*
8 (May 2006)	*P. auriga*	60 ± 27	5.0	*V. ichthyoenteri* *V. harveyi*
9 (August 2006)	*Argyrosomus regius*	10 ± 0.3	80.0	*P. damselae subsp. damselae*

Table 2. Phenotypic features of bacterial isolates collected from fish epizootic outbreaks.

Test	P. damselae subsp. damselae	V. harveyi	V. alginolyticus	V. fischeri	V. splendidus	V. ichthyoenteri
Growth in TCBS	G[a]	Y[b]	Y	G	Y	Y
Swarming	-	+	+	-	-	-
Arginine dehydrolase	+	V[c]	-	-	V	-
Lysine decarboxylase	V	+	+	+	V	+
Ornithine decarboxylase	-	+	+	-	V	+
Urease	V	+	-	+	-	-
Indole production	-	+	V	-	+	+
Acetoine production	V	-	-	-	-	-
Amylase	-	+	+	-	+	-
Gelatinase	-	+	+	-	+	-
Esculin hydrolysis	-	+	-	-	-	-
Nitrate reduction	+	+	+	+	+	-
Citrate utilization	-	+	-	+	+	+
Gas from glucose	+	-	-	-	-	-
Acids from:						
Lactose	-	-	-	V	-	-
Trehalose	V	+	+	-	V	+
Cellobiose	+	+	+	+	+	-
D-galactose	+	+	-	+	-	-
Growth in:						
0% NaCl	-	-	-	-	-	-
1% NaCl	+	+	+	+	V	V
6% NaCl	+	+	+	V	V	+
8% NaCl	-	+	+	-	-	-
Growth at:						
4ºC	V	-	-	-	-	-
35ºC	+	+	+	V	-	-

[a]Green colony; [b]Yellow colony; [c]Variable result.

2.3 Antimicrobial resistance

The antimicrobial resistance pattern to 14 antimicrobials routinely used in aquaculture practice was determined for selected strains of each identified bacterial pathogen. All the bacterial strains tested showed sensitivity to chloramphenicol, enrofloxacine, flumequine and nalidixic acid (Table 3). On the other hand, all the bacterial strains presented resistance to streptomycin, and a variable resistance pattern to the other 9 antimicrobials tested (Table 3). Labella (2010) reported three different resistotype profiles for the bacterial pathogens involved in the epizootic outbreaks. The resistotype I consisted of all the *Photobacterium damselae* subsp. *damselae* strains, and it is characterized for the sensitivity to trimethoprim-sulphametaxazole, oxolinic acid, and nitrofurantoine. The resitotype II grouped the isolates included in the *Vibrio harveyi, V. splendidus, V. fischeri* and *V. alginolyticus* species. This resistotype possessed sensitivity to oxolinic acid, erythromycin, tetracycline, oxytetracycline, nitrofurantoine and novobiocin. Finally, the resistotype III included all the strains of *V. ichthyoenteri*, and presented sensitivity to trimethoprim-sulphametaxazole, tetracycline, oxytetracycline, novobiocin and amoxycillin (Table 3).

Table 3. Antimicrobial resistance patterns of the bacterial isolates involved in epizootic outbreaks.

Antimicrobials (µg/disc)	Resistance profiles		
	I	II	III
Chloramphenicol (30)	S[a]	S	S
Enrofloxacine (5)	S	S	S
Flumequine (30)	S	S	S
Nalidixic acid (30)	S	S	S
Streptomycin (10)	R[b]	R	R
Trimethoprim-sulphametaxazole (1.25 + 23.75)	S	R	S
Oxolinic acid (10)	S	S	R
Erythromycin (15)	R	S	R
Tetracycline (30)	R	S	S
Oxytetracycline (30)	R	S	S
Nitrofurantoine (300)	S	S	R
Novobiocin (30)	R	S	S
Amoxycillin (25)	R	R	S
Ampicillin (10)	V[c]	R	R

[a]Sensitive; [b]Resistant; [c]Variable result.

On the basis of its high prevalence in the epizootic outbreaks recorded, *P. damselae* subsp. *damselae* was considered as the main bacterial pathogen affecting new cultured marine fish species in Southern Spain (Garcia-Rosado et al., 2007).

3. STUDY OF *PHOTOBACTERIUM DAMSELAE* SUBSP. *DAMSELAE*

3.1 Taxonomical position

The taxonomic status of *P. damselae* subsp. *damselae* within the family *Vibrionaceae* has changed repeatedly. *P. damselae* subsp. *damselae* was initially isolated as *Vibrio damselae* from skin ulcers of temperate-water damselfish (Love et al., 1981), and was recognized as an opportunistic pathogen capable of causing disease in a variety of hosts, mainly fish and mammals. MacDonell & Colwell (1985) transferred this species to the new genus *Listonella* based on a review of phylogenetic relationships within the family by 5S rRNA sequence data. Later, Smith et al. (1991) transferred this species to the genus *Photobacterium* based on phenotypic data, which was further supported from the phylogenetic analysis carried out by Ruimy et al. (1994). Similarly, Gauthier et al. (1995) demonstrated that the fish pathogen *Pasteurella piscicida* was closely related to *P. damselae* on the basis of phylogenetic analysis of small-subunit rRNA sequences and DNA-DNA hybridization data. Accordingly, *P. damselae* was proposed to include two subspecies, *P. damselae* subsp. *damselae* and *P. damselae* subsp. *piscicida*. However, Thyssen et al. (1998) have showed that there is no phenotypic evidence that supports the inclusion of *P. damselae* subsp. *piscicida* as a subspecies of *P. damselae*. The distinctive diagnoses of both subspecies of *P. damselae* can be achieved by a multiplex PCR assay, which combines specific primers for 16S rRNA and urease C genes (Osorio et al., 2000b).

A new species, named *P. histaminum*, has been described by Okuzumi et al. (1994) as a halophylic potent histamine-producing bacterium. The new species has been distinguished from other members of the genus based on phenotypic characteristics, 16S rRNA gene sequence and DNA-DNA hybridization. A close physiological similarity between *P. damselae* subsp. *damselae* and *P. histaminum* has been reported (Dalgaard et al., 1997). Kimura et al. (2000) found that the type strain of *P. damselae* subsp. *damselae* has a histamine-producing ability as potent as *P. histaminum*. In addition, the levels of DNA relatedness between both species ranged from 80 to 88%. Regarding to the phenotypic differentiation of these organisms, the authors confirmed a biochemical profile identical for both species, except for the trehalose utilization. For these reasons, Kimura et al. (2000) proposed *P. histaminum* as a later subjective synonym of *P. damselae* subsp. *damselae*.

3.2 Phenotypic and molecular characteristics

According to Bergey's Manual of Systematic Bacteriology (Thyssen & Ollevier, 2005), *P. damselae* subsp. *damselae* belongs to the genus *Photobacterium* included in the family *Vibrionaceae*, displaying morphological characteristics typical of members of the family, appearing as coccobacilli. The flagellate organisms lack of a flagellar sheath, even *P. damselae* subsp. *piscicida* lacks flagella (Baumann & Baumann, 1981).

Labella (2010) carried out a wide study on the phenotypic characteristics of *P. damselae* subsp. *damselae*, resulting positive the following traits: motility, catalase, cytochrome oxidase, growth at the range of 20-35ºC and in 1-6% NaCl, arginine dehydrolase, nitrate reduction, and fermentation of melibiose and maltose. *P. damselae* subsp. *damselae* uses the following compounds as unique carbon source: D-galactose, α-D-glucose, raffinose, turanose, D-ribose, N-acetylgalactosamine, glycogen, methyl α-D-glucoside, dextrin, D- glucose 6 phosphate, glycerol, sorbitol, succinate, D-L-lactic acid, glycil L-glutamic acid, tween 40, tween 80, L-glutamic acid, glycil L-aspartic acid, inosine, L-serine, L-aspartic acid, L-asparagine, L-alanine, uridine, L-alanylglycine and thymine. On the other hand, *P. damselae* subsp. *damselae* strains showed negative results for ornithine decarboxylase, production of H_2S, indole production, alginase, and fermentation of D-mannitol, D-sorbitol, inositol, erytritol, D-adonitol, D-arabitol, dulcitol, raffinose, L-rhamnose and L-arabinose. This subspecies is unable to use D-fucose, α-D-lactose, L-arabinose, gentibiose, melibiose, L- rhamnose, D-mannitol, adonitol, myo-inositol, erytritol, xylitol, arabinitol, acetate, L-glutamate, formate, D-gluconic acid, propionic acid, L-leucine, D-alanine, L-proline, L- threonine, L-ornithine, L-histidine, α-ceto glutaric acid, aconytic acid and ß- hydroxyphenylacetic acid.

The fatty acid profile of *P. damselae* subsp. *damselae* contains high concentrations of $C_{16:1}$, $C_{16:0}$, and $C_{18:1}$ fatty acids, and in lesser extent $C_{12:0}$, $C_{14:0}$, $C1_{4:1}$, $C_{15:0}$, $C_{17:0}$, $C_{17:1}$, and $C_{18:0}$ (Nogi et al., 1998). The electrophoretic analyses carried out revealed similar band pattern for *P. damselae* strains, sharing four major outer membrane proteins (OMP), with molecular masses of 20, 30, 42 and 53 kDa (Magariños et al., 1992). An OMP of 37 kDa (OMP-PD) forms a trimeric structure of approximately 110 kDa that conform an ion channel and acts as a porin in *P. damselae* (Gribun et al., 2004). These results have been confirmed by Western blot performed with anti-OMP polyclonal serum against the monomeric form of OMP-PD. As in the case of the OMP, all *P. damselae* strains showed the same silver-stained lipopolysaccharide (LPS) profile obtained by proteinase K digested whole cell lysates. This profile had a ladder like pattern, typical of smooth type LPS (S-LPS), with low amounts of 2-keto-3-deoxyoctonate (KDO) (Kuwae et al., 1982).

The G+C content of the genomic DNA of *P. damselae* subsp. *damselae* is 40.6-41.4 mol% (Thyssen & Ollevier, 2005). The DNA relatedness of *P. damselae* subsp. *damselae* and other classical *Photobacterium* species, demonstrated by DNA-DNA hybridization, varied between 12 and 37% with *P. leiognathi*, 21 and 30% with *P. phosphoreum*, 19.5% with *P. profundum* and 28% with *P. augustum* (Nogi et al., 1998; Kimura et al., 2000). Plasmids are present in most *P. damselae* subsp. *damselae* strains tested, with sizes ranging from 3.0 kb to higher than 190 kb (Pedersen et al., 1997).

Several studies have demonstrated that strains of *P. damselae* subsp. *damselae* showed a high heterogeneity in biochemical and serological characteristics (Smith et al., 1991; Fouz et al., 1992; Pedersen et al., 1997; 2009; Labella et al., 2006). Botella et al. (2002) established that 11 biochemical features were variable among the 33 *P. damselae* subsp. *damselae* strains tested: acetoin production, luminescence, gas from glucose, lysine decarboxylase, growth at 4º and 40ºC, urease, and utilization of sucrose, D-mannose, D-cellobiose and D-gluconate. Labella et al. (2009) obtained that *P. damselae* subsp. *damselae* strains showed variability for the following tests: acetoin production, ß-galactosidase, lysine decarboxylase, growth at 4º and 40ºC, esculin hydrolysis, acid from mannitol, sorbitol and amygdalin, citrate utilization and assimilation of D-mannose, maltose, malate and N-acetylglucosamine. This variability led the authors to establish 8 different biotypes or phenotypic profiles among the 17 strains tested, isolated from cultured marine fish.

However, the genetic variation of the strains of *P. damselae* subsp. *damselae* has received less attention. Botella et al. (2002), using the amplified fragment length polymorphism (AFLP) technique, demonstrated a high genetic variability among the *P. damselae* subsp. *damselae* strains isolated from gilthead seabream and European seabass cultured in the same geographical area and collected in a short time period (2 years). In fact, the 33 tested strains yielded 24 AFLP profiles, with almost every strain showing a different band pattern.

Takahashi et al. (2008) established, using ribotyping, AFLP and pulsed-field gel electrophoresis (PFGE), that *P. damselae* subsp. *damselae* clinical isolates causing fatal cases in humans had similar genotypes, but they were not clearly distinguishable from environmental isolates (including isolates from fish). Nevertheless, the phenotypic profiles of the clinical isolates were clearly distinct from those showed by environmental isolates. The authors explained the inconsistency between the results obtained from genotypic and phenotypic analysis arguing that the divergence of clinical strains from environmental ones is a recent event, and these phenotypic differences are so small that they are not detected by whole genome typing techniques such as PFGE and AFLP. However, sequencing analysis of the *gyrB*, *toxR* and *ompU* genes showed larger differences between clinical and environmental isolates. Similar results were obtained by Labella (2010) comparing these clinical strains with fish isolates by repetitive extragenic palindromic (REP)-PCR.

Labella et al. (2009) compared three PCR-based techniques [random amplified polymorphic DNA (RAPD), enterobacterial repetitive intergenic consensus (ERIC)-PCR, and REP-PCR] for the analysis of genetic variability within *P. damselae* subsp. *damselae* strains isolated from several fish species in outbreaks occurred in different geographical locations. All the PCR- based typing methods supported the high variability within *P. damselae* subsp. *damselae*, the strains being discriminated into 8-14 genetic groups, depending on the method employed. In addition, no concordance among the genetic assignation of the strains by the different PCR methods was obtained. These results suggest, as concluded also by Botella et al. (2002), that different clonal variants of *P. damselae* subsp. *damselae* potentially pathogenic for several fish species exist, and even can be involved in a single outbreak. On the other hand, and similarly to previous results (Botella et al., 2002), a relationship between the genetic profiles and the origin of isolation or the host fish species could not be established.

3.3 Pathogenicity and virulence factors

P. damselae subsp. *damselae* has been recognized as a bacterial pathogen in a wide variety of aquatic animals including fish, molluscs and crustaceans (Vera et al., 1991; Fouz et al., 1992; Company et al., 1999; Sung et al., 2001; Lozano-Leon et al., 2003; Labella et al., 2006; Wang & Cheng, 2006; Vaseeharan et al., 2007; Han et al. 2009; Kanchanopas-Barnette et al., 2009). This microorganism is an autochthonous inhabitant of aquatic ecosystems, which may survive in seawater and sediment for a long time, maintaining its infectivity and pathogenic properties (Ghinsberg et al., 1995; Fouz et al., 1998; 2000). In addition, *P. damselae* subsp. *damselae* may be a primary pathogen for mammals, including humans (Morris et al., 1982; Clarridge & Zighelboim-Daum, 1985; Fujioka et al., 1988; Perez-Tirse et al., 1993; Yuen et al., 1993; Shin et al., 1996; Fraser et al., 1997; Tang & Wong, 1999; Goodell et al., 2004; Yamame et al., 2004). A comparatively small number of bacterial species belonging to the family *Vibrionaceae* causes diseases in both aquatic animals and humans. However, the fact that an organism provokes disease in an aquatic animal does not necessarily mean that this is the source for human infections (Austin, 2010). Indeed, the origin of some of these bacteria may be the waters in which the aquatic animals are found, and the transmission to humans may be via wound or may be food/water-borne.

The extracellular products (ECPs) are produced by bacterial pathogens to facilitate the uptake of nutrients from the surrounding environment and/or for the successful penetration and survival of pathogens inside the host (Sakai, 1985; Bakopoulos et al., 2003). Main ECP components related to virulence include proteases, haemolysins, and siderophore-mediated iron sequestering systems (Norqvist et al., 1990; Toranzo & Barja, 1993; Balebona et al., 1998a; Rodkhum et al., 2005; Wang et al., 2007). These mechanisms can provoke host tissue destruction and haemorrhages, playing an important role in colonization, invasiveness and dissemination of the bacterial pathogen within the host (Finkelstein et al., 1992; Silva et al., 2003). However, only a few studies have been carried out on the role of ECPs in the pathogenesis of *P. damselae* subsp. *damselae*.

Labella et al. (2010), studying the pathogenicity of *P. damselae* subsp. *damselae* strains isolated from cultured fish, demonstrated that the intraperitoneal inoculation of ECPs from virulent strains (mean LD_{50} of about 1 x 10^5 CFU) was lethal for redbanded seabream at 2 to 4 h post- inoculation, whilst ECP samples from a non-virulent strain (LD_{50} > 10^8 CFU) did not produced toxic effects in fish after a 7d post-inoculation period. The inoculation of heated ECPs (100ºC, 10 min) to fish did not produce deaths, which suggests that the active toxic fraction present in the ECPs is secreted and thermolabile, and it is not associated with the thermorresistant bacterial lipopolysaccharide content. Similar results have been reported for several fish pathogens (Lamas et al., 1994), including *P. damselae* subsp. *damselae* (Fouz et al., 1995). Fish inoculated with heated ECPs (and also with ECPs from non-virulent strains) showed enlarged lymphohaematopoietic organs (Fig. 2), suggesting a stimulation of immune response with cellular accumulation, as also reported for *Aeromonas hydrophila* ECPs (Rey et al., 2009).

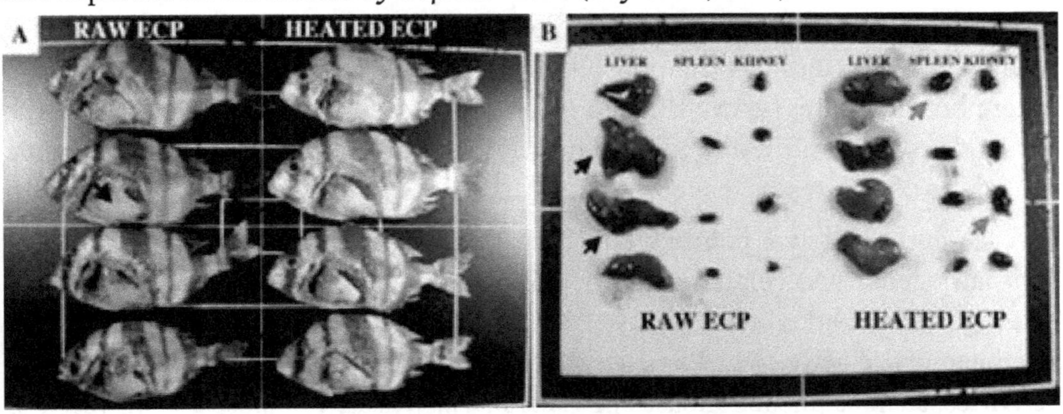

Fig. 2. Effect of the *P. damselae* subsp. *damselae* ECPs (raw and thermally treated at 100ºC for 10 min) on redbanded seabream (*P. auriga*) (10 g weight). (A) Arrows indicate the presence of visceral haemorrhages. (B) Black arrows indicate signs of hepatomegaly, and red arrows show the inflammation of haematopoietic organs (spleen and head kidney).

P. damselae subsp. *damselae* ECPs displayed cytotoxic activity for different fish and mammalian cell lines (Wang et al., 1998; Labella et al., 2010). The cytotoxicity was limited to ECPs from virulent strains, and it was totally lost on heated ECP samples, which suggests the presence of thermolabile cytotoxic components in the raw ECPs (Labella et al., 2010).

The main virulence factor characterized in *P. damselae* subsp. *damselae* is the damselysin, a thermolabile extracellular cytotoxin of 69 kDa, which is a phospholipase D active against the sphingomyelin of the sheep erythrocyte membrane (Kreger, 1984; Kothary & Kreger, 1985; Cutter & Kreger, 1990). The damselysin also presents haemolytic activity against several erythrocyte types, including fish (Kreger et al., 1987). Classically, damselysin production has been related to the pathogenicity of *P. damselae* subsp. *damselae* in diverse animal models (Kothary & Kreger, 1985; Fouz et al., 1993), although Osorio et al. (2000a) demonstrated that the presence of this toxin is not a requisite for the virulence of this bacterial pathogen. Labella et al. (2010) found that 75% of virulent *P. damselae* subsp. *damselae* strains showed phospholipase activity in their ECPs, but the specific 567 bp PCR amplicon corresponding to the phospholipase D (*dly*) gene was detected in only two strains (12.5%). Interestingly, the phospholipase activity in the *dly* + strains remained unaltered after thermal treatment, which differs from the behaviour described for phospholipase toxins in *Vibrionaceae* (Songer, 1987).

Two types of bacterial phospholipases have been described, the extracellular phospholipases (A2, C or D), which are considered as virulence factors (Schmiel & Miller, 1999), and the phospholipases associated with the outer membrane (A type), whose role in pathogenesis had not been established (Dekker, 2000; Snijder & Dijkstra, 2000). Besides the phospholipase D activity associated to damselysin, other phospholipases (extracellular and/or A type) seem to be present in *P. damselae* subsp. *damselae* strains, although they are not directly related to the pathogenic properties of the strains (Labella et al., 2010).

Several authors have pointed out that the pathogenicity of some bacterial fish pathogens was related to their ability to haemolyse the host erythrocytes (Borrego et al., 1991; Fouz et al., 1993; Grizzle & Kiryu, 1993; Pedersen et al., 2009). The haemolytic activity in *Vibrionaceae* can be related to extracellular enzymatic activities, such as phospholipases in *V. parahaemolyticus, V. mimicus, V. harveyi* and *P. damselae* subsp. *piscicida* (Shinoda et al., 1991; Lee et al., 2002; Zhong et al., 2006; Naka et al., 2007) and phospholipase D in *P. damselae* subsp. *damselae* (Kreger et al., 1987), or to the direct action of haemolysins that provoke the pore-structure formation on the erythrocyte membrane (Iida & Honda, 1997; Zhang & Austin, 2005). Labella et al. (2010) reported that all *P. damselae* subsp. *damselae* strains tested produced haemolysis of fish and/or sheep erythrocytes. This ability was exclusively associated with bacterial cultures in 81.25% of the strains. These results could suggest that the haemolysin is associated with the bacterial core or it is an extracellular haemolysin whose activity is inhibited by the enzymatic content of ECPs, as has been described for VTH haemolysin of *V. tubiashii* (Hasegawa & Hase, 2009). As in the case of phospholipases, a correlation between the haemolytic activity of *P. damselae* subsp. *damselae* strains and their virulence properties could not also be established (Labella et al., 2010).

Several extracellular bacterial proteases, mainly metalloproteases and serine-proteases suchas vibriolysins, are considered as virulence factors in numerous bacterial pathogens (Ishihara et al., 2002; Miyoshi et al., 2002; Farto et al., 2006). These proteases provoke tissue damages and degradation of host tissues, favouring the colonization and invasion of pathogens into the host (Miyoshi & Shinoda, 2000). In addition, proteases enable the evasion of the bacteria from several fish defence mechanisms (Vivas et al., 2004). A limited number of enzymatic activities has been detected in *P. damselae* subsp. *damselae* ECPs, including phosphatases, esterases, amylases and glycosidases, but their proteolytic activity was very low, lacking caseinase and gelatinase activities (Toranzo & Barja, 1993; Fouz et al., 1993; Labella et al., 2010). Nevertheless, none of these enzymatic activities could be related with the degree of toxicity, both *in vivo* and *in vitro*, presented by the ECPs (Labella et al., 2010), in contrast to results reported for other fish pathogens such as *Aeromonas* (Esteve et al., 1995). In short, the presence of phospholipases (including damselysin), haemolysins or other enzymatic activities in the ECPs is not directly related to the pathogenicity of *P. damselae* subsp. *damselae*. Labella et al. (2010) hypothesized that another unknown type of toxin, different to the damselysin, could be involved in the toxicity of *P. damselae* subsp. *damselae* ECPs. A neurotoxin possessing an acetylcholine-esterase activity (ictiotoxin) has been described in strains of several species of *Vibrionaceae*, including *P. damselae* subsp. *damselae* (Balebona et al. 1998a; Perez et al. 1998), and may be responsible for several clinical signs observed by these authors.

4. CONCISE SUMMARY

In recent years, Photobacterium damselae subsp. damselae has been consistently isolated from epizootic outbreaks impacting many farmed fish species. This bacterial pathogen has been documented to induce illnesses in humans, thereby qualifying it as a zoonotic agent. The distinctive virulence factor identified in P. damselae subsp. damselae is damselysin, a thermolabile extracellular cytotoxin that functions as a phospholipase D and exhibits haemolytic activity against many types of erythrocytes. Recent findings from our research team indicate that there is no correlation between the presence of the dly gene and the pathogenicity of P. damselae subsp. damselae; thus, other virulence factors may contribute to the pathological damage inflicted by this microorganism on infected fish.

5. REFERENCES

Austin, B. (2010). Vibrios as causal agents of zoonoses. *Veterinary Microbiology* 140: 310-317. Bakopoulos, V., Pearson, M., Volpatti, D., Gousmani, L., Adams, A., Galeotti, M. &

Dimitriadis, G.J. (2003). Investigation of media formulations promoting differential antigen expression by *Photobacterium damselae* ssp. *piscicida* and recognition by sea bass, *Dicentrarchus labrax* (L.), immune sera. *Journal of Fish Diseases* 26: 1-13.

Balebona, M.C., Krovacek, K., Moriñigo, M.A., Mansson, I., Faris, A. & Borrego, J.J. (1998a). Neurotoxic effect on two fish species and a PC12 cell line of the supernate of *Vibrio alginolyticus* and *Vibrio anguillarum*. *Veterinary Microbiology* 63: 61-69.

Balebona, M.C., Zorrilla, I., Moriñigo, M.A. & Borrego, J.J. (1998b). Survey of bacterial pathologies affecting farmed gilt-head sea bream (*Sparus aurata* L.) in southwestern Spain from 1990 to 1996. *Aquaculture* 166: 19-35.

Baumann, P. & Baumann, L. (1981). The marine gram-negative eubacterias: genera *Photobacterium, Beneckea, Alteromonas, Pseudomonas* and *Alcaligenes*, in Starr, M.P., Stolp, H., Trüper, H.G., Balows, A. & Schlegel, H. (ed.), *The Prokaryotes*, Vol. 1, Springer-Verlag, Berlin, pp. 1302-1331.

Borrego, J.J., Moriñigo, M.A., Bosca, M., Castro, D., Martinez-Manzanares, E., Barja, J.L. & Toranzo, A.E. (1991). Virulence properties associated with the plasmid content of environmental isolates of *Aeromonas hydrophila*. *Journal of Medical Microbiology* 35: 264-269.

Botella, S., Pujalte, M.J., Macian, M.C., Ferrus, M.A., Hernandez, J. & Garay, E. (2002). Amplified fragment length polymorphism (AFLP) and biochemical typing of *Photobacterium damselae* subsp. *damselae*. *Journal of Applied Microbiology* 93: 681-688.

Cardenas, S. & Calvo, A. (2003). Reproducción en el mar y en cautividad del pargo común o bocinegro, *Pagrus pagrus* (Pisces: *Sparidae*), *Proceedings of the I Virtual Iberoamerican Congress of Aquaculture*. URL: http//www.civa2003-org.

Cardenas, S. & Manchado, M. (2008). Perspectives for redbanded seabream culture. *Global Aquaculture Advocate* May/June: 56-58.

Clarridge, J.L. & Zighelboim-Daum, S. (1985). Isolation and characterization of two hemolytic phenotypes of *Vibrio damsela* associated with a fatal wound infection. *Journal of Clinical Microbiology* 21: 302-306.

Company, R., Sitja-Bobadilla, A., Pujalte, M.J., Garay, E., Alvarez-Pellitero, P. & Perez- Sanchez, J. (1999). Bacterial and parasitic pathogens in cultured common dentex, *Dentex dentex* L. *Journal of Fish Diseases* 22: 299-309.

Cutter, D.L. & Kreger, A.S. (1990). Cloning and expression of the damselysin gene from *Vibrio damsela. Infection and Immunity* 58: 266-268.

Dalgaard, P., Manfio, G.P. & Goodfellow, M. (1997). Classification of photobacteria associated with spoilage of fish products by numerical taxonomy and pyrolysis mass spectrometry. *Zentralblatt fur Bakteriologie* 285: 157-168.

Dekker, N. (2000). Outer-membrane phospholipase A: known structure, unknown biological function. *Molecular Microbiology* 35: 711-717.

Esteve, C., Amaro, C., Garay, E., Santos, Y. & Toranzo, A.E. (1995). Pathogenicity of live bacteria and extracellular products of motile *Aeromonas* isolated from eels. *Journal of Applied Bacteriology* 78: 555-562.

FAO (2004). *The State of World Fisheries and Aquaculture*, FAO Fisheries Department, Rome.

Farto, R., Armada, S.P., Montes, M., Perez, M.J. & Nieto, T.P. (2006). Presence of a lethal protease in the extracellular products of *Vibrio splendidus-Vibrio lentus* related strains. *Journal of Fish Diseases* 29: 701-707.

Fernandez-Trujillo, A., Porta, J., Borrego, J.J., Alonso, M.C., Alvarez, M.C. & Bejar, J. (2006). Cloning and expression analysis of a Mx cDNA from the Senegalese sole *Solea senegalensis*. *Fish and Shellfish Immunology* 21: 577-582.

Fernandez-Trujillo, M.A., Porta, J., Manchado, M., Borrego, J.J., Alvarez, M.C. & Bejar, J. (2008). c-Lysozyme from Senegalese sole (*Solea senegalensis*): cDNA cloning and expression pattern. *Fish and Shellfish Immunology* 25: 697-700.

Finkelstein, R.A., Boesman-Finkelstein, M., Chang, Y. & Hase, C.C. (1992). *Vibrio cholera* hemagglutinin/protease, colonial variation, virulence, and detachment. *Infection and Immunity* 60: 472-478.

Fouz, B., Larsen, J.L., Nielsen, B., Barja, J.L. & Toranzo, A.E. (1992). Characterization of *Vibrio damsela* strains isolated from turbot *Scophthalmus maximus* in Spain. *Diseases of Aquatic Organisms* 12: 155-166.

Fouz, B., Barja, J.L., Amaro, C., Rivas, C. & Toranzo, A.E. (1993). Toxicity of the extracellular products of *Vibrio damsela* isolated from diseased fish. *Current Microbiology* 27: 341- 347.

Fouz, B., Novoa, B., Toranzo, A.E. & Figueras, A. (1995). Histopathological lesions caused by *Vibrio damsela* in cultured turbot *Scophthalmus maximus* (L.) inoculations with live cells and extracellular products. *Journal of Fish Diseases* 18: 357-364.

Fouz, B., Toranzo, A.E., Marco-Nogales, E. & Amaro, C. (1998). Survival of fish-virulent strains of *Photobacterium damselae* subsp. *damselae* in seawater under starvation conditions. *FEMS Microbiology Letters* 168: 181-186.

Fouz, B., Toranzo, A.E., Millan, M. & Amaro, C. (2000). Evidence that water transmits the disease caused by the fish pathogen *Photobacterium damselae* subsp. *damselae*. *Journal of Applied Microbiology* 88: 531-535.

Fraser, S.L., Purcell, B.K., Delgado, B., Jr., Baker, A.E. & Whelen, A.C. (1997). Rapidly fatal infection due to *Photobacterium (Vibrio) damsela*. *Clinical Infectious Diseases* 25: 935- 936.

Fujioka, R.S., Greco, S.B., Cates, M.B. & Schroeder, J.P. (1988). *Vibrio damsela* from wounds in bottlenose dolphins *Tursiops truncatus*. *Diseases of Aquatic Organisms* 4: 1-8.

Garcia-Rosado, E., Cano, I., Martin-Antonio, B., Labella, A., Manchado, M., Alonso, M.C., Castro, D. & Borrego, J.J. (2007). Co-occurrence of viral and bacterial pathogens in disease outbreaks affecting newly cultured sparid fish. *International Microbiology* 10: 193-199.

Gauthier, G., Lafay, B., Ruimy, R., Breittmayer, V., Nicolas, J.L., Gauthier, M. & Christen, R. (1995). Small-subunit rRNA sequences and whole DNA relatedness concur for the reassignment of *Pasteurella piscicida* (Snieszko et al.) Janssen and Surgalla to the genus *Photobacterium* as *Photobacterium damsela* subsp. *piscicida* comb. nov. *International Journal of Systematic Bacteriology* 45: 139-144.

Ghinsberg, R.C., Drasinover, V., Sheinberg, Y. & Nitzan, Y. (1995). Seasonal distribution of *Aeromonas hydrophila* and *Vibrio* species in Mediterranean coastal water and beaches: a possible health hazard. *Biomedical Letters* 51: 151-159.

Goodell, K.H., Jordan, M.R., Graham, R., Cassidy, C. & Nasraway, S.A. (2004). Rapidly advancing necrotizing fasciitis caused by *Photobacterium (Vibrio) damsela*: a hyperaggresive variant. *Critical Care Medicine* 32: 278-281.

Gribun, A., Katcoff, D.J., Hershkovits, G., Pechatnikov, I. & Nitzan, Y. (2004). Cloning and characterization of the gene encoding for OMP-PD porin: The major *Photobacterium damsela* outer membrane protein. *Current Microbiology* 48: 167-174.

Grizzle, J.M. & Kiryu, Y. (1993). Histopathology of gill, liver and pancreas, and serum enzyme levels of Channel catfish infected with *Aeromonas hydrophila* complex. *Journal of Aquatic Animal Health* 5: 36-50.

Han, E., Gomez, D.K., Kim, J.H., Choresca, C.H., Jr., Shin, S.P., Baeck, G.W. & Park, S.C. (2009). Isolation of *Photobacterium damselae* subsp. *damselae* from zebra shark *Stegostoma fasciatum*. *Korean Journal of Veterinary Research* 49: 35-38.

Hasegawa, H. & Hase, C.C. (2009). The extracellular metalloprotease of *Vibrio tubiashii* directly inhibits its extracellular haemolysin. *Microbiology* 155: 2296-2305.

Iida, T. & Honda, T. (1997). Hemolysins produced by vibrios. *Journal of Toxicology and Toxin Review* 16: 215-227.

Ishihara, M., Kawanishi, A., Watanabe, H., Tomochika, K, Miyoshi, S. & Shinoda, S. (2002). Purification of a serine protease of *Vibrio parahaemolyticus* and its characterization. *Microbiology and Immunology* 46: 298-303.

Kanchanopas-Barnette, P., Labella, A., Alonso, M.C., Manchado, M., Castro, D. & Borrego, J.J. (2009). The first isolation of *Photobacterium damselae* subsp. *damselae* from Asian seabass *Lates calcarifer*. *Fish Pathology* 44: 47-50.

Kimura, B., Hokimoto, S., Takahashi, H. & Fujii, T. (2000). *Photobacterium histaminum* Okuzumi et al. 1994 is a later subjective synonym of *Photobacterium damselae* subsp. *damselae* (Love et al. 1981) Smith et al. 1991. *International Journal of Systematic and Evolutionary Microbiology* 50: 1339-1342.

Kothary, M.H. & Kreger, A.S. (1985). Purification and characterization of an extracellular cytolysin produced by *Vibrio damsela*. *Infection and Immunity* 49: 25-31.

Kreger, A.S. (1984). Cytolytic activity and virulence of *Vibrio damsela*. *Infection and Immunity* 44: 326-331.

Kreger, A.S., Bernheimer, A.W., Etkin, L.A. & Daniel, L.W. (1987). Phospholipase D activity of *Vibrio damsela* cytolysin and its interaction with sheep erythrocytes. *Infection and Immunity* 55: 3209-3212.

Kuwae, T., Sasaki, T. & Kurata, M. (1982). Chemical and biological properties of lypopolysaccharide from a marine bacterium, *Phosphobacterium phosphoreum* PJ-1. *Microbiology and Immunology* 26: 455-466.

Labella, A. (2010). *Aislamiento y Caracterización de Bacterias Potencialmente Patógenas Asociadas a Nuevas Especies de Espáridos Cultivados*, PhD Thesis, University of Malaga, Spain.

Labella, A., Vida, M., Alonso, M.C., Infante, C., Cardenas, S., Lopez-Romalde, S., Manchado,

M. & Borrego, J.J. (2006). First isolation of *Photobacterium damselae* subsp. *damselae* from cultured redbanded seabream, *Pagrus auriga* Valenciennes, in Spain. *Journal of Fish Diseases* 29: 175-179.

Labella, A., Manchado, M., Alonso, M.C., Castro, D., Romalde, J.L. & Borrego, J.J. (2009). Molecular intraspecific characterization of *Photobacterium damselae* ssp. *damselae* affecting cultured marine fish. *Journal of Applied Microbiology* 108: 2122-2132.

Labella, A., Sanchez-Montes, N., Berbel, C., Aparicio, M., Castro, D., Manchado, M. & Borrego, J.J. (2010). Toxicity of *Photobacterium damselae* subsp. *damselae* strains isolated from new cultured marine fish. *Diseases of Aquatic Organisms* 92: 31-40.

Lamas, J., Santos, Y., Bruno, D., Toranzo, A.E. & Anadon, R. (1994). A comparison of pathological changes caused by *Vibrio anguillarum* and its extracellular products in rainbow trout (*Oncorrhynchus mykiss*). *Fish Pathology* 29: 79-89.

Lee, J.H., Ahn, S.H., Kim, S.H., Choi, Y.H., Park, K.J. & Kong, I.S. (2002). Characterization of *Vibrio mimicus* phospholipase A (PhlA) and cytotoxicity on fish cell. *Biochemical and Biophysical Research Communications* 298: 269-276.

Love, M., Teebken-Fisher, D., Hose, J.E., Farmer III, J.J., Hickman, F.W. & Fanning, G.R. (1981). *Vibrio damsela*, a marine bacterium, causes skin ulcers on the damselfish *Chromis punctipinnis. Science* 214: 1139-1140.

Lozano-Leon, A., Osorio, C.R., Nuñez, S., Martinez-Urtaza, J. & Magariños, B. (2003). Occurrence of *Photobacterium damselae* subsp. *damselae* in bivalve molluscs from Northwest Spain. *Bulletin of European Association of Fish Pathologists* 23: 40-44.

MacDonell, M.T. & Colwell, R.R. (1985). Phylogeny of the *Vibrionaceae*, and recommendation for two new genera, *Listonella* and *Shewanella. Systematic and Applied Microbiology* 6: 171-182.

Magariños, B., Romalde, J.L., Bandin, I., Fouz, B. & Toranzo, A.E. (1992). Phenotypic, antigenic and molecular characterization of *Pasteurella piscicida* strains isolated from fish. *Applied and Environmental Microbiology* 58: 3316-3322.

Manchado, M., Ponce, M., Asensio, E., Infante, C., de la Herran, R., Robles, F., Garrido- Ramos, M.A., Ruiz-Rejon, M. & Cardenas, S. (2005). Pagurta, híbrido interespecífico de pargo *Pagrus pagrus* (L., 1758) x hurta *Pagrus auriga* Valenciennes, 1843: caracterización fenotípica y molecular. *Boletin del Instituto Español de Oceanografia* 21: 219-224.

Martin-Antonio, B., Manchado, M., Infante, C., Zerolo, R., Labella, A., Alonso, M.C. & Borrego, J.J. (2007). Intestinal microbiota variation in Senegalese sole (*Solea senegalensis*) under different feeding regimes. *Aquaculture Research* 38: 1213-1222.

Ministerio de Agricultura, Pesca y Alimentación (MAPA) (2008). *La Acuicultura en España*, Junta Asesora de Cultivos Marinos. URL: http://www.mapa.es/es/pesca/pags/jacumar/presentacion/acuicultura_es.htm.

Miyoshi, S. & Shinoda, S. (2000). Microbial metalloproteases and pathogenesis. *Microbes and Infection* 2: 91-98.

Miyoshi, S., Sonoda, Y., Wakiyama, H., Rahman, M.M., Tomochika, K., Shinoda, S., Yamamoto, S. & Tobe, K. (2002). An exocellular thermolysin-like metalloprotease produced by *Vibrio fluvialis*: purification, characterization, and gene cloning. *Microbial Pathogenesis* 33: 127-134.

Morris, J.G., Wilson, R., Jr., Hollis, D.G., Weaver, R.E., Miller, A.R., Tacket, C.O., Hickman, F.W. & Blake, P.A. (1982). Illness caused by *Vibrio damsela* and *Vibrio hollisae. Lancet* 8284: 1294-1297.

NACA/FAO (2001). Aquaculture in the third millennium, *Technical Proceedings of the Conference on Aquaculture in the Third Millennium*, FAO, Bangkok, Thailand, pp. 471- 476.

Naka, H., Hirono, I. & Aoki, T. (2007). Cloning and characterization of *Photobacterium damselae* subsp. *piscicida* phospholipase: an enzyme that shows haemolytic activity. *Journal of Fish Diseases* 30: 681-690.

Nogi, Y., Masui, N. & Kato, C. (1998). *Photobacterium profundum* sp. nov., a new, moderately barophilic bacterial species isolated from a deep-sea sediment. *Extremophiles* 2: 1-7.

Norqvist, A., Norrman, B. & Wolf-Watz, H. (1990). Identification and characterization of a zinc metalloprotease associated with invasion by the fish pathogen *Vibrio anguillarum. Infection and Immunity* 58: 3731-3736.

Okuzumi, M., Hiraishi, A., Kobayashi, T. & Fujii, T. (1994). *Photobacterium histaminum* sp. nov., a histamine-producing marine bacterium. *International Journal of Systematic Bacteriology* 44: 631-636.

Osorio, C.R., Romalde, J.L., Barja, J. & Toranzo, A.E. (2000a). Presence of phospholipase-D (*dly*) gene coding for damselysin production is not a pre-requisite for pathogenicity in *Photobacterium damselae* subsp. *damselae. Microbial Pathogenesis* 28: 119-126.

Osorio, C.R., Toranzo, A.E., Romalde, J.L. & Barja, J. (2000b). Multiplex PCR assay for *ureC* and 16S rRNA genes clearly discriminates between both subspecies of *Photobacterium damselae. Diseases of Aquatic Organisms* 40: 177-183.

Pedersen, K., Dalsgaard, I. & Larsen, J.L. (1997). *Vibrio damsela* associated with diseased fish in Denmark. *Applied and Environmental Microbiology* 63: 3711-3715.

Pedersen, K., Skall, H.F., Lassen-Nielsen, A.M., Bjerrum, L. & Olesen, N.J. (2009). *Photobacterium damselae* subsp. *damselae*, an emerging pathogen in Danish rainbow trout, *Oncorrhynchus mykiss* (Walbaum), mariculture. *Journal of Fish Diseases* 32: 465- 471.

Perez, M.J., Rodriguez, L.A. & Nieto, T.P. (1998). The acetylcholinesterase ichthyotoxin is a common component in the extracellular products of *Vibrionaceae* strains. *Journal of Applied Microbiology* 84: 47-52.

Perez-Tirse, J., Levine, J.F. & Mecca, M. (1993). *Vibrio damsela.* A cause of fulminant septicaemia. *Archives of Internal Medicine* 153: 1838-1840.

Ponce, M., Infante, C., Catanese, G., Cardenas, S. & Manchado, M. (2004). Complete mitocondrial DNA nucleotide sequence of the redbanded seabream *Pagrus auriga*. *European Aquaculture Society Special Publication* 34: 663-664.

Prieto, A., Cañavate, J.P. & Cardenas, S. (2003). Crecimiento de larvas de hurta (*Pagrus auriga*), *Proceedings of the IX National Congress of Aquaculture*, Consejería de Agricultura y Pesca, Junta de Andalucia, Cadiz, Spain, pp. 386-387.

Rey, A., Verjan, N., Ferguson, H.W. & Iregui, C. (2009). Pathogenesis of *Aeromonas hydrophila* strain KJ99 infection and its extracellular products in two species of fish. *Veterinary Record* 164: 493-499.

Rodkhum, C., Hirono, I., Crosa, J.H. & Aoki, T. (2005). Four novel hemolysin genes of *Vibrio anguillarum* and their virulence to rainbow trout. *Microbial Pathogenesis* 39: 109-119.

Ruimy, R., Breittmayer, V., Elbaze, P., Lafay, B., Boussemart, O., Gauthier, M. & Christen, R. (1994). Phylogenetic analysis and assessment of the genera *Vibrio, Photobacterium, Aeromonas* and *Plesiomonas* deduced from small subunit rRNA sequences. *International Journal of Systematic Bacteriology* 44: 416-426.

Sakai, D.K. (1985). Significance of extracellular protease for growth of a heterotrophic bacterium, *Aeromonas salmonicida. Applied and Environmental Microbiology* 50: 1031- 1037.

Schmiel, D.H. & Miller, V.L. (1999). Bacterial phospholipase and pathogenesis. *Microbes and Infection* 1: 1103-1112.

Shin, J.D., Shin, M.G., Suh, S.P., Ryang, D.W., Rew, J.S. & Nolte, F.S. (1996). Primary *Vibrio damsela* septicemia. *Clinical Infectious Diseases* 22: 856-857.

Shinoda, S., Matsuoka, H., Tsuchie, T., Miyoshi, S., Yamamoto, S., Taniguchi, H. & Mizuguchi, Y. (1991). Purification and characterization of a lecithin-dependent haemolysin from *Escherichia coli* transformed by a *Vibrio parahaemolyticus* gene. *Journal of General Microbiology* 137: 2705-2711.

Silva, A.J., Pham, K. & Benitez, J.A. (2003). Haemagglutinin/protease expression and mucin gel penetration in El Tor biotype *Vibrio cholera. Microbiology* 149: 1883-1891.

Smith, S.K., Sutton, D.C., Fuerst, J.A. & Reichelt, J.L. (1991). Evaluation of the genus *Listonella* and reassignment of *Listonella damselae* (Love et al.) MacDonell and Colwell to the genus *Photobacterium* as *Photobacterium damselae* comb. nov. with an emended description. *International Journal of Systematic Bacteriology* 41: 529-534.

Snijder, H.J. & Dijkstra, B.W. (2000). Bacterial phospholipase A: structure and function of an integral membrane phospholipase. *Biochimica et Biophysica Acta* 1488: 91-101.

Songer, J.G. (1987). Bacterial phospholipases and their role in virulence. *Trends in Microbiology* 5: 156-161.

Sung, H.H., Hsu, S.F., Chen, C.K., Ting, Y.Y. & Chao, W.L. (2001). Relationships between disease outbreak in cultured tiger shrimp (*Penaeus monodon*) and the composition of *Vibrio* communities in pond water and shrimp hepatopancreas during cultivation. *Aquaculture* 192: 101-110.

Takahashi, H., Miya, S., Kimura, B., Yamame, K., Arakawa, Y. & Fujii, T. (2008). Difference of genotypic and phenotypic characteristics and pathogenicity potential of *Photobacterium damselae* subsp. *damselae* between clinical and environmental isolates from Japan. *Microbial Pathogenesis* 45: 150-158.

Tang, W.M. & Wong, J.W. (1999). Necrotizing fasciitis caused by *Vibrio damsela. Orthopedics* 22: 443-444.

Thyssen, A. & Ollevier, F. (2005). Genus II. *Photobacterium* Beijerinck 1889, 401AL, in Brenner, D.J., Krieg, N.R. & Staley, J.T. (ed.), *Bergey's Manual of Systematic Bacteriology*, 2nd ed., Vol. 2, Springer, New York, pp. 546-552.

Thyssen, A., Grisez, L., van Houdt, R. & Ollevier, F. (1998). Phenotypic characterization of the marine pathogen *Photobacterium damselae* subsp. *piscicida. International Journal of Systematic Bacteriology* 48: 1145-1151.

Toranzo, A.E. & Barja, J.L. (1993). Virulence factors of bacteria pathogenic for coldwater fish. *Annual Review of Fish Diseases* 3: 5-36.

Toranzo, A.E., Magariños, B. & Romalde, J.L. (2005). A review of the main bacterial fish diseases in mariculture systems. *Aquaculture* 246: 37-61.

Union Europea (2010). *La Política Pesquera Común en Cifras.* http://ec.europa.eu/fisheries/publications/FEP_ES.pdf.

Vaseeharan, B., Sundararaj, S., Murugan, T. & Chen, J.C. (2007). *Photobacterium damselae* subsp. *damselae* associated with diseased black tiger shrimp *Penaeus monodon* Fabricius in India. *Letters in Applied Microbiology* 45: 82-86.

Vera, P., Navas, J.I. & Fouz, B. (1991). First isolation of *Vibrio damsela* from sea bream *Sparus aurata. Bulletin of the European Association of Fish Pathologists* 11: 112-113.

Vivas, J., Razquin, B.E., Lopez-Fierro, P., Naharro, G. & Villena, A. (2004). Correlation between production of acyl homoserine lactones and proteases in an *Aeromonas hydrophila aroA* live vaccine. *Veterinary Microbiology* 101: 107-176.

Wang, F.I. & Chen, J.C. (2006). Effect of salinity on the immune response of tiger shrimp *Penaeus monodon* and its susceptibility to *Photobacterium damselae* subsp. *damselae. Fish and Shellfish Immunology* 20: 671-681.

Wang, X.H., Oon, H.L., Ho, G.W.P., Wong, W.S.F., Lim, T.M. & Leung, K.Y. (1998). Internalization and cytotoxicity are important virulence mechanisms in vibrio-fish epithelial cell interactions. *Microbiology* 144: 2987-3002.

Wang, Q., Liu, Q., Ma, Y., Rui, H. & Zhang, Y. (2007). *LuxO* controls extracellular protease, haemolytic activities and siderophore production in fish pathogen *Vibrio alginolyticus. Journal of Applied Microbiology* 103: 1525-1534.

W.A.S. (2006). *Aqua 2006.* http://www.was.org/Meeting/SessionAbstracts.asp?MeetingCode=AQUA2006& Session=1.

Yamame, K., Asato, J., Kawade, N., Takahashi, H., Kimura, B. & Arakawa, Y. (2004). Two cases of fatal necrotizing fasciitis caused by *Photobacterium damsela* in Japan. *Journal of Clinical Microbiology* 42: 1370-1372.

Yuen, K.Y., Ma, L., Wong, S.S. & Ng, W.F. (1993). Fatal necroziting fasciitis due to *Vibrio damsela. Scandinavian Journal of Infectious Diseases* 25: 659-661.

Zhang, X.H. & Austin, B. (2005). A review. Haemolysins in *Vibrio. Journal of Applied Microbiology* 98: 1011-1019.

Zhong, Y.B., Zhang, X.H., Chen, J.X., Chi, Z.H., Sun, B.G., Li, Y. & Austin, B. (2006). Overexpression, purification, characterization and pathogenicity of *Vibrio harveyi* hemolysin VHH. *Infection and Immunity* 74: 6001-6005.

10 Chapter

BACTERIA ISOLATED FROM AFFECTED WILD AND CULTIVATED MARINE FISH IN GREECE

1. INTRODUCTION: AN OVERVIEW

Global production of sea bream and sea bass aquaculture has been increasing over time. The overall output of gilthead sea bream (Sparus aurata L.) and European sea bass (Dicentrarchus labrax L.) throughout Europe has risen. These marine fish species possess significant economic significance in the aquaculture sector. Specifically, total production increased from 92,310 tonnes in 1999 to 175,196 tonnes in 2006, reflecting an average yearly growth rate of 9.6%. Greece is the leading producer of euryhaline fish (sea bream and sea bass) among Mediterranean nations. In 2008, Greece had its highest production, totaling 130,000 tons, comprising 95,000 tonnes of sea bass and sea bream. The climatic and geomorphological conditions of Greece conducive to the cultivation of euryhaline fish, along with grants and projects provided by the European Union, the decline in fish stocks, and the restrictions imposed on fishing in recent years, have significantly contributed to the development of the aquaculture industry. The sector currently meets the demands of the Greek market, with the majority of production exported to other markets, primarily Italy, Spain, France, England, and Portugal. Presently, new species have been introduced into aquaculture, primarily from the Sparidae family, including Puntazzo puntazzo, Diplodus sargus, Lithognathus mormyrus, Pagrus pagrus, Pagellus erythrinus, and Dentex dentex.

Bacterial infections originating from fish have emerged as significant contributors to economic losses since the inception of marine aquaculture (Kubota & Takakuwa, 1963; Anderson & Conroy, 1970). The advancement of intensive marine aquaculture, characterized by the aggregation of substantial biomass in a limited volume of water, results—under specific conditions (a combination of factors)—in the onset of illnesses that cause population losses. The manifestation of a disease may result in mortality or symptoms, both indicating a divergence from the host's typical structure or function (Hedrick, 1998). The majority of illnesses in farmed fish stem from wild populations. The proximity of farmed and wild fish facilitates the transmission of diseases. The clinical manifestations induced by any pathogen are contingent upon the host type, the age of the fish, and the disease stage (acute, chronic, or subclinical). Furthermore, in certain instances, there exists no association between inside and external injuries. Systemic illnesses, such as pasteurellosis, exhibit significant mortality rates and induce internal damage in infected fish, however these fish often appear healthy. In contrast, several diseases with comparatively low mortality inflict considerable physical harm, such as ulceration, necrosis, and exophthalmos, rendering the fish unsuitable for the market. The diseases and their prevalence Various bacterial infections have been extensively reported in numerous farmed fish species, with the global production of sea bream and sea bass farming experiencing a continuous increase. The occurrence of infections and the variety and quantity of bacterial pathogens have been thoroughly recorded in aquacultured fish species, including salmonids and turbot. Nonetheless, the enumeration and classification of bacteria linked to pathology in both farmed and wild fish species in Greece have been documented intermittently (Athanassopoulou et al., 1999; Bakopoulos et al., 1995; Varvarigos, 1997; Yiagnisis et al., 1999, 2007).

This chapter aims to provide a precise description of the primary bacterial infections identified in both farmed and wild fish in Greece, together with information regarding their prevalence in relation to age and season. This is the inaugural integrative study from Greece encompassing numerous farmed and wild fish species.

1.1 Clinical signs

Signs of diseases include anorexia, lethargies, pale gills, disorientation, darkening in colour, abdominal retention, exophthalmos, abdominal swelling, and external haemorrhages in the head, eyes, skin, gills and at the bases of the fins as well as skin ulcers.

1.2 Post-mortem findings

Visceral petechiation, pale kidneys, enlarged spleen and kidney, liquefactive renal necrosis, lesions in the kidney, tubercles in the spleen, are some of the post-mortem findings.

1.3 Laboratory testing

Diseased fish collected from Greek fish farms. Dead (recently) or moribund fish, showing any symptoms, were transported to the laboratory under refrigeration (2 ºC -8 ºC) and opened aseptically. Sampling was carried out mainly from headkidney, but sometimes from the spleen, liver and the brain (especially in small fry). Tryptic Soy agar (Oxoid), Tryptic Soy Broth (TSB Oxoid) enriched with 2% NaCl (TSAS,

TSBS), Marine agar (MA, Difco) and Thiosulfate-Citrate-Bile salt-Sucrose agar (TCBS, Oxoid) were used as culture media for the bacteria. All inoculated culture media incubated at 22 º C for 2-5 days. A representative number of different types of colonies detected on culture media was collected from plates and streaked on TSAS plates for purity and to carry out identification. Pure cultures of the isolates, obtained by repeated plating on TSAS, were identified by biochemical characterization. Bacteria, identified as *Listonella anguillarum* and *Photobacterium damselae* subsp.*piscicida,* were confirmed serologically by agglutination test (system BIONOR).

2. IDENTIFICATION OF BACTERIA

When identifying bacteria, certain characteristics are selected and used for this purpose. Pure cultures of the isolates, obtained by repeated plating on TSAS are used for identification. Primary identification usually involves simple tests such as morphology, motility, growth on various types of culture media, catalase and oxidase tests. Gram staining reveals the morphology and divides bacteria in two categories - the Gram-positive and the Gram-negative bacteria. For morphological appearance it is preferable to examine young cultures from non-selective media. Bacterial colonies of a single species, when grown on specific media under controlled conditions are described by their characteristic size, shape and pigment. The fermentation of glucose may also be used to distinguish between groups of bacteria. Using these few simple tests it is usually possible to place bacteria, provisionally, in one of the main groups. Then the isolated bacterium is subjected to a battery of tests. Some of them are found in commercial identification systems.

2.1 Growth of bacteria at different temperatures

The growth medium used, for these tests, was the Tryptic Soy Broth with 2% NaCl. The temperatures chosen were 4, 35 and 40 º C for 7, 2 and 1-2 days. The growth of microorganisms at different temperatures is used as an identification key. For example *Vibrio aestuarinus* does not grow at 40 º C but *Vibrio alginolyticus* and *V. parahaemolyticus* can grow.

2.2 Growth of bacteria in different concentration of sodium chloride

The concentrations of sodium chloride were selected 0,3,6,8 and 10%. The growth medium used for these tests was the Tryptic Soy Broth with 2% NaCl. This culture medium is a differential medium because it allows the investigator to distinguish between different types of bacteria based on the trait of NaCl tolerance. Thus a selective, differential medium for the isolation of *Vibrio alginolyticus,* contains a high concentration of salt (10%) that inhibits the growth of *Vibrio harveyi* (it grows until 8% NaCl).

2.3 Using the system API 20E (Biomerieux)

Many commercially available diagnostic kits have been introduced into routine laboratory diagnostics of fish pathogens, such as API 20E, API ZYM, API 20NE, API 50 CH, API Rapid ID 32 (bioMerieux,Marcy-l'Etoile, France), Biolog MicroPlatesGN2, GP2, AN (Biolog, Inc., Hayward, CA, USA), Enterotubes, BBL Crystal E/NF (Becton-Dickinson& Company, Franklin Lakes, NJ, USA), and some others. Of these, API 20E rapid identification system has been the most widely used for identification of fish pathogenic bacteria (Popovic et al., 2007).

The API-20E test kit, a very specific means of bacterial identification (for Enterobacteriaceae and other non-fastidious Gram-negative rods), is a collection of mini test tubes, each with a reagent that test for a different aspect of bacterial metabolism. It provides an easy way to inoculate and read tests. After incubation with an unknown Gram-negative bacterium, the interpretation of positive and negative tests allows for identification to the species level. The preparation of inoculum for the API 20E, was the addition of pure-developed colonies in 5 ml of 2% NaCl. The solutions used for inoculation of cells as indicated respectively by the manufacturer. The inoculation of cells of the tiles and the addition of paraffin oil was made with the help of sterile pipettes volume of 1ml, under sterile conditions. Incubation of plates was at 22-25 º C for 24-48 h. Reading the results was made according to the instructions. It is known that the information from the database of microorganisms API20E identification system is based on results from the study of human clinical strains. For this reason, the identification through the database API 20E system was not used in mind completely. The results of the reactions, however, were considered for identification of bacteria.

2.4 Using the system API Staph (Biomerieux)

For identification of Gram-positive cocci, that give a positive reaction of catalase and negative oxidase reaction, the Api Staph identification system was used (Biomerieux) according to manufacturer's instructions. *Staphylococci* are gram-positive cocci occurring most often in irregular "grape-like" clusters. API-Staph consists of a strip containing dehydrated test substrates in individual microtubes. The tests are reconstituted by adding an aliquot of the API-Staph medium to each tube inoculated with the strain to be studied. The tests included acid production from d-glucose, d-trehalose, d-mannitol, d-mannose, xylose, maltose, lactose, sucrose, *N*-acetylglucosamine, raffinose, d-fructose, d-melibiose, xylitol, and α-methyl-glucosamine; nitrate reduction; and alkaline phosphatase, arginine dihydrolase, urease, and acetoin production. Coagulase production is a significant additional key. It is the ability to clot plasma. A rapid slide coagulase test may be performed. Most *Staphylococcus* species isolated from fish are coagulase negative. The data have been processed using the software program ApiLab Plus.

API Staph (Biomerieux) is an established identification system for *Staphylococci*, used for many years.

2.5 Identification of isolated bacteria: Present work

Pure cultures of the isolates, obtained by repeated plating on TSAS, were identified by biochemical characterization following the criteria described by Austin et al., 1995. Biochemical keys, described by Alsina & Blanch, 1994a, 1994b, were used for identification of *Vibrio* species. These biochemical keys are designed for environmental and clinical isolates and can be used for strains that are Gram negative, giving a positive oxidase reaction, grow in TCBS medium. Some trials were included in the system identification Api 20E. The following biochemical and morphological tests were made: Gram stain, cell morphology, oxidase reaction, catalase reaction, motility, test glucose O-F, growth in TCBS agar, swarming ability of colonies in TSAS, growth in 0, 3,6,8 and 10% NaCl, growth at 4 º C, 35 º C, 40 º C, resistance at 10 and 150 mg O/129, resistance to 10μg ampicillin and tests of API 20E system as the ortho-nitro-phenyl-BD-galaktopyranozidium, hydrolysis of arginine, the decarboxylation of lysine and ornithine, assimilation of citrate, production H_2S, production of urease, deamination of tryptophan, indole production and aketoinis, hydrolysis of gelatin, fermentation of glucose, mannitol, inositol, sorbitol, rhamnose, sucrose, meliviozis, amygdalin and arabinose as carbon sources.

Two thousands one hundred twenty four bacterial isolates obtained from diseased farmed and wild fish species in Greece were identified. During the nine years (1997-2005) a total of 2124 strains of bacteria were isolated from 430 cases, as shown in Table 1.

Table 1. Origin and number of 2124 identified bacterial isolates from diseased farmed and wild fish in Greece (Yiagnisis and Athanassopoulou, 2011).

Fish species	Number of bacterial isolates
Dicentrarchus labrax	1017
Sparus aurata	887
Puntazzo puntazzo	99
Pagellus erythrinus	21
Pagrus pagrus	19
Diplodus sargus	12
Dentex dentex	12
Mugil cephalus	6
Epinephelus marginatus	4
Lithognathus mormyrus	4
Wild fish species	43

Bacteria were identified initially at the genus level and then at the species level.

2.5.1 Bacteria isolated from European sea bass (*Dicentrarchus labrax*)

Figure 1 shows the isolation rates of bacterial groups from diseased farmed European sea bass in Greece from 1997-2005 (Yiagnisis and Athanassopoulou, 2011).

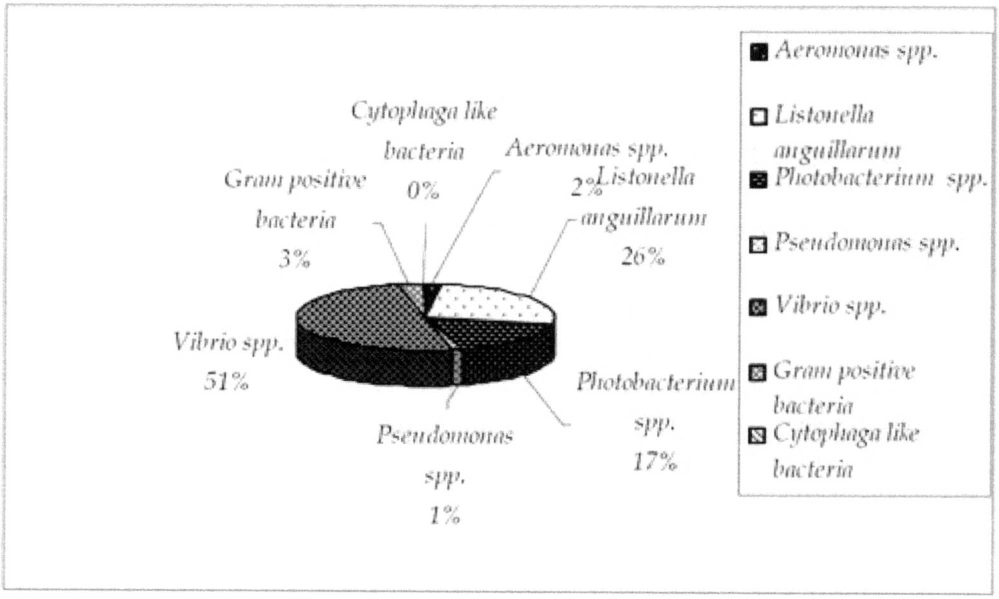

Fig. 1. Bacterial groups isolated from diseased farmed European sea bass in Greece (Yiagnisis and Athanassopoulou, 2011).

Bacteria were identified initially at the genus level and then at the species level.

Different species of *Vibrios* together with *Listonella (Vibrio) anguillarum* constitute 77% of total bacteria isolated from diseased European sea bass. The percentage of isolated *Photobacterium spp.* is 17% while *Photobacterium damselae* subsp. *piscicida* is the main species. The remaining groups of bacteria were isolated at low rates. Table 1 shows the bacteria isolated from sea bass divided into age groups.

Bacterial species most frequently isolated from sea bass are *Listonella anguillarum, Vibrio alginolyticus, Photobacterium damselae* subsp. *piscicida, Vibrio splendidus* II and *Vibrio parahaemolyticus. L. anguillarum* is the most frequent bacterial isolated species from European sea bass. *L. anguillarum* and *Photobacterium damselae* subsp. *piscicida* were isolated mainly from larger fish. *Vibrio splendidus* II has the highest incidence in fry. The majority of bacteria were isolated during spring, as season but on September, as month (Figure 1). As shown in Figure 1, the greater isolation frequency of *Listonella anguillarum* occurred in April while the greater isolation frequency of *Vibrio splendidus* II occurred in May and *Photobacterium damselae* subsp. piscicida and *Vibrio alginolyticus* increased frequencies of isolation were observed in September. For *Vibrio parahaemolyticus* the most observed incidence was in October. It is observed that the bacterium *Vibrio alginolyticus* was isolated simultaneously with *Photobacterium damselae* subsp. *piscicidae.*

Table 1. Number of bacterial isolates from diseased farmed European sea bass in correlation with the age of fish (Yiagnisis and Athanassopoulou, 2011).

Bacterial species	sea bass fry	sea bass	no data	Total
Aeromonas sobria	4			4
Aeromonas spp.	11	1	4	16
Bacillus circulans		2		2
Bacillus pumilus	3			3
Bacillus spp.	4			4
Cytophaga like bacteria	4			4
Flavobacterium spp.		3		3
Listonella anguillarum	58	203		261
Photobacterium damselae subsp.damselae		12		12
Photobacterium damselae subsp.piscicida	61	102		163
Photobacterium phosphoreum	1			1
Pseudomonas fluorescens/putida		8		8
Pseudomonas spp.	6	1		7
Staphylococcus auricularis	2			2
Staph.cohnii	4			4
Staph.epidermidis		6		6
Staph.hominis		3		3
Staph.schleiferi		3		3
Staph.warneri		1		1
Staph.xylosus		4		4
Stenotrofomonas maltophilia		1		1
Vibrio alginolyticus	82	105	4	191
Vibrio anguillarum like	2	3	4	9
Vibrio harveyi	13	8		21
Vibrio costicola	8	4		12
Vibrio fisheri		1		1
Vibrio fluvialis	25			25
Vibrio logei	4			4
Vibrio mediterranei	3			3
Vibrio nereis	4			4
Vibrio nigrapulchritudo	4			4
Vibrio ordalii	3			3
Vibrio parahaemolyticus	32	42		74
Vibrio splendidus II	105	16		121
Vibrio tubiashii	28	1		29
Vibrio vulnificus	4			4
Total	475	530	12	1017

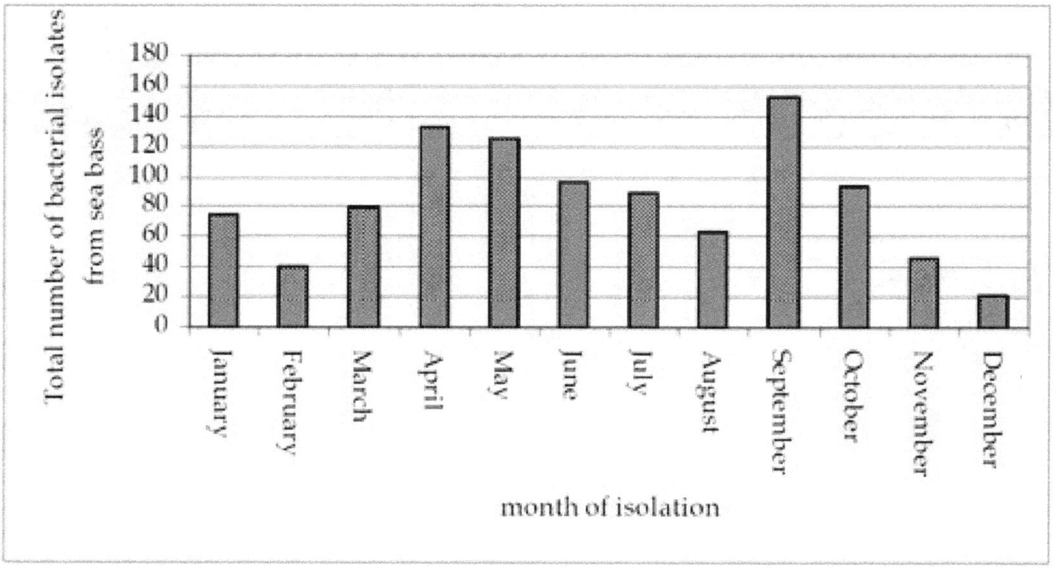

Fig. 2. Total number of bacterial isolates from diseased farmed European sea bass in relation to the month of isolation (Yiagnisis and Athanassopoulou, 2011).

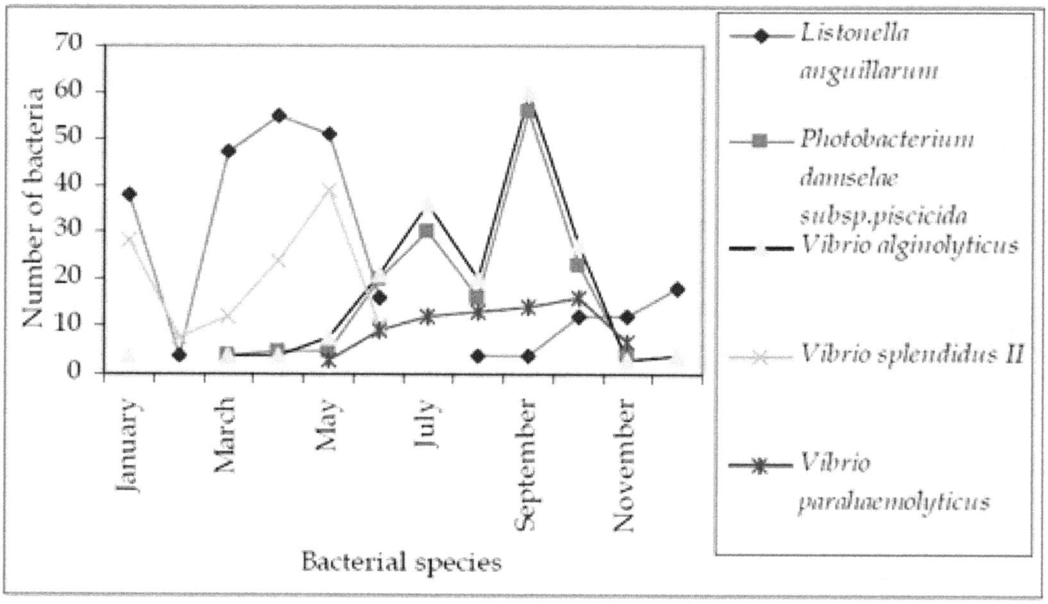

Fig. 3. Number of bacterial species, most frequently isolated from diseased farmed European sea bass in relation to the month of isolation (Yiagnisis and Athanassopoulou, 2011).

Listonella anguillarum was the most frequent species isolated from farmed sea bass during the 9-year sample period.

2.5.2 Bacteria isolated from gilthead sea bream (*Sparus aurata*)

Figure 4 shows the isolation rates of bacterial groups from diseased farmed gilthead sea bream in Greece from 1997-2005 (Yiagnisis and Athanassopoulou, 2011).

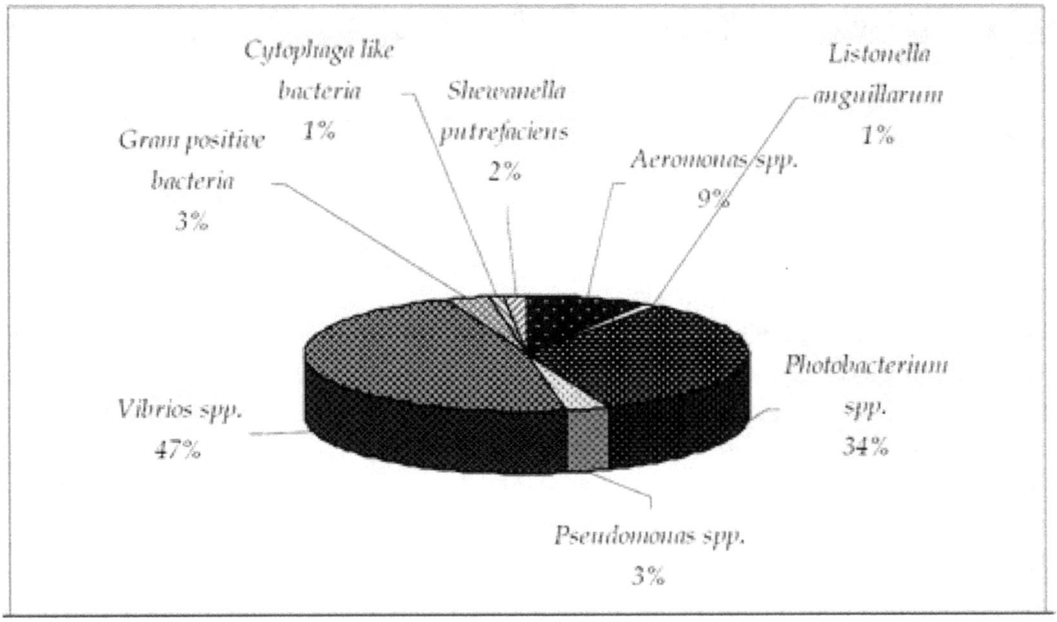

Fig. 4. Rates of isolation of different bacterial groups from diseased farmed gilthead sea bream reared in Greece from 1997 to 2005 (Yiagnisis and Athanassopoulou, 2011).

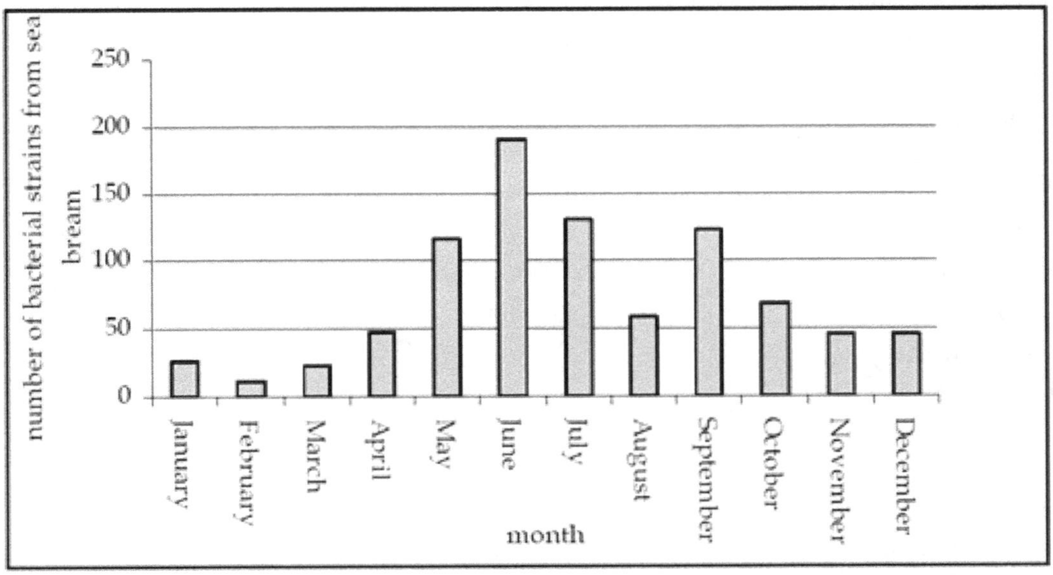

Fig. 5. Number of bacterial strains isolated from diseased farmed gilthead sea bream in correlation to the month of isolation (Yiagnisis and Athanassopoulou, 2011).

Table 3. Number of bacterial isolates from diseased farmed gilthead sea bream in correlation with the age of fish (Yiagnisis and Athanassopoulou, 2011).

Bacterial species	sea bream	fry sea bream	no data	Total
Aeromonas hydro/caviae	7			7
Aeromonas sobria	22			22
Aeromonas spp.	27	8		35
Bacillus spp.		4		4
Cytophaga like bacteria	4	4		8
Listonella anguillarum	8			8
Photobacterium damselae subsp.damselae	56	24		80
Photobacterium damselae subsp.piscicida	139	18		157
Photobacterium spp.		4		4
Pseudomonas cepacia		2		2
Pseudomonas fluorescens/putida	6			6
Pseudomonas spp.	15	2		17
Shewanella putrefaciens	8	4		12
Staph. cobuii	3			3
Staph.lentus		3		3
Staph.schleifer		3		3
Staph.warneri	5			5
Vibrio alginolyticus	117	42	4	163
Vibrio anguillarum like		2	3	5
Vibrio harveyi	20	7		27
Vibrio costicola	60	12		72
Vibrio fisheri	4	12		16
Vibrio fluvialis	15	5		20
Vibrio harveyi	11	4		15
Vibrio mediterranei	4	4		8
Vibrio nereis	4			4
Vibrio nigrapulchritudo	4			4
Vibrio ordalii	2			2
Vibrio parahaemolyticus	50	6		56
Vibrio pelagius		6		6
Vibrio pelagius II	3			3
Vibrio splendidus II	56	8		64
Vibrio tubiashii	8			8
Vibrio vulnificus	31	4		35
Bacillus megaterium	3			3
Grand Total	692	188	7	887

In Figure 4 it is observed that different *Vibrio* species along with *Listonella anguillarum* constitute 48% of total bacteria isolated from diseased sea bream. The percentage of isolated *Photobacterium spp.* is 34% while *Photobacterium damselae* subsp. *piscicida* is the main species. The incidence of *Aeromonas* spp. was 9%, of *Pseudomonas* spp. 3%, of *Shewanella putrefaciens* 2%, of Gram positive cocci 3% and of *Cytophaga* like bacteria 1%. As it is shown in the table 3, the majority (692) of total isolated bacteria (887) belongs to sea bream fry. Bacteria with the highest incidence in the sea bream is *Photobacterium damselae* subsp.*piscicida*, *Vibrio alginolyticus*, *Photobacterium damselae* subsp. *damselae*, *Vibrio costicola*, *Vibrio splendidus* II, *Vibrio parahaemolyticus* and *Vibrio vulnificus*. *Listonella anguillarum* was isolated from fry only, with very low frequency. The majority of bacteria were isolated during summer, especially June (Figure 5).

Figure 6 shows the number and species of bacterial species, most frequently isolated from sea bream in relation to month of isolation.

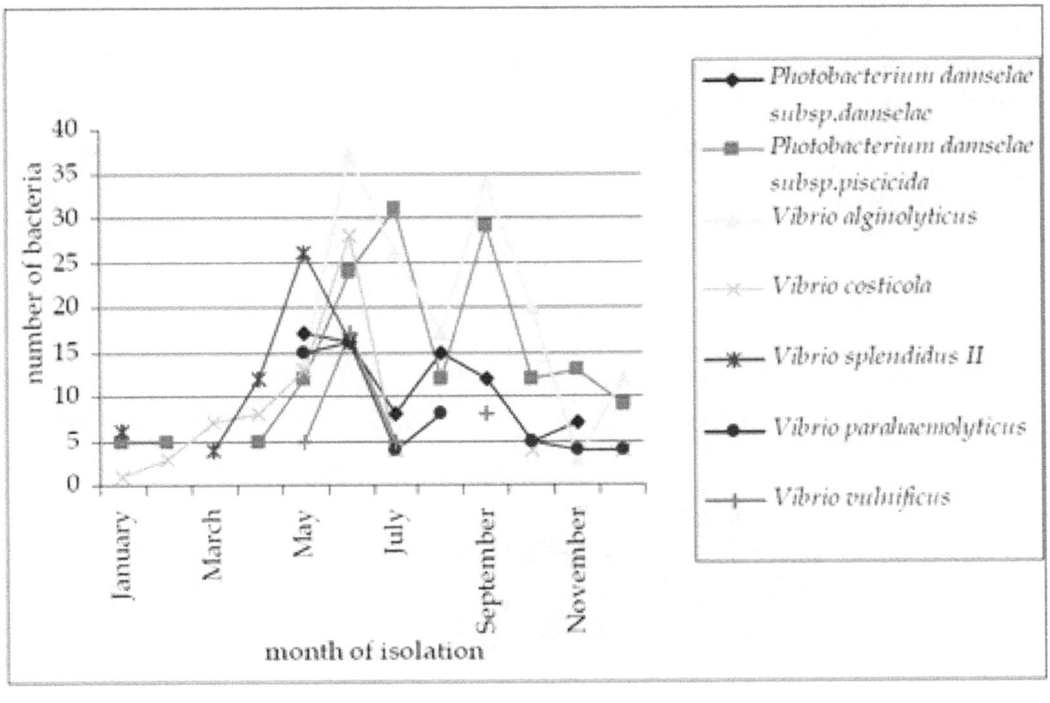

Fig. 6. Number and species of bacterial species, most frequently isolated from diseased farmed gilthead sea bream in relation to month of isolation (Yiagnisis and Athanassopoulou, 2011).

Vibrio alginolyticus, *Photobacterium damselae* subsp. *piscicidae*, *Photobacerium damselae* subsp. *damselae*, *Vibrio vulnificus*, *Vibrio parahaemolyticus* and *Vibio costicola* are most frequently isolated from diseased farmed sea bream in the summer and *Vibrio splendidus* II in May. It is observed that the bacterium *Vibrio alginolyticus* was isolated almost simultaneously with *Photobacterium damselae* subsp. *piscicida*, as in sea bass.

The lower numbers of bacteria from diseased farmed fish were isolated during the winter. These results are in agreement with those of other researchers (Company et al., 1999) where in a bacteriological and parasitological study of farmed dentex conducted in the Mediterranean region, it was reported a relationship between high mortality and high temperature water. These results however are not in agreement with the results of Zorilla et al, 2003, who in a bacteriological study of farmed sea bream (*Sparus aurara*, L.) held in southwest Spain reported the lower numbers of bacterial isolates during summer.

2.5.3 Bacteria isolated from sharpsnout sea bream (*Diplodus puntazzo*)

Figure 7 shows the isolation rates of bacterial groups from diseased farmed sharpsnout sea bream in Greece from 1997-2005. It is observed that different *Vibrio* species along with *Listonella anguillarum* constitute 73% of total bacteria (99) isolated from diseased sharpsnout sea bream. The percentage of isolated *Photobacterium damselae* subsp. *damselae* is 12% while *Vibrio alginolyticus* is the main isolated species. The incidence of *Aeromonas* spp. was 4%, of *Pseudomonas fluorescens/putida* 4% and of *Staphylococcus* spp. 7%. The table 4 shows the bacteria isolated from sharpsnout sea bream divided into age groups. Most of bacterial strains have been isolated from larger fish than fry.

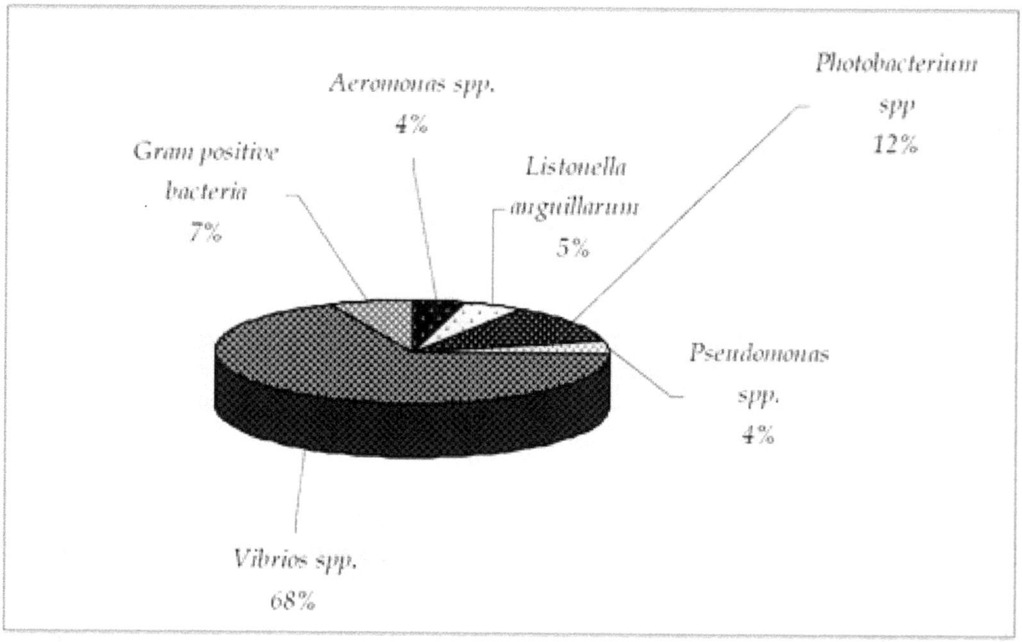

Fig. 7. Bacterial groups isolated from diseased farmed sharpsnout sea bream in Greece (Yiagnisis and Athanassopoulou, 2011).

Bacteria most frequently isolated from farmed diseased sharpsnout sea bream is *Vibrio alginolyticus, Vibrio costicola* and *Photobacterium damselae* subsp. damselae. *Vibrio alginolyticus* is the most often isolated bacterium during the summer and autumn but it is not responsible for great losses. Most of the bacteria from sharpsnout sea bream have been isolated in June, October and November (Fig 8).

Table 4. Number of bacterial isolates from diseased farmed sharpsnout sea bream in correlation with the age of fish (Yiagnisis and Athanassopoulou, 2011).

Bacterial species	sh. sea bream fry	sh. sea bream	Total
Aeromonas caviae		4	4
Listonella anguillarum	5		5
Photobacterium damselae subsp.damselae		12	12
Pseudomonas fluo/putida	4		4
Staphylococcus capitis		4	4
Staphylococcus epidermidis	3		3
Vibrio alginolyticus	14	19	33
Vibrio harveyi		8	8
Vibrio costicola		14	14
Vibrio fisheri		4	4
Vibrio parahaemolyticus	4		4
Vibrio.splendidus II		4	4
Total	30	69	99

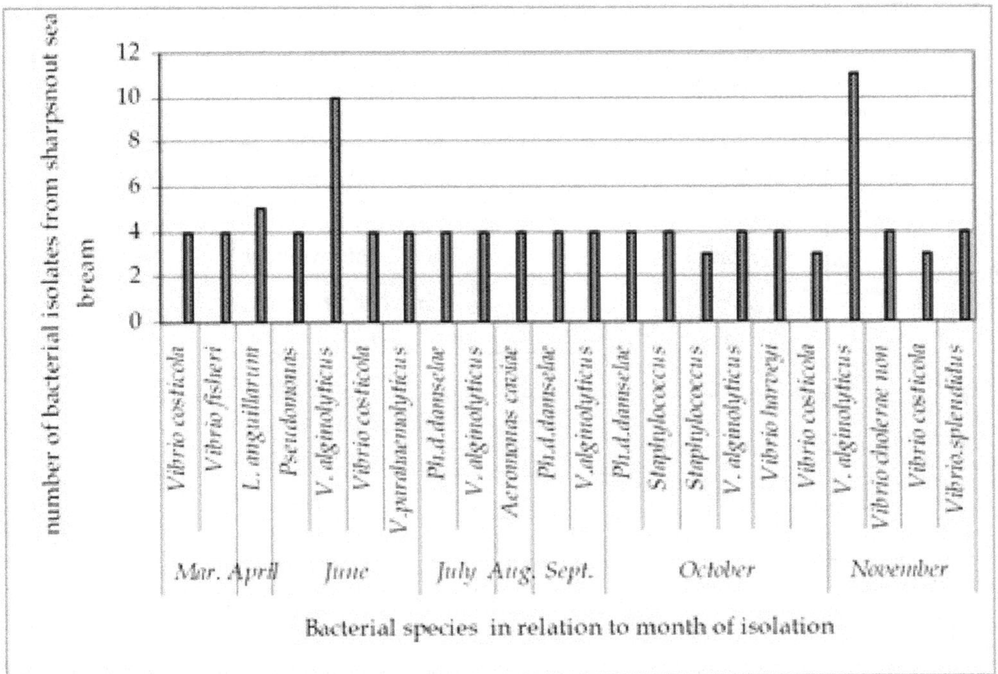

Fig. 8. Number of bacterial species isolated from diseased farmed sharpsnout sea bream in correlation to the month of isolation (Yiagnisis and Athanassopoulou, 2011).

2.5.4 Bacteria isolated from new species of farmed fish

In table 5 it is shown the number of bacterial strains, been isolated, from new species of farmed fish in relation to month of isolation. *Vibrio alginolyticus* is the most commonly isolated bacterial species. ***Listonella anguillarum*** was isolated from *Epinephelus* sp., *Photobacterium damselae* subsp. *piscicida* was isolated from *Mugil cephalus and Pagellus erythrinus. Photobacterium damselae* subsp. *damselae* was isolated from *Pagellus erythrinus*. The species isolated from *Diplodus sargus* are *Vibrio fisheri*, *Vibrio nereis*, and *Vibrio harveyi*. The majority of bacterial strains have been isolated during fall. No bacteria were isolated, from these new species of diseased farmed fish, during the winter.

Table 5. Number of bacterial strains, isolated, from new species of diseased farmed fish in relation to month of isolation (Yiagnisis and Athanassopoulou, 2011).

<table>
<tr><th>New species of farmed fish</th><th>Month of isolation</th><th>Number of bacterial strains</th><th>Bacterial species</th><th>Total bacterial number</th></tr>
<tr><td rowspan="2">Pagrus pagrus</td><td rowspan="2">September</td><td>15</td><td>Vibrio alginolyticus</td><td rowspan="2">19</td></tr>
<tr><td>4</td><td>Vibrio splendidus II</td></tr>
<tr><td rowspan="2">Mugil cephalus</td><td rowspan="2">September and October</td><td>3</td><td>Vibrio alginolyticus</td><td rowspan="2">6</td></tr>
<tr><td>3</td><td>Photobacterium damselae subsp. piscicida</td></tr>
<tr><td rowspan="2">Epinephelus sp.</td><td>March</td><td>2</td><td>Listonella anguillarum</td><td rowspan="2">4</td></tr>
<tr><td>February</td><td>2</td><td>Vibrio anguillarum like</td></tr>
<tr><td rowspan="3">Diplodus sargus</td><td>April</td><td>4</td><td>Vibrio fisheri</td><td rowspan="3">12</td></tr>
<tr><td>March</td><td>4</td><td>Vibrio nereis</td></tr>
<tr><td>April</td><td>4</td><td>Vibrio harveyi</td></tr>
<tr><td rowspan="5">Pagellus erythrinus</td><td rowspan="2">October</td><td>4</td><td>Photobacterium damselae subsp. damselae</td><td rowspan="5">21</td></tr>
<tr><td>4</td><td>Aeromonas hydrophila</td></tr>
<tr><td rowspan="3">November</td><td>5</td><td>Photobacterium damselae subsp. piscicida</td></tr>
<tr><td>4</td><td>Vibrio harveyi</td></tr>
<tr><td>4</td><td>Vibrio costicola</td></tr>
<tr><td>Lithognathus mormyrus</td><td>August</td><td>4</td><td>Vibrio alginolyticus</td><td>4</td></tr>
<tr><td rowspan="3">Dentex dentes</td><td rowspan="2">August</td><td>4</td><td>Aeromonas spp.</td><td rowspan="3">12</td></tr>
<tr><td>4</td><td>Vibrio nereis</td></tr>
<tr><td>February</td><td>4</td><td>Vibrio costicola</td></tr>
</table>

2.5.5 Bacteria isolated from wild fish

In table 6 it is shown the number and the species of bacterial strains, been isolated, from wild fish species in relation to month of isolation. *Listonella anguillarum* (Yiagnisis et al., 2007) and *Photobacterium damselae* subsp.*damselae* are the most frequently isolated species.

Table 6. Number of bacterial strains, isolated, from diseased wild fish species in relation to month of isolation (Yiagnisis and Athanassopoulou, 2011).

Fish species	Month of isolation	Number of bacterial strains	Bacterial species	Total bacterial number
Mugil cephalus	September	1	Photobacterium damselae subsp.*damselae*	3
		1	Vibrio harveyi	
		1	Vibrio alginolyticus	
Atherina boyeri	December	28 (1 case)	Listonella anguillarum	28
Gobius niger	April	1	***Vibrio splendidus*** II	1
Labrus spp	April	1	***Vibrio splendidus*** II	1
Sciaena umbra	October	1	*Pseudomonas* spp.	1
Boops boops	July	9 (1 case)	Photobacterium damselae subsp.*damselae*	9
Total number of bacterial strains	43			

Most of the bacteria, isolated from both farmed and wild fish, were Gram-negative. Specifically 77% (for farmed sea bass), 73% (for farmed sharpsnout sea bream) and 48% (for farmed sea bream) of the isolated bacteria were identified as *Vibrio* species (including *Listonella anguillarum*). For farmed species *Sparus aurata* (sea bream) and *Diplodus puntazzo* (sharpsnout sea bream) the most frequent type was *Vibrio alginolyticus*. This high incidence of Vibrio spp. have was found in previous studies done in Spain by Balebona et al, 1998b and Zorilla et al, 2003. *Vibrio alginolyticus* isolated frequently with other *Vibrio* species and *Photobacterium damselae* subsp. *piscicida*. In fact, from our results it appears that the incidence of *Vibrio alginolyticus* in sea bream and *sea* bass is similar to that of *Photobacterium damselae* subsp. *piscicida*. *V. alginolyticus* was isolated from water of fish farms and live food. Other authors have reported the isolation of this species from water aquaculture (Angulo et al., 1993, Blanch et al., 1997).

3. CONCISE SUMMARY

The European sea bass (Dicentrarchus labrax) is the fish species with the highest number of identified bacterial strains in this investigation. The bacterial species most commonly isolated from sea bass are Listonella anguillarum, Vibrio alginolyticus, Photobacterium damselae subsp. piscicida, Vibrio splendidus II, and Vibrio parahaemolyticus. Listonella anguillarum is the predominant bacterial species identified. L. anguillarum comprises 26% of the total bacteria recovered from infected sea bass, with the highest isolation frequency occurring in April. Photobacterium damselae subsp. piscicida is an obligatory pathogen of fish, also isolated from sea bass fry. Vibrio alginolyticus is an opportunistic vibrio identified from sea bass. In September, there was an increased frequency of isolation for Photobacterium damselae subsp. piscicida and Vibrio alginolyticus. Vibrio splendidus II, another opportunistic vibrio, is predominantly isolated from sea bass fry, with a higher frequency of isolation frequency manifested in May. The predominant bacteria from sea bass were recovered in spring and autumn. The predominant source of total isolated bacteria from sea bream is fry. The predominant bacteria found in sea bream include Photobacterium damselae subsp. piscicida, Vibrio alginolyticus, Photobacterium damselae subsp. damselae, Vibrio costicola, Vibrio splendidus II, Vibrio parahaemolyticus, and Vibrio vulnificus. Listonella anguillarum was only isolated from fry with a frequency of 1%. The predominant bacteria were recovered from sea bream throughout the summer, particularly in June. The highest isolation frequency of Photobacterium damselae subsp. piscicida was observed in July. Vibrio alginolyticus was isolated nearly concurrently with Photobacterium damselae subsp. piscicidae, as observed in sea bass. The proportion of isolated Photobacterium damselae subsp. damselae from sharpsnout sea bream is 12%. Vibrio alginolyticus is the predominant species isolated from sharpsnout sea bream and other novel species of aquacultured fish. The predominant bacterial strains from all three species were isolated in the autumn. The primary pathogens, Listonella anguillarum and Photobacterium damselae subsp. piscicida, can be isolated from wild fish species..

4. REFERENCES

Alsina, M. & Blanch, A. R. (1994a). A set of keys for biochemical identification of environmental *Vibrio* species. *Journal of Applied Bacteriology* Vol.76, pp. 79-85.

Alsina, M. & Blanch, A. R. (1994b). Improvement and update of a set of keys for biochemical identification of environmental *Vibrio* species. *Journal of Applied Bacteriology* Vol.77, pp. 719-721.

Anderson, J. & Conroy, D. (1970). *Vibrio* Disease in Marine Fishes. In: *A Symposium in Marine Fishes and Shellfishes*. Snieszko, F.F. (Ed.), Special Publication No. 5, American Fisheries Society, USA.

Angulo, L., López, J.E &. Vicente, J.A. (1993). Distribución del género *Vibrio* en el agua de entrada de dos piscifactorias de rodaballo (*Scophthalmus maximus*). In: A. Cerviño,

A. Landín, A. de Coo, A. Guerra and M. Torre, Editors, *Proceedings of the IV National Congress of Aquaculture*, Xunta de Galicia, Vilanova de Arousa (Spain) (1993), pp. 611–616.

Athanassopoulou, F., Prapas, A. & Rodger, H. (1999). Diseases of *Puntazzo puntazzo* Cuvier in marine aquaculture systems in Greece. *Journal of Fish Diseases* Vol.22, pp. 215-218.

Austin, B., Alsina, D.A. Austin, A.R. Blanch, F.Grimont, et al. (1995) Identification and typing of *Vibrio anguillarum:* a comparison of different methods. *Systematic and Applied Microbiology,* Vol.18, pp. 285-302.

Bakopoulos, V., Adams, A. &. Richards, R.H. (1995). Some biochemical properties and antibiotic sensitivities of *Pasteurella piscicida* isolated in Greece and comparison with strains from Japan, France and Italy. *Journal of Fish Diseases* Vol.18, pp. 1-7.

Balebona, M.C., Zorrilla, I., Morin˜igo, M.A., Borrego, J.J., 1998b. Survey of bacterial pathologies affecting farmed gilt-head sea bream (*Sparus aurata* L.) in southwestern Spain from 1990 to 1996. *Aquaculture* Vol. 166, pp.19–35.

Blanch A. R., Alsina M., Simón M. & Jofre J. (1997) Determination of bacteria associated with reared turbot (*Scophthalmus maximus*) larvae. *Journal of Applied Microbiology* Vol.82, pp. 729–734.

Company R., Sitjà-Bobadilla A., Pujalte M.J., Garay E., Alvarez-Pellitero P. & Pérez-Sánchez J. (1999) Bacterial and parasitic pathogens in cultured common dentex, *Dentex dentex* L. *Journal of Fish Diseases* Vol.22, pp. 299–309.

Hedrick, R. P. (1998). Relationships of the host, pathogen, and environment: Implications for diseases of cultured and wild populations. *Journal of Aquatic Animal Health* Vol.10, No.2, pp. 107-111.

Kubota, S. & Takakuwa, M. (1963). Studies on the diseases of marine cultured fishes, I. General description and preliminary discussion of fish diseases in the Mie Prefecture, *Journal of the Faculty of Fisheries, Prefectural University of Mie-Tsu* 6 (1963), pp. 107–124 (In Japanese) (Translation of the Fisheries Research Board of Canada, Biological Station, Nanaimo, BC, 1966).

Popovic, N.T. Coz-Rakovac, R. & Strunjak-Perovic, I., 2007. Commercial phenotypic tests (API 20E) in diagnosis of fish bacteria: a review, *Veterinarni Medicina* 52, pp. 49–53.

Varvarigos, P. (1997): Marine fish diseases in Greece Fish Farmer November / December, pp. 33-34.

Yiagnisis, M., Christofilogiannis, P., Rigos, G., Koutsodimou, M., Andriopoulou, A., Anastasopoulou, G., Nengas, I. & Alexis, M. (1999). Review of bacterial isolates in Greek mariculture during the period of 1995-1998. *Abstract book p-060. 9th International Conference EAFP "Diseases of Fish and Shellfish"*, 19-24 September 1999, Rhodos, Greece.

Yiagnisis, M., Vatsos, I.N., Kyriakou, C. & Alexis, M. (2007). First report of *Listonella anguillarum* isolation from diseased big scale sand smelt, *Atherina boyeri* Risso 1810, in Limnos, Greece. *Bulletin of the European Association of Fish Pathologist* Vol.27, No.2, pp. 61-69.

Zorrilla, M., Chabrillon, A. S., Rosales, P. D. Manzanares, E. M., Balebona, M. C. & Morinigo, M. A. (2003): Bacteria recovered from diseased cultured gilthead sea bream, *Sparus aurata* L. in southwestern Spain.Aquaculture, Vol.218, No.1-4, 27 March 2003, pp. 11–20.

11 Chapter

PHARMACEUTICALS FOR THE IMMUNE SYSTEM IN FISH: IMMUNE FUNCTION, IMMUNOASSAY, PHARMACEUTICALS

1. INTRODUCTION: AN OVERVIEW

Fish constitute a diverse assemblage of organisms, encompassing agnathans (hagfishes and lampreys), chondrichthyans (sharks and rays), and teleosts (bony fish). Similar to all vertebrates, fish possess cellular and humoral immune responses, as well as central organs primarily involved in immune defense. Fish and mammals have both similarities and variances in immunological function (Cabezas, 2006; Nelson, 1994; Tort et al., 2003; Zapata et al., 1996). The fish defense system is fundamentally analogous to that of mammals. Teleost fish possess phagocytic cells akin to macrophages, neutrophils, and natural killer (NK) cells, in addition to T and B lymphocytes, for their cellular defense mechanisms. Teleosts possess several humoral defense components, including complement (both classical and alternative routes), lysozyme, natural hemolysin, transferrin, and C-reactive protein (CRP). The presence of cytokines, including interferon, interleukin 2 (IL-2), and macrophage activating factors (MAF), has been documented (Secombes et al., 1996; Sakai, 1999). Conversely, the anatomy of the immune system markedly differs between fish and mammals. It is evident that fish do not possess bone marrow or lymph nodes. The head kidney functions as a principal lymphoid organ, alongside the thymus and spleen (Press & Evensen, 1999). Gut-associated lymphoid tissues, sometimes referred to as lymphoid organs, have demonstrated their role in generating immunological responses in carp (Joosten et al., 1996). Certain teleosts, like plaice, exhibit a lymphatic system distinct from the blood circulatory system, although the presence of such a system has been disputed in other species (Hølvold, 2007).

The health of fish is contingent upon the interaction between key components of the fish and their surrounding environment (Figure 1). The tolerance of these numerous elements is contingent upon the host and, in many instances, the husbandry procedures. The environment is arguably the most vital element of the fish health matrix, as ambient quality affects the fish's physiological health, cultured species, feeding practices, development rate, and capacity to sustain both natural and learned resistance and immunity. The overall physiological condition of the fish host is influenced by husbandry practices, environmental quality, nutritional state, and pathogens, all of which affect the host's inherent resistance and acquired immunity. It is widely acknowledged that fish subjected to these stressors exhibit increased vulnerability to infection. (Magnadóttir, 2010; Plumb & Hanson, 2011).

(modified from Magnadóttir, 2006 and Plumb & Hanson, 2011).

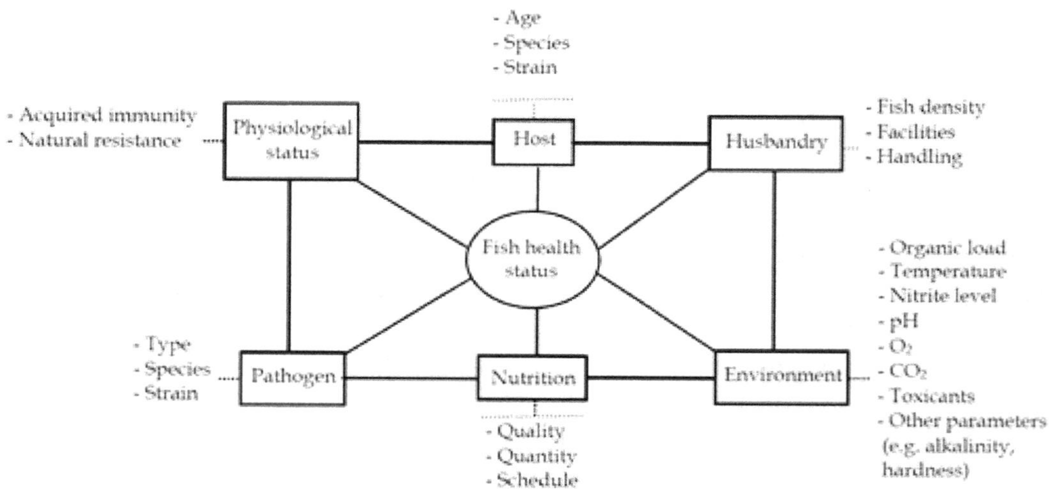

Fig. 1. The relationship of various factors in fish health status.

In addition, in the Food and Drug Administration (FDA) and the European Union (EU) member states, although a limited number of antimicrobial agents are licensed for use in fin fish culture, various drugs such as chemotherapeutics have been used to an increasing levels treat bacterial infections in cultured fish in the last decades years. However, the incidence of drug-resistant (including multiple and cross-resistance) bacteria has become a major problem in fish culture and public health (Alderman & Hasting, 1998; Aoki, 1992; Horsberg, 2003). Vaccination is a useful prophylaxis for infectious diseases of fish and is already commercially available for bacterial infections such as vibriosis, enteric red mouth disease (ERD) and furunculosis and some viral infection such as infectious pancreatic necrosis (IPN). Vaccination may be the most effective method of controlling fish disease. Furthermore, the development of vaccines against intracellular pathogens such as *Renibacterium salmoninarum* has not so far been successful. Therefore, the immediate control of all fish diseases using only vaccines is impossible. Immunostimulants such as synthetic chemicals, bacterial derivatives, polysaccharides or animal and plant extracts increase resistance to infectious disease, not by enhancing specific immune responses, but by enhancing non-specific immune defence mechanisms. Although, there is no memory component and the response is likely to be of short duration. Use of these immunostimulants is an effective means of increasing the immunocompetency and disease resistance of fish. Research into fish immunostimulants is developing and many agents are currently in use in the aquaculture industry (Klesius et al., 2001; Sakai, 1999; Subasinghe, 2009). Moreover, the incorporation of diverse dietary additives such as vitamins, carotenoids, probiotics, prebiotics, synbiotics, and herbal treatments into fish feed has been evaluated in fish. The overall results have been advantageous, including the reduction of stress responses, enhancement of innate parameters, and improvement of disease resistance. (Austin & Brunt, 2009; Hoffmann, 2009; Magnadóttir, 2010; Nayak, 2010).

2. IMMUNE SYSTEM COMPONENTS

2.1 Tissues and cells

Types of immune organs vary between different types of fish. In the jawless fish (hagfishes and lampreys), true lymphoid organs are absent. Instead, these fish rely on region of lymphoid tissue within other organs to produce their immune cells (Zapata et al., 1996). However, genetic differences may be small and some molecular and cellular agents similar, the anatomical and functional organisation such as the structure and form of the immune system (Press & Evensen, 1999; Randeli et al., 2008). The immune system of fish has cellular and humoral immune responses, and organs whose main function is involved in immune defence (Jimeno, 2008). Most of the generative and secondary lymphoid organs present in mammals are also found in fish, except for the lymphatic nodules and the bone marrow (Alvarez-Pellitero, 2008; Jimeno, 2008; Press & Evensen, 1999; Zapata et al., 1996). Instead, the anterior part of kidney usually called head kidney, aglomerular, assumes hemopoietic functions (Jimeno, 2008; Meseguer et al., 1995; Tort et al., 2003), and unlike higher vertebrates is the principal immune organ responsible for phagocytosis (Danneving et al., 1994; Galindo-Villegas & Hosokowa, 2004), antigen processing activity and formation of IgM and immune memory through melanomacrophagic centres (Tort et al., 2003). The most important immunecompetent organs and tissue of fish include the *kidney* (anterior/or head and posterior/or caudal), *thymus*, *spleen*, *liver*, and *mucosa-associated lymphoid tissues* (Figure 2) (Press & Evensen, 1999; Shoemarker et al., 2001). In fish, myelopoiesis generally occurs in the head kidney and/or spleen, whereas thymus, kidney and spleen are the major lymphoid organs (Zapata et al., 2006). Next to the thymus as the primary T cell organ head kidney is considered the primary B cell organ. Also, head kidney and spleen present macrophage aggregates, also known as melano-macrphage centres (Alvarez-Pellitero, 2008).

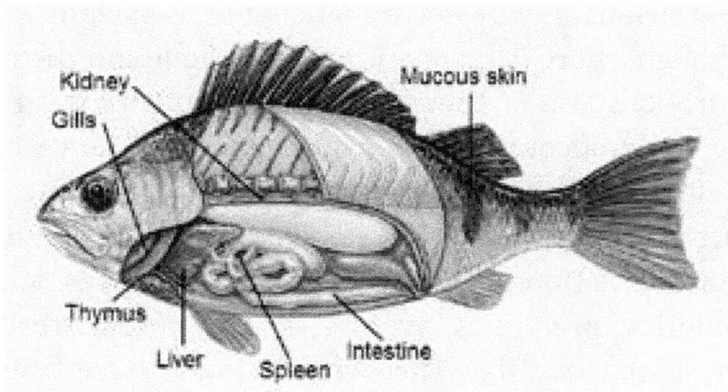

(modified from http://www.dkimages.com/discover/previews/1171/10686362.JPG).

Fig. 2. Immune structures in teleost fish.

The kidney often referred to as the head kidney tissue is important in hematopoiesis and immunity in fish. And it is predominantly a lympho-myeloid compartment (Press & Evensen, 1999). Early in development, the entire kidney is involved in production of immune cells and the early immune response. As the fish mature, blood flow through the kidney is slow, and exposure to antigens occurs. There appears to be a concentration of melanomacrophage centers are aggregates of reticular cells, macrophages, lymphocytes and plasma cells; they may be involved in antigen trapping and may play a role in immunologic memory (Galindo-Villegas & Hosokowa, 2004; Press et al., 1996; Secombes et al., 1982). The head kidney or anterior kidney (pronephros), the active immune part, is formed with two Y-arms, which penetrate underneath the gills. In addition, this structure of the kidney has a unique feature, and it is a well innervated organ, and the kidney is also an important endocrine organ, homologous to mammalian adrenal glands, releasing corticosteroids and other hormones. Thus, the kidney is a valuable organ with key regulatory functions and the central organ for immune-endocrine interactions and even neuroimmuno-endocrine connections (Press & Evensen, 1999; Tort et al., 2003).

The thymus is a paired bilateral organ situated beneath the pharyngeal epitelium in the dorso-lateral region of the gill chamber. But it seems that the size of the thymus varies with seasonal changes and hormonal cycles (Galindo-Villegas & Hosokowa, 2004; Meseguer et al., 1995; Press & Evensen, 1999; Zapata et al., 1996). The thymus appears to have no executive function. It is regarded, as a primary lymphoid organ where the pool of virgin lymphocytes in the circulation and other lymphoid organs. However, much of the data supporting this is indirect evidence obtained either by immunizing with T-dependent antigens (Ellsaesser et al., 1988) or by using monoclonal antibodies as cell surface markers (Passer et al., 1996) and functional *in vitro* assay. In addition, trout-labeled blood lymphocytes migrate through the thymus before reaching the spleen and kidney (Tatner & Findlay, 1991). It suggest that teleost thymus, despite its striking morphology, has the same function as in higher vertebrates, that is, it is the main source of immunocomponent T cells (Zapata et al., 1996), and research shows that the thymus is responsible for the development of T-lymphocytes, as in other jawed vertebrates (Alvarez-Pellitero, 2008; Galindo-Villegas & Hosokowa, 2004). In general, the available data support a correlation between the histological maturation of the teleost thymus, appearance of the lymphocytes in peripheral lymphoid organs, and development of the cell-mediated immune response (Zapata et al., 1996).

The spleen is the major peripheral and a secondary lymphoid organ in fish which contains fewer haemopoietic and lymphoid cells than the kidney, being composed mainly of blood held in sinuses, and it is believed to be involved in immune reactivity and blood cell formation (Galindo-Villegas & Hosokowa, 2004; Manning, 1994; Zapata et al., 1996). Most fish spleen is not distinctly organized into red and white pulp, as in mammals, but white and red pulp is identifiable. It contains different sized lymphocytes, numerous developing and mature plasma cells, and macrophages in a supporting network of fibroblastic reticular cells. Lymphocyte and macrophage are present in the spleen of fish, contained in specialized capillary walls, termed ellipsoids. In addition, ellipsoids appear to have a specialised function for plasma filtration and particularly immune complex. Most macrophage is arranged in malanomacrophage centers, and it is defined that they are primarily responsible for the breakdown of erythrocytes. These centers may retain antigens as immune complexes for long periods. Although the lymphoid tissue is poorly developed in the teleost spleen, after antigenic stimulation, increased amount of lymphoid tissue does appear, and indirectly suggesting the presence of T-like and B-like cells in this group fish (Espenes et al., 1995; Galindo-Villegas & Hosokowa, 2004; Zapata et al., 1996). The spleen of teleosts has also been implicated in the clearance of blood-borne antigens and immune complexes in splenic ellipsoids and also has a role in the antigen presentation and the initiation of the adaptive immune response (Alvarez-Pellitero, 2008; Chaves-Pozo et al., 2005; Whyte, 2007).

The liver is included under this chapter, because in mammals, it is responsible for production of components of the complement cascade and acute phase proteins (such as CRP), which are important in the natural resistance of the animal, defined that the liver of fish plays a similar role (Fletcher, 1981). On the contrary, research to support this claim is lacking (Galindo-Villegas & Hosokowa, 2004; Shoemarker et al., 2001).

The mucosa-associated lymphoid tissues in fish are distributed around the intestine referred to as the gut, skin and gills, thus complementing the physical and chemical protection provided by the structure (Jimeno, 2008; Press & Evensen, 1999; Tort et al., 2003). Teleost lack organized mucosa-associated lymphoid tissues such as Peyer's patches of mammals, though there is evidence that skin, gills and intestine contains populations of leucocytes (Jimeno, 2008; Press & Evensen, 1999) and innate and adaptive immunity act in case of attack of microorganisms (Ellis, 2001; Schluter et al., 1999). This equipment is completed with immunocompetent cells such as leucocytes and intraepithelial plasmatic cells (Dorin et al., 1994; Moore et al., 1998; Tort et al., 2003). Recently, several additional defences have been discovered in fish mucous membranes (Bols et al., 2001), such as the production of nitric oxide by the gill as well as antibacterial peptides and proteins by skin (Campos-Perez et al., 2000; Galindo-Villegas & Hosokowa, 2004; Ebran et al, 1999; Tort et al., 2003). Not only the mucous membranes of these tissues are an important physical barrier in fish, but also contain several components with a role in the host-parasite interaction, and release antimicrobial agents or proteins. Besides that among the epidermal secretions, complement, lysozyme, lectins (or pentraxins), alkaline phosphatase and esterase, trypsin (or trypsin- like), natural antibodies or immunoglobulins are often found, although their amount and activity depend on the species, and hemolysine are among the substances present with biostatic or biocidal activities (Alexander & Ingram, 1992; Alvarez-Pellitero, 2008; Aranishi & Mano, 2000; Arason, 1996; Balfry & Higgs, 2001; Ellis, 1999; Galindo-Villegas & Hosokowa, 2004; Jones, 2001; Fast et al., 2002; Magnadóttir, 2006; Palaksha et al., 2008; Shoemarker et al., 2001; Tort et al., 2003). Most research on the presence of immunoglobulin or antibody in the mucus suggests that mucus immunoglobulin is not a result of the transduction of immunoglobulin from the serum (Shoemarker et al., 2001). Mucous or goblet cells secrete mucus, which has at least three different types of defensive roles: (1) Mucus interrupts establishment of microbes by being continually sloughed off. (2) If establishment is accomplished, mucus acts as a barrier to be crossed. (3) The mucus on skin, and presumably the other surfaces, contains a variety of humoral factors with antimicrobial properties (Galindo-Villegas & Hosokowa, 2004; Tort et al., 2003).

All multicellular organisms possess a selection of cells and molecules that interact in order to ensure production from pathogens (Abbas & Lichtmann, 2006). This collection of highly specialised components makes up the immune system, and poses a physiological defence against microbe invasion (Jimeno, 2008). Fish immune cells show the same main features as those of other vertebrates, and lymphoid and myeloid cell families have been defined. Key cell types involved in non-specific cellular defence responses of teleost fish include the phagocytic cells monocytes/macrophages, non-specific cytotoxic cells (or NK cells), thrombocytes, granulocytes (or neutrophils) and lymphocytes (Table 1) (Buonocore & Scapigliati, 2009; Hamerman et al., 2005; Hølvold, 2007; Magnadóttir, 2006; Jimeno, 2008; Shoemarker et al., 2001).

Epithelial and antigen presenting cell also participate in the innate defence in fish, and some teleost have been reported to have both acidophilic and basophilic granulocytes in the peripheral blood in addition to the neutrophils. Furthermore, recently it has been observed that basophilic granular cells (acidophilic/eosinophilic granule cells or mast cells) of fish Perciformes order, the largest and most evolutionarily advanced order of teleosts, are endowed with histamine (Garcia-Ayala & Chaves-Pozo, 2009; Jimeno, 2008; Magnadóttir, 2006; Murelo et al., 2007; Whyte, 2007). Mononuclear cells in fish include the macrophages (and/or tissue macrophages) and monocytes. These cells are probably the single most important cell in the immune response in fish. Not only are they important in the production of cytokines, but they also are the primary cells involved in phagocytosis and the killing of pathogens upon first recognition and subsequent infection (Buonocore & Scapigliati, 2009; Cabezas, 2006; Clem et al., 1985; Garcia-Ayala & Chaves-Pozo, 2009; Secombes et al., 2001; Shoemarker et al., 2001). Macrophages also play major roles as being the primary antigen-presenting cell in teleost, thus linking the non-specific and acquired immune response (Balfry & Higgs, 2001; Galindo-Villegas & Hosokowa, 2004; Jimeno, 2008; Shoemarker et al., 2001; Vallejo et al., 1992). Thrombocytes are thought to be a nucleated version of the mammalian platelet. These cells are involved in blood clotting and have recently been thought to have phagocytic properties (Balfry & Higgs, 2001; Secombes, 1996).

Table 1. Non-specific immune cells in fish and their functional characteristics and mode of action.

Cellular components	Functional characteristics and mode of action
Monocytes/Macrophages	Phagocytosis, and phagocyte activation, cytokine production, intracellular killing, antigen processing and presentation, Secretion of growth factors and enzymes to remodel injured tissue, T-lymphocyte stimulation.
Granulocytes (or Neutrophils)	Phagocytosis, secretion and phagocyte activation, cytokine production, extracellular killing, inflammation.
Non-specific cytotoxic cells (or natural killer cells)	Recognition and target cell lysis, induce apoptosis of infected cells, Synthesize and secrete **interferon-gamma** (IFN-γ).

(modified from Hølvold, 2007; Shoemarker et al., 2001).

Fish possess polymorph nuclear cells, or granulocytes (especially neutrophils, and eosinophils, and basophils), that contain granules, the contents of which are released upon stimulation (Balfry & Higgs, 2001). These cells are highly mobile cell, phagocytic, produce reactive oxygen species, traveling via the blood and lymphatic systems to sites of infection and injure, thereby playing a vital role in the inflammatory response. Also, neutrophils are the primary cells involved in the initial stages of inflammation in fish, between 12 to 24 hours, and the function of the granulocytes may be cytokine production to recruit immune cells to the area of damage or infection (Galindo-Villegas & Hosokowa, 2004; Manning, 1994; Shoemarker et al., 2001). However, eosinophilic granular cells found in the stratum granuloma of the gut, gills and skin, and surrounding major blood vessels, are not considered to be eosinophils but rather mast cells (Vallejo & Ellis, 1989; Reite, 1998; Galindo- Villegas & Hosokowa, 2004). Cells mediating the lytic cycle to occur and destroy tumour target cells lines following receptor binding in fish have been denominated non-specific cytotoxic cells (Galindo-Villegas & Hosokowa, 2004), and are similar to (or closely related in function) the mammalian NK cells (Shoemarker et al., 2001). These cells capable of be important in protozoan parasites (Evans & Gratzek, 1989; Evans & Jaso-Friedman, 1992), and viral immunity of fish (Hogan et al., 1996), and are found in the blood, lymphoid tissue, and gut of fish (Balfry & Higgs, 2001). Lymphocytes are the cells responsible for the specificity of the specific immune response. The two different classes of lymphocytes (T and B) are the acknowledged cellular pillars of adaptive immunity, and can be distinguished by their cell surface markers and subsequent function (Balfry & Higgs, 2001; Garcia-Ayala & Chaves-Pozo, 2009; Pancer & Cooper, 2006). T lymphocytes recognize antigen that is presented by antigen-presenting cells such as macrophages, and are primarily responsible for cell-mediated immunity. These cells are also important sources of cytokines, which are particularly important in the inflammatory response (Balfry & Higgs, 2001). On the other hand, B lymphocytes are responsible for humoral immunity, and recognize antigen and produce specific antibodies to that antigen. T and B cells can be worked together and with other types of cells to mediate effective adaptive immunity (Garcia-Ayala & Chaves-Pozo, 2009; Jimeno, 2008; Miller et al., 1998; Pancer & Cooper, 2006). Interestingly, B cells from rainbow trout have high phagocytic capacity, suggesting a transitional period in B lymphocyte evolution during which a cell type important in innate immunity and phagocytosis evolved into a highly specialized component of the adaptive arm of the immune response in higher vertebrates (Jimeno, 2008; Li et al., 2006).

2.2 Humoral molecules

The classification of humoral parameters is commonly based on their pattern recognition specificities or effector function. Most non-specific humoral molecules involved in the natural resistance of fish are presented with composition and mode of action in Table 2 (Magnadóttir, 2006; Shoemarker et al., 2001). These components are act in several ways to kill and/or prevent the growth and spread of pathogens. Other acts as agglutinins (aggregate cells) or precipitins (aggregate molecules). There are also opsonins that bind with the pathogen and, in doing so, facilities its uptake and removal by phagocytic cells. In addition, some of these substances have important role in the inflammatory immune response, such as opsonins, anaphylatoxins, neutrophil, and macrophage chemo-attractants. Briefly, these factors involve various lytic substances/or hydrolase enzymes (lyzosyme, cathepsine L and B, chitinase, chitobiase, trypsin-like), agglutinins /or precipitins (CRP, serum amyloid P (SAP), lectins, α- and natural precipitins, natural antibodies, natural hemagglutinins), enzyme inhibitors (α_2-macroglobulin, serine-/cysteine-/ and metal- proteinase inhibitors) and pathogen growth inhibitors (interferon (IFN), myxovirus (Mx)- protein, transferrin, ceruloplasmin, metallothionein). Antimicrobial peptides such as cathelicidins (CATH-1, -2), defesins (DB-1, -2, -3), hepsidins (hepsidinLEAP-1, -2), piscidins (e.g. pleurocidin, epinecidin-1, dicentracin), ribosomal proteins, histone derivates (e.g. parasin, histon H2B, SAMP H1, oncorhyncins, hipposin), which widespread in nature as defence mechanism in plant and animals are also substances that have been identified in the tissue such as mucus, liver, skin and gills of some teleost species, including halibut and flounder (Alvarez-Pellitero, 2008; Aoki et al., 2008; Aranishi & Mano, 2000; Balfry & Higgs, 2001; Buonocore & Scapigliati, 2009; Cole et al., 1997; Ellis, 1999; Ellis, 2001; Galindo-Villegas & Hosokowa, 2004; Hølvold, 2007; Lemaître et al, 1996; Magnadóttir, 2006; Rodriguez-Tovar et al., 2011; Shoemarker et al., 2001; Smith & Fernandes, 2009; Smith et al., 2000; Tort et al., 2003; Whyte, 2007; Yano, 1996).

In addition, in teleost fish, evaluating the complement system as a humoral component is an essential part of the innate immune systems, and can be activated through the two /or three pathways of complement; (1) the classical pathway such as specific immunoglobulin or IgM is triggered by binding of antibody to the cell surface but can also be activated by acute phase proteins such as ligand-bound CRP or directly by viruses, bacteria and virus-infected cells, (2) the alternative pathway such as bacteria cell wall and viral components or surface molecules of parasites is independent of antibody and activated directly by foreign microorganisms, (3) the lectin pathway is elicited by binding of a protein complex consisting mannose-binding lectins to mannans on bacterial cell surfaces. All three pathways converge to the lytic pathway, leading to opsonisation or direct killing of the microorganism (Aoki et al., 2008; Balfry & Higgs, 2001; Ellis, 1999; Ellis, 2001; Galindo-Villegas & Hosokowa, 2004; Holand & Lambris, 2002; Nakao et al., 2003; Randelli et al., 2008; Shoemarker et al., 2001; Tort et al., 2003; Whyte, 2007; Yano, 1996).

Table 2. Non-specific humoral molecules and their composition and mode of action in fish.

Humoral components	Composition	Mode of action
Antibacterial peptides (*e.g. histone H2B, cecropin P1, pleurocidin, parasin, hipposin, SAMP H1*)	Protein	Constitutive and inducible innate defence mechanism, active against bacteria, defence before development of the specific immune response in the larval fish

Humoral components	Composition	Mode of action
Antiproteases (*e.g. α1-anti-protease, α2-anti-plasmin, α2-macroglobulin*)	----	Restricts the ability of bacteria to invade and growth *in vivo*, active against bacteria
Ceruloplasmin	Protein	Copper binding
Complement system (*e.g. C3, C4, C5, C7, C8, C9 and their isoforms, B- and D-factors*)	Protein	Promote binding of microbes to phagocytes, promote inflammation at the of complement activation, cause osmotic lysis or apoptotic death
Interferons (IFNs) /Myxovirus (Mx)-proteins (*e.g. IFN-αβ, IFN-γ*)	Glycoprotein /or Protein	Aid in resistance to viral infection, inhibit virus replication, inducible IFN-stimulated genes
Lectins (*e.g. legume and cereal lectins, mannose-binding lectin, C-type lectins, intelectin, cod, ladder lectin*)	Glycoprotein and/or specific sugar binding protein	Induce precipitation and agglutination reactions, recognition, promote binding of different carbohydrates in the presence of Ca+2 ions, active complement system, opsonin activity and phagocytosis
Lytic enzymes (*e.g. lysozyme, chitinase, chitobiase*)	Catalytic proteins lysozyme, complement components	Change the surface charge of microbes to facilitate phagocytosis, haemolytic and antibacterial and/or antivirucidal, antiparasitical effects, opsonic activity, inactivation of bacterial endotoxin(s)
Natural antibodies	----	Recognition and removal of senescent and apoptotic cells and other self-antigens, control and coordinate the innate and acquired immune response, activity against haptenated proteins
Pentaxins (*e.g. C-reactive protein, serum amyloid P*)	Protein	Opsonisation or activation of complement, promote binding of polysaccharide structures in the presence of Ca+2 ions, induce cytokine release, coast microbes for phagocytosis by macrophage
Proteases (*e.g. cathepsine L and B, trypsin-like*),	----	Defence against bacteria, activity against Vibrio anguillarum
Transferrin/Lactoferrin	Glycoprotein	Iron binding, acts as growth inhibitors of bacteria, activates macrophage

(modified from Hølvold, 2007; Shoemarker et al., 2001).

2.3 Cytokines and chemokines

The initiation, maintenance, and amplification of the immune response are regulated by soluble mediators named cytokines. Cytokines are the soluble messengers of the immune system and have the capacity to regulate many different cells in an autocrine, paracrine, and endocrine fashion, and can also be immune effectors (King et al., 2001). In the last few years, much interest has been generated in the study of fish cytokines and chemokines and significant progress, and has been made in isolating these molecules from fish. In recent years, various cytokines have been described in fish, but the major drawback in identifying fish cytokines is the low sequence identity compared to their mammalian counterparts. The low sequence identities also limit the detection of proteins of fish cytokines by using the antibodies of human cytokines (Plouffe et al., 2006). Most of these have been identified in biological assays on the basis of their functional similarity to mammalian cytokine activities. Some have also been detected through their cross-reactivity with mammalian cytokines (Manning & Nakanishi, 1996).

The predominant pro-inflammatory cytokines are interleukins (ILs) (especially IL-1β and IL-6) and tumour necrosis factor-alfa (TNF-α) (Balfry & Higgs, 2001; Bird et al., 2005; Corripio- Miyar et al., 2006; Garcia-Ayala & Chaves-Pozo, 2009; Hølvold, 2007; Jimeno, 2008; King et al., 2001; Magnadóttir, 2010; Randelli et al., 2008; Savan et al., 2005; Tort et al., 2003). These cytokines have a number of systemic effects, including body temperature elevation neutrophil mobilization, and stimulation of acute phase protein production in the liver (Balfry & Higgs, 2001; King et al., 2001; Randelli et al., 2008). Additional several cytokine /or cytokine homologues found in fish include IL-2, IL-4, IL-10, IL-11, IL-12, IL-15, IL-18, IL21, IL22, IL-26 and IFN-γ, (Balfry & Higgs, 2001; Bei et al., 2006; Bird et al., 2004; Buonocore & Scapigliati, 2009; Corripio-Miyar et al., 2006; Garcia-Ayala & Chaves-Pozo, 2009; Hølvold, 2007; Igawa et al., 2006; Inoue et al., 2005; Jimeno, 2008; King et al., 2001; Li et al., 2007; Magnadóttir, 2010; Randelli et al., 2008; Tort et al., 2003; Wang et al., 2005; Whyte, 2007; Yoshiura et al., 2003; Zou et al., 2004), and others cytokines in some fish species include transforming growth factor-β family such as TGF-β_1, -β_2, -β_3, -βA, and –βB, macrophage- migration inhibition factor (MIF), macrophage-colony stimulating (M-CSF or CSF-1; such as CSF-1R or sCSF-1R), chemotactic factor (CF) and plateled activating factor (PAF). However, no antibody markers are at present available for fish TGF-β, M-CFS and PAF (Belosevic et al., 2006; Garcia-Ayala & Chaves-Pozo, 2009; Klesius et al., 2010; Manning & Nakanishi, 1996; Randelli et al., 2008; Tafalla et al., 2003). On the other hand, orthologous cytokines in teleost fish have been classed as Class I, Class II, chemokines, TNF superfamily and IL-1 family (Table 3) (Alvarez-Pellitero, 2008; Aoki et al., 2008; Lutfalla et al., 2003).

IL-1β has been identified in 13 different species of teleost, and is produced by macrophage and also by a variety of other cells such as neutrophilic granulocytes. These ones are play a role in immune regulation through stimulation of T cells which is analogous to mammalian IL-1β. In addition, it is an important mediator of inflammation in response to infection and it has been reported in the trout to directly affect hypothalamic-pituitary-interrenal axis function, stimulating cortisol secretion. Another potentially important cytokines, TNF-α has been cloned in various fish. Besides, TNF-like protein activity has been shown to induce apoptosis, and to enhance neutrophil migration and macrophage respiratory burst activity. The number of studies in fish have provided indirect evidence suggesting that TNF-α is an important macrophage activating factor (MAF) produced by leukocytes. In some fish species, homologous MAF containing supernatants have been shown to induce a typical activated-macrophage response, evidence by increases in phagocytosis and nitric oxide production (Balfry & Higgs, 2001; Garcia-Ayala & Chaves-Pozo, 2009; Holland et al., 2002; Hølvold, 2007; Tort et al., 2003; Whyte, 2007). In addition, TNF-α has been shown increase chemotaxis of rainbow trout anterior kidney leukocytes and induces the expression of a number of genes in the immune response including IL-1β, IL-8 and cyclooxygenase-2 (COX- 2) (Zou et al., 2003). Other vital cytokines, IFNs are secreted proteins, are also pH-resistant cytokines which are produced by many cell types in response to a viral infection (within 2 days in rainbow trout injected viral haemorrhagic septicemia virus), and occurs in very young fish. In isolated Atlantic salmon macrophage stimulated with polyinosinic polycytidylic acid (poly I:C), peak IFN production occurred within 24 h and peak Mx protein production after 48 hours (Ellis, 2001; Nygaard et al., 2000). Therefore, IFN-mediated antiviral defence mechanisms are able to response during the early stages of a viral infection, which is mediated by the innate non-specific IFN responses while long-term protection is mediated by the specific immune response (Galindo-Villegas & Hosokowa, 2004; Ellis, 2001).

Table 3. Cytokines of teleost fish, and their function/or structure and members.

Class	Function /or Structure	Members
Cytokine class I	Involved in expansion and differentiation of cells. Have a 4-α helix bundle structure	IL-a and -b, IL-1 1-a and -b*, epo, GCSF-a and -b*, leptin, PRL, GH, M17*, M17 homologue (MSH)*
Cytokine class I	Involved in minimizing damage to host after insult. Contain more than 4-α helices.	IFN-α1, IFN-α2, IFN-γ, IL-10, IL-20, IL-24
Chemokines	Regulate cell migration under both inflammatory and homeostasis. Small proteins with 4 conserved Cys residues.	CXC (CXCL8-like, CXC-10, -12, - 13, - 14), CC (CCL19/21/25, CCL20, CCL27/28, CCL17/22,MIP, MCP)
TNF super family	Involved in inflammation and lymphoid organ development. Compact trimmers as membrane bound or soluble proteins.	Lymphotoxin-β,lymphotixin-β, TNF-α
IL-1 family	Involved in pro-inflammatory responses. Fold rich in β-strands.	IL-1α, IL-1β, IL-18

*: Only found in fish. (modified from Aoki et al., 2008).

Chemokines are known as second-order /or chemotactic cytokines, are a superfamily of small secreted cytokines that direct migration of immune cells to sites of infection, produced by different cell types that have, among other function, chemoattractant properties stimulating the recruitment activation and adhesion of cells to sites of infection injury (Alvarez-Pellitero, 2008; Aoki et al., 2008; Ellis, 2001; Hølvold, 2007). Different chemokines have been characterized in some fish species such as rainbow trout, carp, catfish, flounder and Atlantic halibut, including members of the first two conserved cysteines in their sequence: CXC, CC, C and CX_3C class /or family. Although, the CC chemokines represent the largest subfamily of chemokines, IL-8 was the first known chemokines, and other chemokines such as CXCL8 (or IL-8), γIP-10, CK-1 and CK-2 belongs to the subfamily. Chemokines play a key role in the movement if immune effector cells to sites of infection and it is becoming increasingly clear that their function is also necessary to translate an innate immune response into an acquired adaptive immune (Alvarez-Pellitero, 2008; Aoki et al., 2008; Hølvold, 2007; Peatman & Liu, 2007; Whyte, 2007).

3. FISH IMMUNE SYSTEM DESCRIPTION

In this section, since complexity and due this component of the immune system including innate (non-specific) and acquired (specific / or adaptive) immune systems in fish is out of the scopes of this chapter, will not be described in detail, but will be briefly mentioned here- in. Hereof, components of these systems and its mode of action were given in detail at Section 2.

The classical division of the immune system is into the innate and the adaptive systems. Despite the fact that dividing immune system into the innate and the acquired immunity is a common practice, recent studies in both fish and mammalian immunology demonstrate that these are combined systems rather than independent systems. Thus, the innate immune response is also important in activating the acquired immune response (Figure 3) (Fearon & Locksley, 1996; Jimeno, 2008; Medzhitov, 2007; Shoemarker et al., 2001).

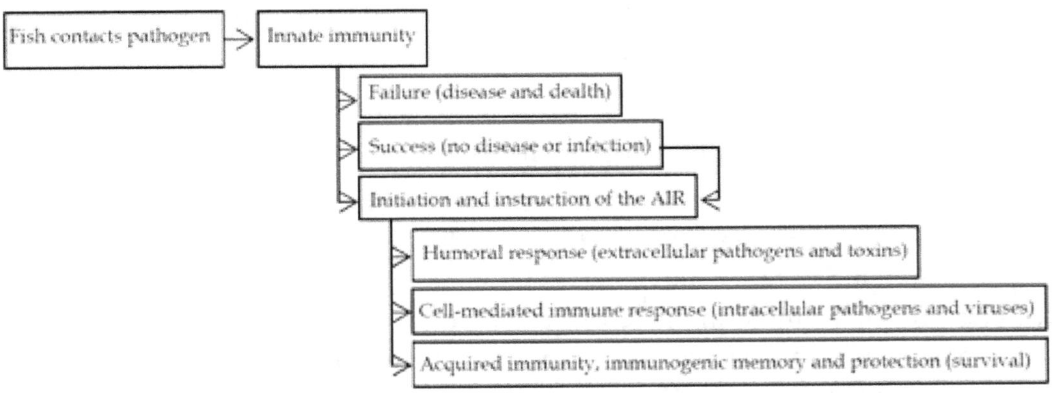

AIR: Acquired immune response. (modified from Shoemarker et al., 2001).

Fig. 3. Schematic representation of the response of a fish following an encounter with a pathogen.

3.1 Innate (non-specific) immune system

The innate immune system is of prime importance in the immune defence of fish. It is commonly divided into 3 compartments: (1) physiochemical barriers and/or the epithelial and/or mucosal barrier such as scales, epithelial surface (on gills, skin and gut) with secreted mucus, (2) the humoral parameters such as cell secretions of complement, CRP, IFN, lysozyme, transferrin, lectins, antimicrobial peptides, and (3) the cellular components such as non-specific cytotoxic cells (or NK cells), monocytes/macrophages, thrombocytes, granulocytes (or neutrophils), lymphocytes (see Section 2) (Buonocore & Scapigliati, 2009; Jansson, 2002; Magnadóttir, 2010; Rodriguez-Tovar et al., 2011). The general term for these innate parameters is pattern recognition proteins or receptors. These parameters recognize pathogen associated molecular patterns (PAMPs) associated with microbes and also inhered danger signals from malignant tissue or apoptotic cells. Typical PAMP are polysaccharides and glycoproteins like bacterial lipopolysaccharide, fragellins, teichoic acid and peptidoglycans, bacterial CpG and virus associated double-stranded RNA (Alvarez- Pellitero, 2008; Cabezas, 2006; Ellis, 2001; Hølvold, 2007; Jimeno, 2008; Magnadóttir, 2010; Medzhitov & Janeway, 2002; Whyte, 2007). However, under normal conditions the fish maintains a healthy state by defending itself against the potential invaders by a complex system of innate defence mechanisms. These mechanisms are both constitutive and responsive and provide protection by preventing the attachment, invasion or multiplication of microbes on or in the tissue. Immune systems effecting drugs such as immunostimulants, probiotics, prebiotics and synbiotics should act through the enhancement of the innate immune response (Austin & Brunt, 2009; Galindo-Villegas & Hosokowa, 2004; Hoffmann, 2009; Magnadóttir, 2006; Nayak, 2010).

The production or expression of both humoral and cellular innate parameters is commonly amplified or up-regulated during immune response, but there is believed to be no memory. This mean that a second encounter with the same pathogen will not result in enhance response as is seen in acquired immune response (Magnadóttir, 2010).

3.2 Acquired (specific) immune system

If a pathogen evades or overwhelms the innate defence mechanism of the lost, causing the foreign antigen to persist beyond the first several days of infection, an acquired immune system components is initiated. In addition, the antigen-specific lymphocytes of acquired immune response are capable of swift clonal expansion and of a more rapid and effective immune response on subsequent exposures to the pathogen (King et al., 2001). However, activation of the acquired immune system is relatively slow, requiring specific receptor selection, cellular proliferation and protein synthesis but it is long lasting (Magnadóttir, 2010).

In contrast to the innate immune systems components, the acquired immune system produces effector cells (T- and B-lymphocytes) and molecules (immunoglobulins (Igs)/or specific antibodies), which are highly specific to the antigen of the invading microbe. The B-cells, similar to the B1-subset of mammalian B-cells, are involved in the humoral response while the T-cells are responsible for the cell-mediated response (Galindo- Villegas & Hosokowa, 2004; Jansson, 2002; King et al., 2001; Magnadóttir, 2010). Furthermore, the other key elements in the evolution of the acquired immune system are the appearance of the thymus, the recombination activation gene (RAG; especially RAG 1 and 2 genes) enzymes, which through gene rearrangement generate the great diversity of the Ig superfamily (T- and B-cell receptors) and major histocompatibility complex (MHC). On the other hand, the key humoral parameter of the acquired system is the Igs (antibodies), expressed either as B-lymphocytes receptor or secreted in plasma. The trigger for activation of the acquired immune system, the activation and proliferation of lymphocytes, take place in organized lymphoid tissue. Following activation by a specific antigen, either in soluble form or in association with the MHC marker on antigen presenting cells, the B-cells proliferate and differentiate into long lasting memory cells and plasma cells, which secrete the specific antibody. Also, T-cells, using a specific receptor, recognise pathogen only in association with the MHC marker on antigen presenting cells (Alvarez-Pellitero, 2008; Buonocore & Scapigliati, 2009; Galindo-Villegas & Hosokowa, 2004; Jansson, 2002; King et al., 2001; Magnadóttir, 2010; Rodriguez-Tovar et al., 2011). Effectively only one functional Ig class, a tetrameric IgM, is demonstrated in teleost fish, and these molecule is also made up of eight heavy (mu)-/and light (lambda)- chains. This is in contrast to the pentameric Ig classes and sub-classes mammals on the basis of heavy chain molecular weight and on their surface and secrete-antibodies only of the Ig class. Other Ig-like molecules have been described in some fish species, which may increase the diversity of the B-cell recognition capacity (Lorenzen, 1993; Magnadóttir, 2010; Randelli et al., 2008; Shoemarker et al., 2001; Wilson et al., 1997). Resistance to and recovery from first infection are a results of complex interactions between innate and acquired defence mechanism (Lorenzen, 1993). A summary of innate and acquired immune systems in fish is shown in Figure 4 (Jimeno, 2008).

Briefly, the immune reaction in fish is influenced by endogen rhythms and environmental parameters, of which temperature is by far the most important. Another important factor is nutrition, which may be subject to enormous variation within and between wild populations (Lorenzen, 1993). The immunosuppressive effects of population and stress resulting in higher disease susceptibility are well known. Choosing a universal trait or an innate component that could act as a biomarker for adverse conditions in aquaculture is however problematic. This is because of the variable effects on innate an acquired parameters depending on the type and duration of adverse conditions and on the fish species (Magnadóttir, 2006; Ortuño et al., 2001). The innate and acquired immune systems are given activity/or factor, cells involved and cellular markers in Table 4 (Jansson, 2002; Randelli et al., 2008).

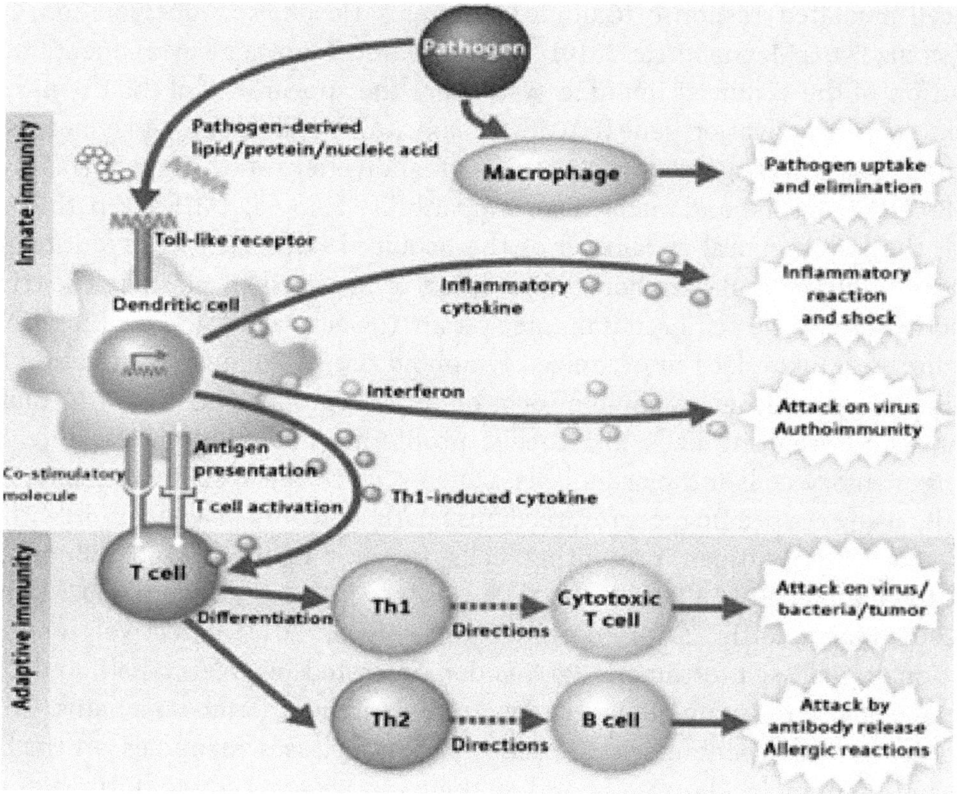

Th1: T-helper 1, Th2: T-helper 2. (Jimeno, 2008).

Fig. 4. Cross-talk between innate and acquired immune systems.

Table 4. The innate and acquired immune systems activity and/or factors and cellular markers.

Activity/Factor	Cell involved	cDNA sequence coding for	Cellular marker
Innate immunity			
Phagocytosis	Mononuclear phagocytes B-cells	-	mAb to M□, and IgM, neutrophils, pAb to granulocytes, granulin
ROS species	Mononuclear phagocytes	iNOS	NBT, no antibodies
Complement, APR	Hepatocytes	C3, C4, C5, C7, C8, CRP, SAP	pAb to C3

Activity/Factor	Cell involved	cDNA sequence coding for	Cellular marker
Antibacterial	Various types	Families of peptides	None
Antiviral	Leucocytes, fibroblasts	IFN-1, IFN, Mx-protein	None
Enzymes	Various types	Lysozyme, caspases, proteases	None
Inflammation, cytokines, monokines	Leucocytes	TNF-α, COX-2, PLA2, TLRs, ILs (1, 6, 12, 14, 16, 17, 18, 20, 21, 22), >16 chemokines	pAb for IL-1, pAb and mAb for TNF-α
Non-specific killing	Leucocytes	NCCRP-1	mAb to 5C6
Acquired immunity			
Memory, specific antibody,	B-cells	IgM, IgD, IgT, RAGs	mAb to IgM, B-cells
Memory, cellular recognition,	T-cells	TcR-α, -, β -γ, -δ, CD3, RAGs	DLT15, WCL38
Specific killing	T-cells	CD8-α, CD8-β, MHCI	None
Helper activity	T-cells	CD4, MHCII	None
	Th1 /or Th2	IFN-γ, IL-2 /or IL-4, IL-10	None
	Leucocytes	ILs (7, 15, 21, 22, 26), LtB	None

APR: Acute phase response, CD: Cell-differentiation cluster; COX-2: Cyclooxygenase 2, CRP: C-reactive protein, iNOS,: Inducible nitric oxide synthase, IFN: Interferon, Ig: Immunoglobulin, IL: Interleukin, LtB: Lymphotoxin B, M: Macrophage, MHC: Major histocompatibility complex, NCCRP-1: Non-specific cytotoxic cells receptor protein-1, NBT: Nitroblue tetrazolium, PLA2: Phospholypase A2, RAGs: Recombinase-activating genes, ROS: Reactive oxygen species, SAP: Serum amyloid P, TcR: T-cell receptor, Th1: T-helper 1, Th2: T-helper 2, TLRs: Toll-like receptors, TNF: Tumor necrosis factor, mAb: Monoclonal antibodies, pAb: Polyclonal antibodies. (modified from Randelli et al., 2008).

4. IMMUNOASSAY

Diagnostics is the determination of the cause of a disease or clinical pathology. The techniques used range from gross observation to highly technical biomolecular-based tools. Pathogen screening is another health management technique, which focuses on detection of pathogens in sub-clinical, or apparently healthy, hosts. Schematic representation of the diagnosis using a stepwise clinical approach is presented in Figure 5 (King et al., 2001; Subasinghe, 2009).

In recent years, fish immunological research has been mainly focused on two aspects: (1) Firstly, comparative and development studies have contributed to a better understanding of the characterize, the structural and functional evolution of the immune system mechanisms and pathways from invertebrate, through fish to mammals, (2) The second aspect, and one that has received the major funding, is the requirement of the fish farming industries, and also has understated how the fish immune system responds the foreign agents. The word-wide growth in aquaculture in the past 2-3 decades has demanded the development of a comprehensive knowledge of the immune system of the commercially important fish species, and also has understated how the fish immune system responds the foreign agents. The purpose has been twofold: to secure to optimum activity of the natural immune defence of the fish through cultural conditions and the choice of fish stock (or by breeding to produce stock of fish with superior disease resistance), and also to develop and improve prophylactic measure such as vaccination, immunostimulants and probiotics (Alvarez-Pellitero, 2008; Galindo-Villegas & Hosokowa, 2004; Ellis, 2001; Magnadóttir, 2010).

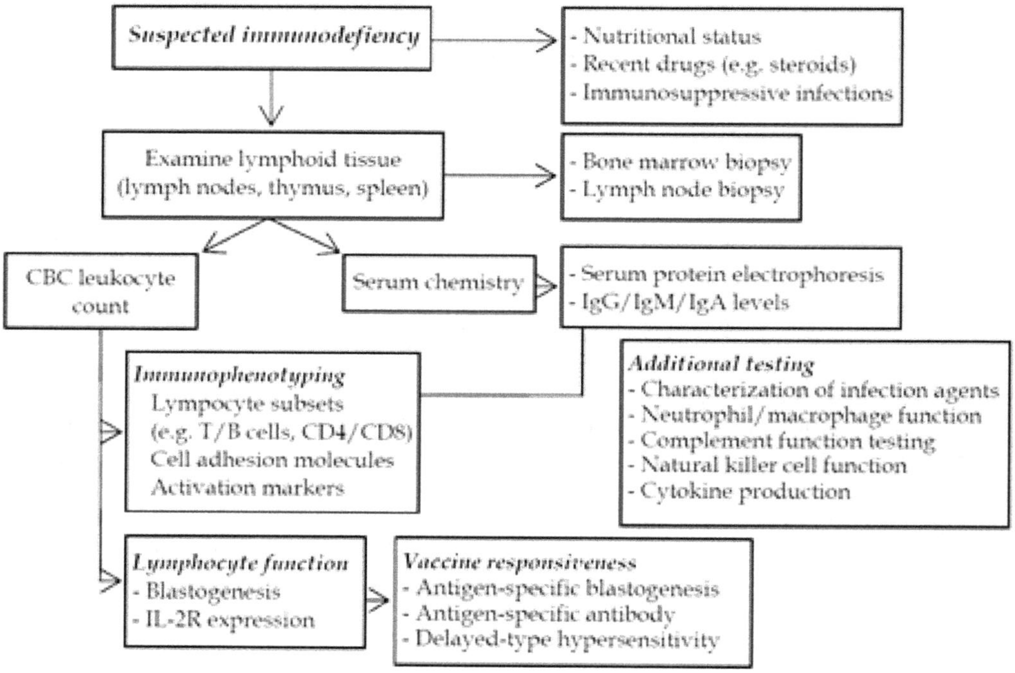

CBC: Complete blood count, Ig: Immunoglobulin, IL: Interleukin. (modified from King et al., 2001).

Fig. 5. Schematic representation clinical evaluation of the immune system.

A variety of technologies have already made an impact in reducing disease risk and many novel methods will contribute in the future (Adams & Thompson, 2006; Adams & Thompson, 2008). Improved nutrition, use of probiotics, improved disease resistance, quality control of water, seed and feed, use of immunostimulants, rapid detection of pathogens and the use of affordable vaccines have all assisted in health control in aquaculture. The success of vaccination in reducing the risk of furunculosis in salmon is an excellent example of technology having made a significant impact. This is turn led to a reduction of the use of antibiotics that has been sustained, and productivity has increased as a result of vaccination (Gudding et al., 1999; Adams et al., 2008).

Table 5. Hematological, innate and acquired immune response assays.

Hematological/physiological assays-blood samples Specific immune response assays	**Specific immune response assays**
Hematocrit: Percent of red blood cell pack	*Scale rejection*: Transplantation indicator
Leukocrit: Percent of white blood cell pack	*Delayed hypersensitivity*: Allergenic reactions
Cell counts and differentials: Numbers of cells and types	*Trypan blue*: Killer cell activity
Lysozyme levels: Enzyme level in blood	*Chromium release*: Killer cell activity
Serum immunoglobulin level: Specific and nonspecific antibody	*Melanomacrophage centers*: Antigen processing cells, Antigen accumulation: Concentration in spleen or kidney areas, Cell aggregates:
Serum protein level: Total protein in serum	Increase in numbers of melanomacrophage cells *Passive hemolytic plaque assay* (*Jerne assay*): Antibody- producing cells
Innate defensive mechanism or acquired immune response assays	*Assays measuring serum antibody levels*
(These assays can be used for either response)	**Immunolectrophoresis /or immunoassay and DNA-based diagnosis**
Phagocytosis: Percents and indexes; engulfment by phagocytic cells/or phagocytic activity: By incubating blood with a killed bacterial culture	*CF*: Complement fixation
Bactericidal activity: By incubating macrophages with a live bacterial culture	*DBH or WB*: Dot blot hybridization or Western blot
Rosette-forming cells: Adherence of particles around lymphocpes	*ELISA*: Enzyme-linked immunosorbent assay
Glass or plastic adherence: Stickiness of phagocytic cells	*FAT*: Fluorescent antibody assays (or technique) *IFAT*: Indirect-FAT
Pinocytosis: Engulfment of fluids by phagocytic cells	*QFAT*: Quantitative-FAT
Neutrophil activation: Myelo-peroxidase production and NBT dye reduction by oxidative burst e.g. oxidative radicals, Chemilurninescence: light detection from oxidative burst	*ISH*: In situ hybridization
Blastogenesis: Mitosis of lymphocytes cells;	*LAMP*: Loop-mediated isothermal amplification
Agglutination / or Hemagglutination	*PCR*: Polymerase chain reaction
Preciptinin (*Ouchterlony gel*): Measures soluble antigens in gels	*QPCR*: Quantitative (or real-time)-PCR
Immunoelectrophoresis: For defining blood or antigenic components	*rcb-PCR*: Reserve cross blot-PCR
	RT-PCR: Reverse transcriptase-PCR
	RIA: Radioimmunoassay
	SNT or VNT: Serum- /or virus- neutralization test
	Lateral-flow immunoassays
	Multiplex assays (e.g.*Protein array system, Micro- arrays*)

(modified from Anderson, 1996).

Many of the assays for detecting the changes in the protective mechanism of the fish due to immunomodulations are divided from those used in fish disease diagnostics and immunization programs. Although, most used tests in the last decades, most used assays for fish immunomodulation diagnosis are given as list in Table 5 (Adams & Thompson, 2008; Anderson, 1996; Brown-Treves, 2000, Jeney & Anderson, 1993; King et al., 2001; Lorenzen, 1993; Roque et al., 2009; Plumb & Hanson, 2011; Subasinghe, 2009). A large number of methods have been developed for immunodiagnostics and these are used routinely in many laboratories for the detection of fish and shellfish pathogens. These tools include both immunoassay and DNA-based diagnostic methods such as enzyme-linked immunosorbent assay (ELISA), radioimmunoassay (RIA), polymerase chain reaction (PCR), quantitative (or real-time)-PCR (QPCR), reverse transcriptase-PCR (RT-PCR), fluorescent antibody assays (FAT), indirect-IFAT, quantitative-FAT (QFAT), immunohistochemistry (IHC), *in situ* hybridization (ISH) and blot (dot-blot/dip-stick/western blot) amplification techniques (Adams et al., 2008; King et al., 2001; Plumb & Hanson, 2011; Roque et al., 2009; Subasinghe, 2009). However, with the development of Rapid Kits (immunochromatography/lateral flow) which are simple to use, sensitive and inexpensive (Adams & Thompson, 2008).

These molecular-based techniques (immunoassay and nucleic acid assay) provide quick results, adaptable to field situation, with high sensitivity and specificity, at relativity low cost, and can be easily applied to a large number of samples, and are also particularly valuable for infections which are difficult to detect such as sub-clinical infections using standard histology and tissue-culture procedures such as histopathology, bacteriology, virology, parasitology and mycology. They can be used for non-lethal sampling, and are valuable for monitoring challenge experiments under controlled laboratory conditions. Further development of this technology is likely to enhance more rapid detection and diagnosis of disease, which is crucial for early and effective control emergent disease situations (Adams & Thompson, 2008; Subasinghe, 2009). Although, modern immunoassays are very sensitive, sometimes their result may not be easy to analyse. This is partly because the blood chemistry and/or immune parameters of fish is highly depended on environmental conditions, nutrition, and other factors such as degree of antigen purity, genetic make-up, maternal effects, age and sexual maturation. There are also differences in sensitivities and specificities for each method and in the type of samples that can be used such as formalin fixed, fresh, tissue, blood, water. Further limitations of some immunoassays are that they can be lengthy assay to perform, required cell culture expertise, specific reagent and equipment, and requiring up to 7 to 14 days before they can be evaluated. In addition, non-specific reactions in immunoassays may vary by an order of magnitude between fish caught at the same time and palace, and may eventually obscure specific antibody activity (Table 6)

(Adams et al., 2008; Adams & Thompson, 2008; King et al., 2001; Lorenzen, 1993; Magnadóttir, 2010; Vatsos et al., 2003).

Any antibody-based test is only as good as the antibody used in it, and a standard protocol and reliable source of standard specific antibody is crucial. Antibody probes can be produced in a number of ways, including polyclonal antibodies (prepared in animal species, and can also be very useful tools for the detection of pathogens), monoclonal antibodies (prepared using hybridoma technology), phage display antibodies or antibody fragments. However, serum contains many different types of antibodies and mixed populations of antibodies can create problems in some immunological techniques (Adams, 2004; Adams & Thompson, 2006; Adams et al., 2008), some of which are now commercial available. Although some antibody-based methods can be very sensitive and carrier status can be detected, such technology can be limited in sensitivity when environmental samples are used, such as water samples, and molecular methods are ideal in this situation. (Adams et al., 2008; Zhang et al., 2004; Zhang et al., 2006)

Molecular technologies are also widely used for the detection of fish pathogens (Adams & Thompson, 2006; Cunningham, 2004; Wilson & Carson, 2003). They have been successfully utilized for the detection and identification of low levels of aquatic pathogens. In addition, molecular methods can be used for the identification to pathogens to species level and in epidemiology for the identification of individual strains and differentiating closely related strains. The DNA-based methods such as PCR are extremely sensitive. However, false positive and false negative results can cause problems due to contamination or inhibition. Real-time PCR (closed tube to reduce contamination) and Nucleic Acid Sequence Based Amplification (NASBA) are alternatives that reduce this risk and offer high sample throughput. Some of the most common PCR-based technologies used for the detection of pathogens are nested PCR, random amplification of polymorphic-DNA (RAPD), reverse transcriptase-PCR (RT-PCR), reverse cross blot-PCR (rcb-PCR) and RT-PCR enzyme hybridisation assay. *In situ* hybridisation is also widely used in the detection of shrimp

Table 6. Used methods, advantages and disadvantages to diagnose fish disease.

Method	Advantage	Disadvantage
Conventional methods		
Culture	Useful because the pathogen is isolated and the etiological agent can be confirmed	Labour intensive, can be expensive, not always possible to confirm identity of etiological agent
Histopathology	Useful for assisting in the diagnosis of disease, particularly where the causative agents of new diseases have not yet been identified	Labour intensive; skilled personnel required, not always possible to identify agent

Microscopy	It is an important tool in many of the methods shown in this Table. Many different types of microscopes are now available	Can be labour intensive; skilled personnel required; can be expensive if using confocal microscope or TEM. Not always possible to identify agent
Biochemical analysis	Useful for identifying bacteria with characteristic biochemical profiles; commercial kits available for this purpose	Can be labour intensive; skilled personnel required. Not always possible to identify agent
Molecular methods		
PCR	Very sensitive, can be automated to analyse large sample numbers	Only detects presence of DNA of pathogen, not the whole organism. False positive and negative results can occur
Nested-PCR	Extremely sensitive method, more sensitive and specific than one-round PCR	Takes longer than the one-round PCR. False positive and negative results can occur
RT-PCR	Can detect live pathogens (e.g. detects RNA)	Care needed to ensure RNA is not degraded
Random amplified polymorphic DNA	Useful method for determining the identity of microorganisms at a strain level, assessing the genetic relationship of samples or analysing mixed pathogen populations in samples	Can be labour intensive. Skilled personnel required
Reverse cross blot-PCR	Useful for distinguishing closely related species	Expensive. Labour intensive. Skilled personnel required
RT-PCR enzyme hybridisation assay	Can detect live pathogens. Large sample numbers can be analysed	Labour intensive. Skilled personnel required
In situ hybridisation	Detects DNA or RNA of pathogen, therefore there is no need for antibodies to detect protein	Labour intensive. Skilled personnel required. Expensive, sometimes difficult to see pathology in tissue sections after procedure
LAMP	Fast, with results obtained in a couple of hours. Suitable for field application. Does not require skilled operator. Results easy to interpret. Sensitive	Complex to set up initially
Quantitative-PCR	Allows quantification of DNA that can be related to pathogen level in infected tissue. Extremely sensitive	Labour intensive. Requires specialised equipment. Skilled personnel required. Expensive
Immunological methods		
Agglutination	Simple method, no requirement for specialised equipment	Not very sensitive in comparison to other immunological methods

ELISA-detection of pathogen	Versatile method that can be used to identify pathogens or antibodies depending on how assay is set up. Microassay–therefore small amounts of reagent needed. Quantitative; can be automated to analyse large sample numbers. Sensitive	Standardised reagents and specialised equipment needed. Need careful selection of controls and a skilled operator
Immuno-histochemistry	An extension of histopathology–the pathology can be observed around the infected tissue as the slide is counterstained. Can be amplified to increase sensitivity	Need formalin-fixed, wax embedded tissue sections, therefore procedure is labour intensive. Need a skilled operator to analyse results
Western blot	Particularly useful for serology to identify pathogen-specific proteins	Standardised reagents and specialised equipment needed. Need careful selection of controls and a skilled operator
Dot blot	Versatile method which can be used to identify pathogens or antibodies depending on how assay is set up. Microassay–therefore only small amounts of reagent needed. Protein not denatured in process unlike Western blotting	Standardised reagents need to be available to perform analysis. Need a skilled operator
FAT/IFAT	Fast method if performed directly on infected tissue smears, takes longer if fixed tissue sections are used (e.g need to process infected tissue). Sensitive. Useful for detection of viruses	Need a skilled operator to analyse results, auto-fluorescence on tissue sections can interfere with interpretation of results. Requires specialised equipment
Serology-ELISA detection of fish antibodies	Non-destructive sampling method, uses ELISA format therefore can screen large numbers of samples	Indirectly detects the presence of the pathogen. Most suitable for viral infections as antibodies against Gram- negative bacteria may cross-react in assay. In order to perform the assay a specific anti-fish species antibody is required. Needs careful interpretation
Rapid kits	Fast (results obtained in minutes), inexpensive, suitable for field application. Easy to interpret results. Sensitive	Designed to be used with fresh tissue. Using frozen or fixed tissue may affect sensitivity of results

Multiplex methods		
Protein array system (Luminex)	Versatile method that can be used to identify pathogens or antibodies depending on how assay is set up. Can detect proteins or DNA. Microassay–therefore only small amounts of reagent needed. Quantitative. Can measure several pathogens or analytes simultaneously. Sensitive	Labour intensive. Needs a skilled operator. Expensive. Standardised reagents need to be available to perform analysis. Requires specialised equipment
Multiplex-PCR assays	Can detect more than one pathogen with the assay. Sensitive	Difficult to standardise. Expensive
Micro-arrays	Can detect more than one pathogen with the assay. Allows up and down regulation of genes to be examined. Very sensitive	Needs a skilled operator, very expensive, labour intensive, designated software needed to analyse results. Requires specialised equipment

ELISA: Enzyme-linked immunosorbent assay, FAT: Fluorescent antibody assays (or technique), IFAT: Indirect FAT, LAMP: Loop-mediated isothermal amplification, PCR: Polymerase chain reaction, RT- PCR: Reverse transcriptase-PCR, TEM: Transmission electron microscopy (modified from Adams & Thompson, 2008).

viruses and confirmation of mollusc parasites. Colony hybridisation has also been used successfully for the rapid identification of *Vibrio anguillarum* in fish (Powell & Loutit, 2004), and has the advantage of detecting both pathogenic and environmental strains (Adams et al., 2008).

Serology is an alternative approach to pathogen detection, and can also be applied to the detection of pathogen-specific antibodies in fish. The ELISA is well suited to large scale screening and this can be performed in any species of fish when an anti-fish species antibody is available (Adams et al., 2008). A number of new technologies are being developed for the rapid detection of pathogens and monitoring host responses. These include immunochromatography, such as lateral flow technology, and multiplex testing using the Bio-Plex Protein Array System or microarray technologies (Adams and Thompson, 2006). Lateral Flow is simple methodology enabling accurate (high sensitivity, specificity), simple, easy to use (2 steps, no instrument required) testing that is also economic (time/labor saving). The Protein Array System (Luminex) theoretically offers simultaneous quantitative analysis of up to 100 different biomolecules from a single drop of sample in an integrated, 96-well formatted system, mainly focusing on the detection of cytokines. Therefore, it can be used in molecular and immunodiagnostics to detect pathogens directly from tissue samples or culture, or it can be used in serology to measure fish antibodies (Adams et al., 2008; Adams & Thompson, 2008; Dupont, 2005; Giavedoni, 2005).

5. IMMUNOSUPPRESSION

Aquatic environment of fish is in close contact with numerous pollutants. Aquatic pollutants such as heavy metals, aromatic hydrocarbons, pesticides and mycotoxins modulate the immune system of fish, thus increasing the host susceptibility to infectious pathogens. Pollutants in the water which may be particulate or soluble can also be natural source such as metals showing the seasonal increase in lakes as well as drugs used in the prevention or treatment of disease such as corticosteroid hormones, used drugs in terrestrial animal health in aquaculture such as florfenicol, oxolinic acid, and oxytetracycline (Table 7). Immunosuppressive effects of these compounds may occur at high concentrations or long-term exposures (Anderson, 1996; Bols et al., 2001; Brown-Treves, 2000; Duffy et al., 2002; El-Gohary et al., 2005; Enis-Yonar et al., 2011; Kusher & Crim, 1991; Lumlertdacha & Lovell, 1995; Lundén & Bylund, 2002; Lundén et al., 1998; Lundén et al., 1999; Manning, 2001; Manning, 2010).

Table 7. Nonspecific defense mechanisms and specific immune response assays /or parameters in fish effected by presence of some immunosuppressive compounds.

Substances	Parameters	Fish species
Metals and organometallies		
Aluminum	Reduced chemiluminescence	Rainbow trout
Arsenic	Phagocytosis elevated or lowered	Rainbow trout
Cadmium	Elevated serum antibody	Rainbow trout
	Chemiluminescence reduced	Rainbow trout
	Lymphocyte number and mitogenic response reduced	Goldfish
	Antibody-binding lymphocyte reduced	Bluegill
Chromium	Serum antibody reduced	Brown trout, carp
Copper	Chemiluminescence reduced	Rainbow trout
	Susceptibility to lHNV increased	Rainbow trout
	Leukocyte respiratory burst activity inhibed	Rainbow trout
	Serum antibody reduced	Brown trout
	Antibody-producing cells reduced	Rainbow trout
	Susceptibility to *Vibrio anguillarum* increased	Eel
Lead	Serum antibody reduced	Brown trout
Mercury	Lymphocyte numbers reduced	Barb
Nikel	Serum antibody reduced	Brown trout
Zinc	Serum antibody reduced	Brown trout
	Phagocytosis decreased	Rainbow trout
Aromatic hydrocarbones		
Benzidine	Non-specific agglutination rise	Estuarine fish
PAHs	Macrophage activity reduced	Spot, Hogchoker
	Melanomacrophage numbers reduced	Flounder

PCBs Benzo[a]pyrene PCB 126 Aroclor 1254 Aroclor 1232 Aroclor 254/1260	Phagocytic capacity reduced Antibody-producing cells reduced Non-specific cytotoxic cell activity reduced Antibody-producing cells reduced Susceptibility to disease increased Susceptibility to disease increased	Rainbow trout Medaka Catfish Coho salmon Channel catfish Rainbow trout
Phenols Hydroquinone	Antibody- producing cells reduced Non-specific cytotoxic cell activity reduced	Rainbow trout Carp
TCDD	Mitogenic response partially suppressed Susceptibility to IHNV	Rainbow trout
Pesticides		
Bayluscide	Serum African antibody reduced	Catfish
Dichlorvos	Lysozyme activity reduced	Carp
DDT	Antibody-producing cell, serum antibody reduced	Goldfish
Endrin	Phagocytic, antibody-producing cell activities reduced	Rainbow trout
Malathion	Lymphocyte number reduced	Channel catfish
Metrifonate	Phagocytic, neutrophilic and lysozyme activity reduced , antibody-producing cell reduced,	Cichlid fish
Methyl bromide	Thymic necrosis	Medaka
Tributyltin	Chemiluminescence reduced	Oyster, Hogchoker
Trichlorophon	Phagocytic, neutrophilic, lysozyme activity reduced	Carp
Mycotoxins		
Aflatoxin-B1	B-cell memory loss, neutrophilic activity reduced	Rainbow trout
Fumonisin-B1	Antibody-producing cells reduced	Catfish
Antibiotics		
Florfenicol	Chemiluminescence reduced Phagocytic cells counts reduced after 5-6 weeks	Rainbow trout
Oxolinic acid	Antibody-producing cells reduced	Rainbow trout
Oxytetracycline	Mitogenic response reduced, Antibody-producing cells reduced, phagocytic activity reduced	Carp Rainbow trout
Other compounds		
Cortisol/ Kenalog-40	Antibody-producing cells reduced	Rainbow trout
Hydrocortisone	Phagocytic activity reduced	Striped bass

DDT: Dichloro-diphenyl-trichloroethane, PAHs: Polynuclear aromatic hydrocarbons, PCBs: Polychlorinated biphenyls, TCDD: 2,3,7,8-tetrachlorodibenzo-p-dioxin. IHNV: Infectious hematopoietic necrosis virus. (modified from Anderson, 1996).

6. IMMUNOMODULATION

Immunomodulators present in the diets stimulate the innate immune systems, while antigenic substance such as bacterins and vaccines initiate the more prolonged process of antibody production and acquired immune systems. Prophylactic and therapeutics administration of immunomodulators will need to be adapted to each cultured fish species in anticipation of recognize pathogens, under known environmental conditions (Gannam & Schrock, 2001). Prophylactic and therapeutic compounds and/or drugs against infections are rarely successful or limited effects; currently there are no approved some drugs for the control and treatment fish disease in the aquaculture industry. For example, several substances, such as fumagilin and albendazole have been used in fish with potential value in controlling microsporidian infections. However, other drugs, like sulphaquinoxaline, amprolium and metronidazole have been ineffective to control the disease (Berker & Speare, 2007; Dykova, 2006; Rodriguez-Tovar et al., 2011). Most of similar drugs have ambiguous result and it is has been reported that high concentrations and prolonged treatment of infections with some drugs might cause side-effects. More promising results have been achieved by using immune-prophylactic control components such as probiotics (e.g. basillus P64, yeasts and lactic acid bacteria), prebiotics (e.g. fructo-galacto-, transgalacto- oligosaccharide), vaccination (e.g. Vibrio spp., Yersinia ruckerii) and immunostimulants (e.g. β-glucan, chitosan and levamisole) (Austin & Brunt, 2009; Hoffmann, 2009; Magnadóttir, 2010; Nayak, 2010; Rodriguez-Tovar et al., 2011). On the other hand, in recent years, organically produced aquatic products are increasingly available to consumers and, in particular, sea bass and sea bream from certificated fish farms (Perdikaris & Paschos, 2010). The initial legislative framework for organic aquaculture in the European Union (EU) was the Directives (EC) 834/07 and (EC) 889/08 (EU, 2007; 2008).

7. IMMUNOSTIMULANTS

Various chemotherapeutic compounds have been extensively used to treat bacterial infections in cultured for about the last 20-30 years. However, the incidence of drug-resistant bacteria has become a major problem in fish culture (see Chapter 11: Section 5.2). Although, vaccination is a useful prophylaxis for infectious disease, and is also already commercially available for bacterial infections such as vibriosis, redmouth disease and for viral infections such as infectious pancreatic necrosis, the development of vaccines against intracellular pathogens such as *Renibacterium salmoninarum* has not so for been unmitigated successful. Therefore, the immediate control of all fish disease using only vaccines is impossible. Even thought, use of immunostimulants, in addition to chemotherapeutic drugs and vaccines, has been widely accepted by the aquaculture industry, many question about the efficacy of immunostimulants from users still continue such as whether this components can protect against infections disease (Table 8). Also, the biological activities of the immunostimulants may be so multiple and potent that some of them may be more harmful than beneficial (Dalmo, 2002; Sakai, 1999).

By definition, an immunostimulant is a naturally occurring compound that molecules that modulates the immune system by increase the host's resistance against disease that in most circumstances are caused by pathogens (Bricknell & Dalmo, 2005). However, synthetic chemicals such as isoprinosine, bestatin, levamisole, muramyl dipeptide and FK-565 well- known as lactoyl tetrapeptide are known to possess immunostimulatory properties. It is important to note the use of the term "modulate", as a substance with the potential immunostimulatory properties may lead to a down regulation of the immune response if administered in excess amounts or long-term usage. Hence, administration of an immunestimulant prior to an infection may elevate the defence barriers of the animal and thus provide protection against an otherwise severe or lethal infection. Also, immunostimulants enhance individual components of innate immune response, but this does not always translate into increased survival. An important point to have in mind is that not by enhancing acquired immune response. Therefore, there is no memory component and the response is likely to be of short duration (Gannam & Schrock, 2001; Hølvold, 2007; Maqsood et al., 2011; Raa, 2000; Sakai, 1999).

Table 8. A comparison of characteristics of chemotherapeutics, vaccines and immunostimulants (Sakai, 1999).

	Chemotherapeutics	Vaccines	Immunostimulants
When	Therapeutically	Prophylactically	Prophylactically
Efficacy	Excellent	Excellent	Good
Spectrum of activity	Middle	Limited	Wide
Duration	Short	Long	Short

A division of immunostimulants depended on which effects they include such as anti- bacterial, -viral, –fungal and –parasitic effects may be helpful but hard to accomplish. Some immunostimulants may induce both antibacterial and antiparasitic effects, whereas other may help the organism to fight virus and fungus. Generally, immunostimulants used in fish and shrimp in many countries can be divided into two main groups as biological substances and synthetic chemicals depending on their sources (Table 9) (Anas et al., 2005; Brown-Treves, 2000; Dügenci et al., 2003; Galindo-Villegas & Hosokowa, 2004; Gannam & Schrock, 2001; Gildberg et al., 1996; Glina et al., 2009; Jiye et al., 2009; Lauridsen & Buchmann, 2010; Maqsood et al., 2011; Noga, 2010; Paulsen et al., 2003; Perera & Pathiratne, 2008; Petersen et al., 2004; Raa, 2000; Sakai, 1999). But, some immunostimulants may be included in different subgroups by some researchers, such as schizophyllan and scleroglucan. These substances may be included in bacterial derivatives-subgroups as various β-glucan products from *Schizopyllum commune* and *S. glucanicum*, respectively, or may be included in polysaccharides-subgroups as polysaccharides containing sugars.

7.1 Dose, timing, administration-route and -period of immunostimulants

The effect of timing the administration on immunostimulant function is a very important issue. Usually, the most effective timing of antibiotics is upon the occurrence of disease, and they cannot often be used prophylactically due to risk of fostering the development of drug- resistant bacteria. Researchers proposed that immunostimulants may improve health and performance of fish and shrimp in aquaculture, if used prior to: (1) before the outbreak of disease to reduce disease-related losses, (2) situations known to result in stress and impaired general performance of animals such as handling, change of temperature and environment, weaning of larvae to artificial feeds, (3) expected increased exposure to pathogenic micro- organisms and parasites such as spring and autumn blooms in the marine environment, high stocking density, (4) developmental phases when animals are particularlysusceptible to infectious agents such as the larvae phase of shrimp and marine fish, smoltification in salmon, sexual maturation (Raa, 2000; Sakai, 1999). In addition, the effects of immunostimulants may also be different dependent on the administration route, the dose used, the duration of the treatment and growth period. Immunostimulants does not show a linear dose/effect relationship; instead they most often show a distinct maximum at a certain intermediate concentration and even a complete absence of effect or toxicity, at high concentration. The explanations for these phenomena are still speculative and include competition for receptors (analogous to substrate inhibition of enzyme), over stimulation resulting in exhaustion and fatigue of the immune system (Bright-Singh & Philip, 2002).

It is reported that oral administration of an immunostimulant such as lipopolysaccharide is increased larval growth. This may be important in the intensive production of fish larvae and juveniles. In spite of advantages and limitations, the basic methodologies adopted are injection, immersion and oral (Table 10). Injection and immersion methods are suitable only for intensive aquaculture and both require the fish to be handled or at least confined in a small space during the procedures (Dalmo, 2002; Guttvik et al. 2002; Raa, 2000). By injections of immunostimulants enhances the function of leucocytes and protection against pathogens. However, this method is labour intensive, relatively time- consuming and becomes impractical when fish weight less than 15 gram. By immersion, efficacies had been confirmed by several researchers (Anderson et al., 1995; Baba et al., 1993; Jeney & Anderson, 1993; Perera & Pathiratne, 2008), although, since dilution, exposure time and levels efficacy are not well defined, caution must be taken in account by applying this methods. Oral administration is only method economically suited to extensive aquaculture, is non-stressful and allows mass administration regardless of fish size, but of course may be administration only in artificial diet (Galindo-Villegas & Hosokowa, 2004; Noga, 2010).

Table 9. Groups, substances and examples of immunostimulants evaluated in many countries that have been tried to increase disease protection in fish species and/or shrimps.

Groups	Substances	Compounds
Biological substances	Animal compounds	EF-203 (Chicken egg), Ete (Tunicate, *Ecteinascida turbinata*), Hde (Abalone, *Haliotis discus hannai*), cod milt, firefly squid and acid-peptide fractions (fish protein hydrolysate)
	Plant extracts	Glycyrrhizin (Licorice, saponin in *Glycyrrhiza glabra*), quil-A saponin, ergosan (*Laminaria digitata*), C-UP III (a Chinese herb mix), laminaran (Seaweed), spirulina (*Spirulina plantensis*) *Quillaja saponica* (Soap tree), leaf extract (*Ocimum sanctum*), scutellaria extract (*Scutellaria baicalensis*), astragalus extract (*Astragalus membranaceus*), ganoderma extract (*Ganoderma lucidum*), lonicera extract (*Lonicera japonica*), phyllanthus extract (*Phyllanthus emblica*), azadirachta extract (*Azadirachta indica*), solanum extract (*Solanum trilobatum*), mistletoe (*Viscum album*), nettle (*Urtica dioica*), ginger (*Zingiber officinale*) and chevimmun (*Echinacea anguistifolia-Baptista tinctoria-Eupatorium perfoliatum*)
	Bacterial and yeast derivatives	ß-glucans (from bacteria and mycelial fungi; MacroGard, VitaStim, SSG, Eco-Activa, Betafectin, Vetregard, Dinamune, Aquatim, AquaStim, Curdlan, Krestin), ascogen (Aquagen), peptidoglycan (*Brevibacterium lactofermentum*; *Vibrio sp.*), pDNA (*Escherichia coli*), lipopolysaccharide, fragellins (recombinant- *Borrelia*), *Vibrio anguillarum* cells, *Clostridium butyricum* cells, *Achromobacter stenohalis* cells and streptococcal components (*Bordetella pertuosis, Brucella abortus, Bacillus subtilis, Klebsiella pneumonia*)
	Cytokines	Interferon, interleukin-2, tumor necrosis factor
	Hormones	Growth hormone, prolactin, melanin stimulating hormone, ß-endorphin and melanin concentrating hormone
	Nutritional factors	Vitamin–A, -C, -E, carbohydrate (Acemannan), soybean protein, trace elements (zinc, iron, copper, selenium) and nucleotides
	Polysaccharides	Chitin, chitosan, lentinan, schizophyllan, sclerotium, scleroglucan, protein-bound polysaccharide (PS-K), oligosaccharide and polyglucose
	Others	Lactoferrin
Synthetic chemicals		Avridine, bestatin, DW-2929, ISK, KLP-602, FK-156 (lactoyl tetrapeptide), FK-565, fluro-quindone, Freund's complete adjuvants, imiquimod, isoprinosine, levamisole, muramyl dipeptide and sodium alginate

(modified from Galindo-Villegas & Hosokowa, 2004; Sakai, 1999).

Table 10. Administration methods, advantages and limitations of immunostimulants in aquaculture.

Route	Dose	Exposure time	Advantages	Limitations
Injection	Variable	1 or 2 doses	Allows use of adjuvants, Most potent immunization route, most cost effective method for large fish	Only for intensive aquaculture, fish must be >10~15 g, stressful (anesthesia, handling), labour hard
Immersion	2 - 10 mg/L	10 min to hours	Allows mass immuno-stimulation of small (<5 g) fish, most cost effective method for small fish.	Only for intensive aquaculture, dip rise handling stress, potency not as high as injection route
Oral	0.01–4%	Some days or longer	Only not-stressful method, Allows mass immuno- stimulation of fish any size, no extra labour cost	Poor potency, requires large amounts of immunostimulation to achieve protection, suitable only for fish fed artificial diet

(Galindo-Villegas & Hosokowa, 2004).

The effects of immunostimulants were dose and/or application time, route, and period related. For example, low-dose glucan content being beneficial whereas high-dose glucan content had limited effects (Ai et al., 2007). Peptidoglycan is not influence the high-dose (0.1%) in shrimp diets, and not effect after 60 days of oral administration in rainbow trout growth. On the other hand, Ete exerted a protective effect in eels injected intra-peritoneal 2 days after challenge with *A. hydrophila*. However, the protection was not seen when Ete was administered intra-peritoneal 2 days before or concurrently with the bacteria. The adjuvant effects of glucan against *A. salmonicida* vaccine oral delivery (7 days administration) and immersion (15 min). No adjuvant effects were seen with the immersion treatment, although the fish administered glucan orally showed enhanced vaccine effects (Sakai, 1999). The number of NBT-positive cell in catfish increased following oral administration of glucan and oligosaccharide over 30 days, but not over 45 days (Yoshida et al., 1995). The effects of long- term oral administration of immunostimulants are still unclear. However, the dilution, the effective administration period and the levels of efficacy require more complete investigation for each immunostimulants.

7.2 *In vivo* and *in vitro* effects of immunostimulants

The benefit of immunostimulants is considerable. They have the potential to elevate the innate defence mechanisms of fish prior to exposure to a pathogen, or improve survival following exposure to a specific pathogen when treated with an immunostimulant. There are two main procedures for evaluating the efficacy of an immunostimulant; (1) *in vivo* such as protection test against fish pathogen, (2) *in vitro* such as the measurements of the efficiency of cellular and humoral immune mechanism (Bricknell & Dalmo, 2005; Maqsood et al., 2011). *In vivo* evaluation should be based at least on the following parameters: phagocytosis, antibody production, free radical production, lysozyme activity, natural cytotoxic activity, complement activity, mitogen activity, macrophage activating factor (MAF), nitroblue tetrazolium reaction (NBT), etc. The evaluation of an immunostimulant by the *in vitro* methods which test the effects of that substance on the immune system is to be preferred in preliminary studies (Table 11 and Figure 6) (Aly & Mohamed, 2010; Anas et al., 2005; Barman et al., 2011; Brown-Treves, 2000; Dügenci et al., 2003; Galina et al., 2009; Galindo-Villegas & Hosokowa, 2004, Gannam & Schrock, 2001; Magnadóttir, 2010; Maqsood et al., 2011; Noga, 2010; Paulsen et al., 2003; Peddie et al., 2002; Raa, 2000; Sakai, 1999; Yin et al., 2009). Nevertheless, if possible *in vitro* test should be performed together wit *in vivo* experiments in order to elucidate the basic mechanisms responsible for the protection (Bricknell & Dalmo, 2005; Brown, 2006; Brown-Treves, 2000; Dalmo, 2002; Sakai, 1999). Used aquaculture potential immunostimulants with *in vivo* and/or *in vitro* effects and administration route are given at Table 12, as well as with doses at Table 13.

A number of studies have been reported that potential immunostimulants may be showed at *in vivo* and/or *in vitro* effects in fish species in different countries and various periods (Aly & Mohamed, 2010; Anas et al., 2005; Dugenci et al., 2003; Galindo-Villegas et al., 1996; Galindo-Villegas et al., 2006; Gildberg et al., 1996; Ispir & Dorucu, 2005; Kunttu et al., 2009; Lauridsen & Buchmann, 2010; Ortuño et al., 2002; Peddie et al., 2002; Perera & Pathiratne, 2008; Sakai et al., 1995; Seker et al., 2011; Soltani et al., 2010; Yin et al., 2009; Zhao et al., 2010).

Table 11. The *in vivo* and *in vitro* responses seen in fish treated with immunostimulants.

In vivo effects	*In vitro* effects
Increased survival after challenges with bacteria, antiparasitic effects including reduced settlement of sea lice, improved resistance to viral infection and increased interferon levels	Increased macrophage activity including: - Phagocytosis, free radical production, enzyme activity, migration activity, production of cytokines, nitric oxide production, bacterial killing, antibody production, respiratory burst, MAF, NBT
Growth enhancement	Increased cytotoxicity
Increased antibody production following vaccination	Increased lysozyme activity
Increased lysozyme levels	Increased cytokine induction Increased oxygen radical induction Increased cell proliferation

MAF: Macrophage activating factor, NBT: Nitroblue tetrazolium reaction. (modified from Bricknell & Dalmo, 2005).

Table 12. Potential immunostimulants, *in vivo* and/or *in vitro* effects of this immunostimulants with administration route.

Immunostimulant	Species	Route	*In vivo* or *in vitro* effects	Resistance
Synthetic chemicals				
FK-565	Trout	ip	phagocytosis ↑	*A. salmonicida* ↑
		in vitro	antibody ↑	-
Freund's adjuvants	Trout	ip	-	*V. anguillarum* ↑, *Y. ruckeri* ↑
	Yellowtail	ip	-	*P. piscicida* →
Levamisole	Carp	ip/oral	phagocytosis ↑ / NBT →	-
		im	phagocytosis ↑, CL ↑	*A. hydrophila* ↑
	Trout	*in vitro*	phagocytosis ↑, NBT →	-
		ip	phagocytosis ↑, CL ↑, complement ↑	*V. anguillarum* ↑
		im	-	*A. salmonicida* ↑
Muramyl dipeptide	Trout	ip	phagocytosis ↑, CL ↑	*V. anguillarum* ↑
Bacterial and yeast derivatives				
Achromobacter stenohalis cells	Char	ip	CL ↑, complement ↑	*A. salmonicida* ↑
Clostridium butyricum cells	Trout	oral	phagocytosis ↑, NBT ↑	*V. anguillarum* ↑
Glucan	Trout	im / ip	phagocytosis ↑ / NBT ↑	-
Lipopolysaccharide	Plaice	ip	macrophage migration ↑	-
	Red sea bream		phagocytosis ↑	
	Goldfish	*in vitro*	macrophage activating factor ↑	
	Salmon		phagocytosis ↑, NBT ↑	
	Catfish		IL-1 ↑	
Peptidoglucan	Trout	oral	-	*V. anguillarum* ↑
	Shrimp		-	YHB ↑
	Yellow tail		phagocytosis ↑	*E. seriolicida* ↑
	J. flounder		lysozyme ↑, phagocytosis ↑	*E. tarta* ↑

Vibrio bacteria	Trout	im	-	*A. salmonicida* ↑, *E. seriolicida* ↑
	Prawn/ shrimp	ip,im,oral	-	*Vibrio sp.* ↑
VitaStim	Coho	ip, oral	-	*A. salmonicida* ↑
	Chinook	oral / im	-	*A. salmonicida* ↑/→
	Catfish	oral	antibody ↑	*E. ictaluri* →
Yeast glucan	Salmon	oral	-	*V. anguillarum* ↑, *A. salmonicida* ↑
		ip	complement ↑, lysozyme ↑, antibody ↑, NBT ↑, killing →	*A. salmonicida* →, *Y. ruckeri* ↑
	Catfish	ip	phagocytosis ↑, NBT ↑, antibody ↑, killing ↑	*E. ictaluri* ↑
		oral	CL ↑, migration ↑	*E. ictaluri* →
	Shrimp	im,*in vitro*	phenoloxidase ↑, lysozyme →, NBT ↑, CL ↑	-
	Trout	ip	lysozyme ↑, killing ↑, $O2^-$ ↑, NBT →	*V. anguillarum* →
	Turbot	oral	lysozyme ↑, complement →, CL ↑	*V. anguillarum* ↑
Yeast glucan + Vit-C	Trout	oral	lysozyme →, complement ↑, CL ↑	-

↑:Increase, ↓:Decrease, →:No change, Brook: Brook trout, Chinook: Chinook salmon, CL: Chemiluminescent response, Coho: Coho salmon, Con-A: Concanavalin-A, IL-1: Interleukin-1 production, im: immersion, ip: Intraperitoneal injection, J. flounder: Japanese flounder, Killing: Bactericidal activity of macrophage, LPS: Lipopolysaccharide, NBT: Nitroblue tetrazolium reaction, NK: Natural killer cell, $O2^-$: Production of superoxide anion, Prawn: Kuruma prawn, PS-K: Protein-bound polysaccharide, Salmon: Atlantic salmon, Shrimp: Black tiger shrimp, Trout: Rainbow trout, Turbot: Scophthalmus maximus, Vit-C: Vitamin-C, *A. hydrophila: Aeromonas hydrophila, A. salmonicida: Aeromonas salmonicida, E. ictaluri: Edwardsiella ictaluri, E. seriolicida: Enterococcus seriolicida, E. tarda: Edwardsiella tarda, HRV: Hirame rhabdo virus, P. piscicida: Pasteurella piscicida, R. salmoninarum: Renibacterium salmoninarum, V. anguillarum: Vibrio anguillarum, Y. ruckeri: Yersinia ruckeri, YHB: Yellow-head baculo virus.* (modified from Gannam & Schrock, 2001; Sakai, 1999).

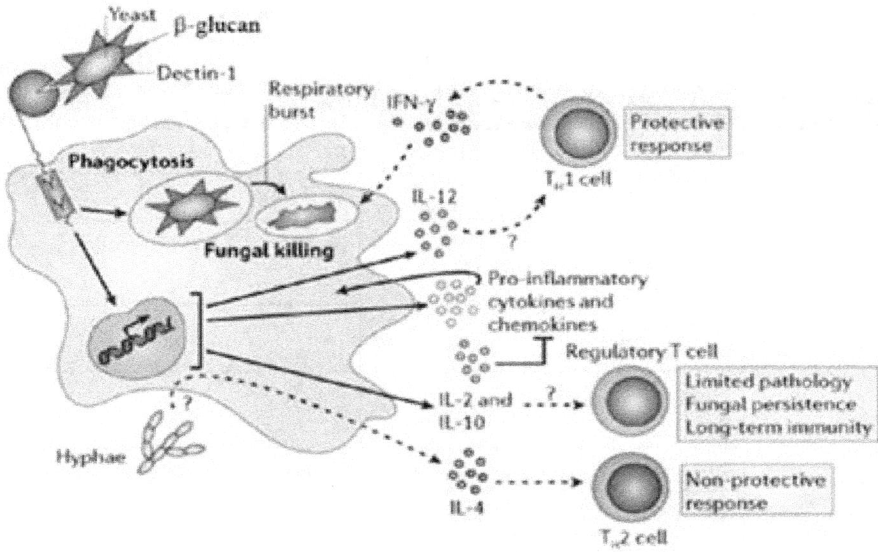

Fig. 6. Recognition of β-glucans on fungal particles induces several dectin-1-mediated cellular responses, which might contribute to anti-fungal immunity *in vivo***.**

Recognition of β-glucans on fungal particles induces several dectin-1-mediated cellular responses, which might contribute to anti-fungal immunity in vivo. These include fungal uptake and killing and the production of pro-inflammatory cytokines and chemokines, such as tumour-necrosis factor (TNF) and CXC-chemokine ligand 2, in collaboration with the Toll-like receptors (TLRs), which is likely to lead to cellular recruitment and activation. Dectin-1-mediated recognition also stimulates the production of interleukin-12 (IL-12), which might result in a protective T-helper 1 (TH1)-cell response and the production of interferon-γ (IFN-γ), thereby activating the fungicidal activities of phagocytes. In dendritic cells, β-glucan recognition by dectin-1 can also induce the production of IL-10 and IL-2), which could potentially contribute to the development of regulatory T cells, thereby limiting inflammatory pathology and promoting fungal persistence and long-term immunity, as proposed previously. IL-10 would also inhibit the production of pro-inflammatory cytokines and chemokines. Fungi might also mask their β-glucan, by conversion from yeast to hyphal forms. This could result in the induction of non-protective TH2-cell immune responses, mediated by IL-4; this could be the result of preventing recognition by dectin-1 although the pathways leading to this response are unknown. Although dectin-1 is described here as having a central role in the generation of protective immune responses, it should be noted that many other opsonic and non-opsonic receptors such as the mannose receptor, complement receptor 3, dendritic-cell-specific ICAM3-grabbing non-integrin and TLRs also contribute to this process. (modified from Brown, 2006).

Table 13. Doses and effects of immunostimulants as nutritional factors and nucleotides in some fish species.

Immunostimulant	Species	Dose - Route	*In vivo* or *in vitro* effects
β-glucan	Atlantic salmon	15 mg/kg, inj	ROS ↑, lysozyme ↑
		150 mg/kg, oral	acid phosphatase↑
		1 ml/fish, inj	lysozyme ↑, complement ↑, antibody ↑
	Coho salmon	5 and 15 mg/kg, inj	→
	Japanese flounder	3 g/kg, oral	NBT ↑
	Sea bass	2% wet bw, oral	humoral activation ↑
	Dab	0.5 μg/kg, iv	ROS ↑
	Dentex	0.5%, oral	→
	Turbot	0.5 – 500 μg/kg, iv	ROS ↑
		2 g/kg	leukocyte number ↑
	Yellow tail	2 - 10 mg/ml, inj	phagocytic activity ↑
	Rainbow trout	88 and 350 μg/g, oral	lysozyme ↑, ROS ↑, complement bacteriolytic activity ↑
β-glucan + LPS	Atlantic salmon	1 – 250 + 10 μg/ml	lysozyme ↑
β-glucan + Mannose	Japanese flounder	1%, oral	NBT ↑, lysozyme ↑
	Pink snapper	0.1 - 1% w/w, oral	ROS ↑, macrophage activation ↑
β-glucan + FKC + Quillaja saponica	Pink snapper	122 + 34 + 5 mg/kg	agglutination titers ↑
CFA	Sockey salmon	5 mg/kg, inj	antibody ↑
Chitosan	M. rosenbergi larva	0.25 - 1% v/v, 150 ml *in vitro*	Antibacterial activity ↑
Chitin	Gilhead sea bream	0.1 ml/fish, inj	→
		1 mg/fih, inj	humoral and cellular activation ↑
		25 – 100 mg/kg	NCCs ↑, ROS ↑, phagocytic activity ↑
EF-203	Rainbow trout	30 μg/kg, oral	NBT ↑, phagocytic activity ↑
Ergosan	Rainbow trout	1 mg/fish	complement ↑
Fungi	Sockey salmon	10 g/kg, oral	→

Glycyrrhizin	Yellow tail	0 – 50 mg/kg, oral	complement ↑
Laminaran	Blue gourami	20 mg/kg, inj	CL ↑
Levamisole	Atlantic salmon	2.5 mg/L, bath	ROS ↑, phagocytic activity ↑, lysozyme ↑
	Coho salmon	5 mg/kg, inj	→
	Sockey salmon	125 – 500 μg/ml, oral	phagocytosis ↑, complement ↑, ROS ↑, lymphokine ↑
		0.5 – 500 μg/ml, iv	ROS ↑
		75 – 300 mg/kg, oral	NCCs ↑
	Japanese flounder	125 – 500 mg/kg, oral	phagocytic activity ↑, NBT ↑, lysozyme ↑
	Rainbow trout	10 and 50 μg/ml, in vitro	phagocytic activity ↑, NBT ↑
		5 mg/kg, inj	NBT ↑, lysozyme ↑, phagocytic activity ↑, killing ↑
LPS	Red sea bream	1 ml/fish, inj	phagocytic activity ↑
MCFA	Coho salmon	5 mg/kg, inj	→
Microsporidian	Flounder	106 spores, inj	antibody ↑
Myxsosporean	Sea bass	multiple, iv	ROS ↑
Peptidoglycan	Japanese flounder	1.5 – 4.5 g/kg, oral	phagocytosis ↑, complement ↑, MAF ↑, ROS ↑
	Yellow tail	0.2 mg/kg, oral	phagocytic activity ↑
Quillaja	Yellow tail	0.5 – 50 mg/kg	chemotaxis ↑
Yeast	Gilhead sea bream	1 – 10 g/kg	cellular response ↑
Yeast	Sea cucumber	5%, oral	phagocytic activity ↑, phagocytic index ↑, lysozyme ↑
Wy-18, 251	Coho salmon	10 mg/kg, inj	→

↑: Increase, ↓: Decrease, →: No change, μg: Microgram, inj: Injection, iv: Intra venous, bw: Body weight, CFA: Complete Freund's adjuvant, CL: Chemiluminescent response, FKC: Formalin-killed *Edwardsiella tarda* cells, L: Litre, LPS: Lipopolysaccharide, Killing: Bactericidal activity of neutrophil and monocytes, MAF: Macrophage activating factor, M. rosenbergi: Macrobrachium rosenbergi, MCFA: Medium-chain fatty acid(s), NBT: Nitroblue tetrazolium reaction, NCCs: Non-specific cytotoxic cells, ROS: Reactive oxygen species, Turbot: Scophthalmus maximus, Wy-18, 251: an analog of levamisole at structure 3-(p- chlorophenyl) thiazolol [3,2-a] benzimidazole-2-acetic acid. (modified from Galindo-Villegas & Hosokowa 2004).

Improvements in the health status of fish can certainly be achieved by balancing the diets with regard to nutritional factors (see Table 9), in particular lipids such as fatty acids, essential oils, and anti-oxidative vitamins or minerals such as vitamin-C, -E and selenium, and also minerals iron and fluoride, but this is primarily a result of an input of substrates and cofactors in a complex metabolic system. These compounds were identified as micronutrients that could affect disease resistance. This is unlike immunostimulants, which interact directly with the cells of the immune system and make them more active, because they enhance in immune system by providing substrate and co-factors necessary for the immune system to work properly. Nevertheless, some nutritional factors are so intimately interwoven with the biochemical processes of the immune system that significant health benefits can be obtained by adjusting the concentration of such factors beyond the concentration range sufficient to avoid deficiency symptoms or below a certain concentration range (Balfry & Higgs, 2001; Gannam & Schrock, 2001; Lim et al., 2001a, 2001b; Raa, 2000; Soltani et al., 2010). The modulatory effects of dietary nutritional factors on macrophage-, haemolytic-, lysozyme- and complement activation, lymphocyte proliferation, macrophage phagocytic response as well as oxidative burst, pinocytosis and bactericidal activity have been reported in aquaculture (Balfry & Higgs, 2001; Galindo-Villegas & Hosokowa, 2004; Gannam & Schrock, 2001, Sakai et al., 1999).

Other a type of nutritional factors as immunostimulants in aquaculture, nucleotides have essential physiological and biochemical functions including encoding and deciphering genetic information, mediating energy metabolism and cell signaling as well as serving as components of co-enzymes, allosteric effectors and cellular agonists. Also, these compounds have traditionally been considered to be non-essential nutrients. Nucleotides consist of a purine or a pyrimidine base, a ribose or 2'-deoxyribose sugar and one or more phosphate groups. The term nucleotide in this context refers not only to a specific form of the compounds but also to all forms that contain purine or pyrimidine bases.

In recent years, world-wide heightened attention on nucleotide supplementation for fishes was aroused by the reports of some researches, indicating that dietary supplementation of nucleotides enhanced resistance of salmonids to viral, bacterial and parasitic infections as well as improved efficacy of vaccination and osmoregulation capacity (Burrells et al., 2001a, 2001b; Grimble & Westwood, 2000; Li & Gatlin, 2006). The modulatory effects of dietary nucleotides on lymphocyte maturation, activation and proliferation, macrophage phagocytosis, immunoglobulin responses as well as genetic expression of certain cytokines have been reported in humans and animals including some fish species such as hybrid tilapia, rainbow trout, Coho salmon, Atlantic salmon and common carp (Gil, 2002; Li & Gatlin, 2006). To date, research pertaining to nucleotide nutrition in fishes has shown rather consistent and encouraging beneficial results in fish health management, although most of the suggested explanations remain hypothetical and systematic research on fishes is far from complete. Because increasing concerns of antibiotic use have resulted in a ban on sub- therapeutic antibiotic usage in some countries, research on immune nutrition for aquatic animals is becoming increasingly important. Also, research on nucleotide nutrition in fish is needed to provide insights concerning interactions between nutrition and physiological responses as well as provide practical solutions to reduce basic risks from infectious diseases for the aquaculture industry (Burrells et al., 2001a, 2001b; Li & Gatlin, 2006). In aquaculture, used immunostimulants as nutritional factors and nucleotides with dose, administration route and effects are given at Table 14 (Galindo-Villegas & Hosokowa, 2004; Li & Gatlin, 2006; Sakai et al., 1999).

Table 14. Doses and effects of immunostimulants as nutritional factors and nucleotides in fish species.

Immunostimulant	Species	Dose - Period	*In vivo* or *in vitro* effects / Resistance
Nutritional factors			
Vitamin-C	Catfish	150 mg/kg, 14 w	*E. tarta* ↑
		1000 mg/kg, 7 w	neutrophil → (phagocytosis)
		3000 mg/kg	complement ↑, phagocytic index ↑ antibody → macrophage killing → *E. tarta* ↑
		4000 mg/kg, 9 w	complement →, antibody →, *E. tarta* ↑
	Red sea bream	1000 mg/kg	complement →
		10 000 mg/kg	phagocytic activity ↑
	Atlantic salmon	2750 mg/kg, 26 w	complement ↑, NBT →, phagocytosis →, MAF ↑, *A. salmonicida* ↑
		2980 mg/kg	antibody ↑
		3170 mg/kg, 16 w	NBT ↑, killing ↑, migration ↑, antibody ↑, *A. salmonicida* ↑
		4000 mg/kg, 52 d	*V. salmonicida* ↑
		5000 mg/kg	antibody →
	Coho salmon	3000 mg/kg	phagocytosis ↑, ROS ↑, complement ↑
	J. flounder	6100 mg/kg	NBT ↑
	Trout	244 mg/kg, 16 w	proliferation ↑, NBT →, MAF ↑
		550 mg/kg, 10 d	IHNV ↑
		2000 mg/kg, 4 w/12 w	*I. mutifiliis* ↑ / *V. anguillarum* ↑
		2000 mg/kg, 127 d	phagocytic index ↑, lysozyme ↑→
	Turbot	300 – 200 mg/kg	phagocytic activity ↑, lysozyme ↑
	Sockey salmon	> requirement	→
	Yellow tail	122 – 6100 mg/kg	phagocytic activity ↑, lysozyme ↑

Vit-C + Vit-E	Gilhead sea bream	2900 + 1200 mg/kg	lysozyme ↑, NCCs ↑
		Many concentration, in vitro	migration ↑, phagocytic activity ↑, ROS (mix) ↑
Vit-C + Yeast glucan	Trout	oral	lysozyme →, complement ↑, CL ↑
Vitamin-E	Catfish	2500 mg/kg, 180 d	phagocytic index ↑, antibody →
	Chinook	300 mg/kg and > requirement	→
	J. flounder	600 mg/kg	phagocytic activity ↑, lysozyme ↑
	Atlantic salmon	low levels	IHNV ↓
		> requirement	→
		800 mg/kg, 20 w	NBT →, complement ↑, lysozyme ↓, *A. salmonicida* ↑
	Trout	500 mg/kg, 12 w	phagocytosis ↑
		low levels	antibody ↓
	Turbot	500 mg/kg	phagocytic activity ↑
Vit-E + Selenium	Catfish	240 + 0.8 mg/kg, 120 d	NBT ↑
Vitamin-A (retinol)	Atlantic salmon	oral	anti-protease activity ↑, migration ↑
	Gilhead sea bream	50 -300 mg/kg	ROS ↑
α-tocopherol	Gilhead sea bream	600 – 1800 mg/kg	complement ↑
α-tocopherol acetate	Yellow tail	119 – 5950 mg/kg	phagocytic activity ↑, lysozyme ↑

Arginine	J. flounder	150 mg/kg	NBT ↑, lysozyme ↑
Ascorbate 2-monophospate	Atlantic salmon	20 – 1000 mg/kg	→
Ascorbil 2-sulfate	Atlantic salmon	4770 mg/kg	→
		82, 44, 3170 mg/kg, 23 w	antibody ↑
		1000 mg/kg	ROS ↑, lymphocyte number ↑
		2750 mg/kg	complement ↑
		4000 mg/kg	lysozyme ↑
Axtahantin	J. flounder	100 mg/kg	chemotaxis ↑, NBT ↑

<table>
<tr><td>Essential oil</td><td>Common carp</td><td colspan="2">30, 60, 120 ppm diet, 1% bw, 8 d</td><td>antibody ↑, bactericidal activity ↑</td></tr>
<tr><td>Protein hydrosilate</td><td>Atlantic salmon</td><td colspan="2">1 – 25 in vitro</td><td>ROS ↑</td></tr>
<tr><td>Soybean protein</td><td>Trout</td><td colspan="2">oral</td><td>phagocytosis ↑, NBT ↑, killing ↑</td></tr>
<tr><td colspan="5">Nucleotides</td></tr>
<tr><td>Ascogen P [1]</td><td>Hybrid striped bass</td><td colspan="2">5 g/kg, fixed ration approaching satiation daily</td><td>neutrophil oxidative radical ↑, survival after challenge with Streptoccus iniae ↑</td></tr>
<tr><td rowspan="3">Ascogen S [2]</td><td rowspan="2">Hybrid tilapia</td><td colspan="2">2 and 5 g/kg, 16 w</td><td>growth ↑, survival ↑</td></tr>
<tr><td colspan="2">5 g/kg, 120 d</td><td>antibody after vaccination ↑
lymphocyte mitogenic response ↑</td></tr>
<tr><td>Trout</td><td colspan="2">0.62, 2.5 and 5 g/kg, diet at 1% bw/d, 37 d</td><td>growth ↑</td></tr>
<tr><td rowspan="8">Optimun [2]</td><td rowspan="2">Trout</td><td rowspan="8">2 g/kg, containing 0.03% nucleotide</td><td>2% bw/d, 3 w</td><td>survival after challenge with, V. anguillarum ↑</td></tr>
<tr><td>1% bw/d, 2 w</td><td>survival after challenge with infectious salmon anaemia virus ↑</td></tr>
<tr><td>Coho salmon</td><td>2% bw/d, 3 w</td><td>survival after challenge with Piscirickettsia salmonis ↑</td></tr>
<tr><td rowspan="4">Atlantic salmon</td><td>2% bw/d, 3 w</td><td>sea lice infection ↓</td></tr>
<tr><td>1.5% bw/d, 3 w [a] and 5 w [b]</td><td>antibody ↑, mortality ↓</td></tr>
<tr><td>1.5% bw/d, 8 w</td><td>plasma chloride ↓, growth ↑</td></tr>
<tr><td>10 w</td><td>intestinal fold ↑</td></tr>
<tr><td>Turbot</td><td>to hand saniation daily</td><td>Altered immunogene expression in various tissues</td></tr>
<tr><td>Ribonuclease-digested yeast RNA [3]</td><td>Common carp</td><td colspan="2">15 mg/fish, by intubation, 3 d</td><td>phagocytosis ↑, complement ↑, lysozyme ↑, respiratory burst ↑, A. hydrophila infection ↓</td></tr>
</table>

1: Canadian Biosystem Inc. Calgary-Canada, 2: Chemoforma Augst-Swithzerland, 3: Amano Siyaku Co-op Tokyo, a: Before vaccination, b: Post-vaccination, bw: Body weight, d: Day, w: Week, ↑: Increase, ↓: Decrease, →: No change, Chinook: Chinook salmon, CL: Chemiluminescent response, J. flounder: Japanese flounder, Killing: Bactericidal activity of macrophage, NCCs: Non-specific cytotoxic cells, NBT: Nitroblue tetrazolium reaction, NT: Nucletoide, ROS: Reactive oxygen species, Trout: Rainbow trout, Turbot: Scophthalmus maximus, Vit-C: Vitamin-C, Vit-E: Vitamin-E, *A. hydrophila: Aeromonas hydrophila, A. salmonicida: Aeromonas salmonicida, E. tarda: Edwardsiella tarda, V. anguillarum: Vibrio anguillarum.* (modified from Galindo-Villegas & Hosokowa, 2004; Li & Gatlin, 2006; Sakai et al., 1999).

7.3 Risks and benefits using immunostimulants

Immunostimulants are more widely applied both within the aquaculture sector and in traditional animal husbandry. There are many examples of successful use of immunostimulants to improve fish welfare, and also *in vivo* or *in vitro* effects of immune system (see Table 12, Table 13 and Table 14). One of the earliest applications of immunostimulants in fish was the use of glucans in salmon diets. These diets were considered to be effective in managing disease outbreaks after stressful events such as grading and there was believed to be some benefit in reducing sea lice settlement; allowing the stock to go longer between anti-sea lice treatments. Certainly, the use of in-diet immunomodulators has become widely accepted in aquaculture with commercially available diets supplemented with nucleotides which have been demonstrated to reduce sea lice settlement and provide better protection against *A. salmonicida* and *V. anguillarum* infection (Bricknell & Dalmo, 2005; Burrells et al., 2001a; 2001b). Immunostimulants can provide particular benefits when used in order to: (1) reduce mortality due to opportunistic pathogens, (2) prevent virus disease such as Vitamin-C on infectious hematopoietic necrosis (IHN) virus and yeast glucan on yellow-head baculovirus, (3) enhance disease resistance of farmed fish and shrimp, (4) reduce mortality of juvenile fish especially in fry and larval fish, (5) enhance the efficacy of antimicrobial as adjoint substances, if used in combination with curative antimicrobial drugs at an early phase of disease development, or prior to anticipated disease outbreak, (6) enhance the resistance to parasites or microsporidias, such as Vitamin-C on *Ichthyophthirius multifiliis,* lactoferrin on *Cryptocaryon irritans,* or glucans and chitin on *Loma salmonea*, (7) enhance the efficacy of vaccines, (8) improve fish welfare against stress (e.g grading, sea transfer, vaccination and environmental change), such as glucans may be helped reduce the negative effects of stress on the innate immune response, soybean lecithin may be provided higher tolerance for increased water temperature, and vitamin-E may be protected the complement system against stress-related reduction of activity, (9) promoting a greater and more effective sustained immune response to those infectious agents producing subclinical disease without risks of toxicity, carcinogenicity or tissue residues, (10) maintaining immune surveillance at heightened level to ensure early recognition and elimination of neoplastic changes in tissues, and (11) selectively stimulating the relevant components of the immune system or non-specific immune mechanism that preferentially confer protection against micro-organisms, such as via interferon release, especially for those infectious agents for which no vaccines currently exists (Ai et al., 2007; Bricknell & Dalmo, 2005; Cerezuela et al., 2009; Gannam & Schrock, 2001; Maqsood et al., 2011; Raa, 2000; Rodriguez-Tovar et al., 2011; Yin et al., 2009).

Naturally, there is a risk that use of immunostimulants in aquaculture may cause unforeseen problems. Continual feeding of immunostimulants has generally been abandoned, in adult fish, in favor of pulse feeding. There are two possible outcomes of continuous feeding of an immunostimulant; (1) although, it is a very rare occurrence, the immunostimulant up-regulates the immune system to heightened levels and this is maintained until the immunostimulant is withdrawn, (2) the most obvious contra-indication as it would be in larval fish, continual exposure to an immunostimulant can induce tolerance. This is caused by the immune system of the host becoming de-sensitized to the immunostimulant and the immunostimulant response is lost, or in extreme circumstances the continued expose to an immunostimulant causes the immune response to become suppressed, giving a lower level of innate defences whilst exposure to that particular immunostimulant is maintained (Bricknell & Dalmo, 2005; Sakai et al., 1999). Besides, no research has yet been performed concerning the influence of immunostimulants at some stage such as maturation and spawning of fish. The immune systems become suppressed by sex hormones, testosterone and estradiol-17β, at these stages. Although the use of immunostimulants could cause recovery of the immune systems suppressed by sex hormones, they may disturb sexual maturation and other essential functions associated with spawning, or may include sterility through polyploidy (Cuesta et al., 2007; Magnadóttir, 2010; Piferrer et al., 2009). On the other hand, the mere deleterious side-effects of immunostimulants have not been completely investigated.

8. CONCISE SUMMARY

Significant advancements have been achieved in recent years regarding the immunological regulation of fish infections, benefiting the expanding aquaculture sector globally and enhancing comprehension of fundamental immunological phenomena. There are primarily three approaches for the control of fish diseases: immunization, chemotherapeutics, and immunostimulants. Moreover, new research on probiotics, prebiotics, and synbiotics has demonstrated beneficial health effects in various fish species. Immunostimulants and vaccinations are utilized in conjunction to avert infectious disorders. Immunostimulants may be utilized in the treatment of certain infectious disorders; however, they may not be as efficient as various chemotherapeutics. Antibiotic-resistant microorganisms jeopardize the treatment of fish diseases with chemotherapeutics. Immunostimulants may mitigate the limits of chemotherapeutics. Immunostimulants are considered to be safer than chemotherapeutics and possess a broader spectrum of efficacy compared to vaccinations. The integration of vaccination and immunostimulant delivery may enhance the efficacy of vaccinations.. In addition, continued pressure on the use of antimicrobials associated with food residue and environmental issue will encourage the use of immunostimulants. However, precautions must be observed concerning issues such as tolerance and adverse side effects, including immunosuppression resulting from excessive doses of immunostimulants or undesirable effects stemming from chronic usage of such chemicals. The current understanding of possible immunostimulants remains unclear in various aspects, particularly concerning the paths and processes via which these chemicals can target certain cells.

9. REFERENCES

Abbas, A. K. & Lichtmann, A. H. (2006). *Basic Immunology: Functions and disorders of the immune system, 2nd Edition*, W.B. Saunders Company, ISBN 1416029745, Philadelphia, USA

Adams, A. & Thompson, K.D. (1990). Development of an ELISA for the detection of *Aeromonas salmonicida* in fish tissue. *Journal of Aquatic Animal Health*, Vol. 2, pp. 281- 288

Adams, A. & Thompson, K.D. (2008). Recent applications of biotechnology to novel diagnostics for aquatic animals. *Revue Scientifique et Technique-Office International des Epizooties*, Vol. 27, No. 1, pp. 197-209

Adams, A. (2004). Immunodiagnostics in aquaculture. *Bulletin of the European Association of Fish Pathologists*, Vol.24, pp. 33-37

Adams, A.; Aoki, T.; Berthe, C.J.; Grisez, L. & Karunasagar, I. (2008). Recent technological advancements on aquatic animal health and their contributions toward reducing disease risks–a review, In: *Diseases in Asian Aquaculture VI. Fish Health Section*, M.G. Bondad-Reantaso, C.V Mohan, M. Crumlish & R.P. Subasinghe (Eds.), 71-88. Asian Fisheries Society, Manila, Philippines

Ai, Q.; Mai, K.; Zhang, L.; Tan, B.; Zhang, W.; Xu, W. & Li, H. (2007). Effects of dietary [beta]-1, 3 glucan on innate immune response of large yellow croaker, *Pseudosciaena crocea. Fish and Shellfish Immunology*, Vol. 22, pp. 394-402

Alderman, D.J. & Hastings, T. S. (2003). Antibiotic use in aquaculture: development of antibiotic resistance-potential for consumer health risks. *International Journal of Food Science and Technology*, Vol. 33, No. 2, pp. 139–155

Alexander, J.B. & Ingram, G.A. (1992). Noncellular nonspecific defence mechanisms of fish. *Annual Review of Fish Diseases*, Vol. 2, pp. 249-279

Alvarez-Pellitero, P. (2008). Fish immunity and parasite infections: from innate immunity to immunoprophylactic prospects. *Veterinary Immunology and Immunopathology*, Vol. 126, pp. 171–198

Aly, S.M & Mohamed, M.F. (2010). *Echinacea purpurea* and *Allium sativum* as immunostimulants in fish culture using Nile tilapia (*Oreochromis niloticus*). *Journal of Animal Physiology and Animal Nutrition*, Vol. 94, No. 5, pp. e31-e39

Anas, A.; Paul, S.; Jayaprakash, N.S.; Philip R.; Bright-Singh I. S. (2005). Antimicrobial activity of chitosan against vibrios from freshwater prawn *Macrobrachium rosenbergii* larval rearing systems. *Disease of Aquatic Organisms,* Vol. 67, pp. 177-179

Anderson, D.P.; Siwicki, A.K. & Rumsey, G. L. (1995). Injection or immersion delivery of selected immunostimulants to trout demonstrate enhancement of nonspecific defense mechanisms and protective immunity. In: *Diseases in Asian Aquaculture, Vol. 11, Fish Health Section*, M. Shariff, R.P. Subasinghe & J.R. Arthur, (Eds.), 413- 426, Asian Fisheries Society, Manila, Philippines

Anderson, D.P. (1996). Environmental factors in fish health: Immunological aspects. In: *The Fish Immune System: Organism, Pathogen, and Environment*, G. Iwama & T. Nakanishi (Eds.), 289-310, Academic Press, ISBN 0-12-350439-2, San Diego, California, USA

Aoki, T.; Takano, T.; Santos, M.D. & Kondo, H. (2008). Molecular innate immunity in teleost fish: Review and future perspectives. In: *Fisheries for Global Welfare and Environmental, Memorial book of the 5th Word Fisheries Congress,* K. Tsukamoto, T. Kawamura, T. Takeuchi, T.D. Beard, Jr. & M.J. Kaiser (Eds.), 263-276, ISBN 978-4- 88704-144-8, Terrapub, Setagaya-ku, Japan

Aranishi, F. & Mano, N. (2000). Antibacterial cathepsins in different types of ambicoloured Japanese flounder skin. *Fish and Shellfish Immunology,* Vol. 10, pp. 87-89

Arason, G. (1996). Lectin as defence molecules in vertebrates and invertebrates. *Fish and Shellfish Immunology,* Vol. 6, pp. 277-289

Austin, B. & Brunt, J.W. (2009). The use of probiotics in aquaculture, Chapter 7. In: *Aquaculture Microbiology and Biotechnology: Volume 1*, D. Montet & R.C. Ray (Eds), 185-207, Science Publishers, ISBN 978-1-57808-574-3, Enfield, New Hampshire, USA

Baba, T.; Watase, Y. & Yoshinaga, Y. (1993). Activation of mononuclear phagocyte function by levamisole immersion in carp. *Nippon Suisan Gakkaishi,* Vol. 59, pp. 301-307

Balfry, S.K. & Higgs, D.A. (2001). Influence of dietary lipid composition on the immune system and disease resistance of fish, Chapter 11, In: *Nutrition and Fish Health,* L. Chhorn & C.D. Webster (Eds.), 213-234, The Haworth Press, Inc., ISBN 1-56022-887- 3, Binghamton, New York, USA

Barman, D.; Kumar, V.; Roy, S.; Singh, A.S.; Majumder, D.; Kumar, A. & Singh, A.A. (1991). The role of immunostimulants in Indian aquaculture. Cited 21.05.2011. Available from http://aquafind.com/ articles/Immunostimulants-In-Aquaculture.php

Becker, J.A. & Speare D.J. (2007). Transmission of the microsporidian gill parasite, *Loma salmonae. Animal Health Research Reviews*, Vol. 8, pp. 59-68

Bei, J.X.; Suetake, H.; Araki, K.; Kikuchi, K.; Yoshiura, Y.; Lin, H R. & Suzuki, Y. (2006). Two interleukin (IL)-15 homologues in fish from two distinct origins. *Moleculer Immunology*, Vol. 43, pp. 860-869

Belosevic, M.; Hanington, P.C. & Barreda, D.R. (2006). Development of goldfish macrophages in vitro. *Fish and Shellfish Immunology*, Vol. 20, pp. 152-171

Bird, S.; Zou, J.; Kono, T.; Sakai, M.; Dijkstra, J.M. & Secombes, C. (2004). Characterization and expression analysis of interleukin 2 (IL-2) and IL-21 homologues in the Japanese pufferfish, *Fugu rubries*, following their discovery by synteny. *Immunogenetics*, Vol. 56, pp. 909-923

Bird, S.; Zou, J.; Savan, R.; Kono, T.; Sakai, M.; Woo, J. & Scombes, C. (2005). Characterisation and expression analysis of an interleukin 6 homologue in the Japanese pufferfish, *Fugu rubripes. Developmental and Comperative Immunology*, Vol. 29, pp.775-789

Bols, N.C.; Brubacher, J.L.; Ganassin, R.C. & Lee, L.E.J. (2001). Ecotoxicology and innate immunity in fish. *Developmental and Comparative Immunology*, Vol. 25, pp. 853-873

Bounocore, F. & Scapigliati, G. (2009). Immune defence mechanism in the sea bass *Dicentrarchus labrax* L., Chapter 6, In: *Fish Defenses, Volume 1: Immunology*, G. Zaccone, J. Meseguer, A. García-Ayala, B.G. Kapoor (Eds.), 185-219, Science Publishers, ISBN 978-1-57808-327-5, Enfield, New Hampshire, USA

Bricknell, I. & Dalmo, R.A. (2005). The use of immunostimulants in fish larval aquaculture. *Fish Shellfish Immunology*, Vol. 19, pp. 457-472

Bright-Singh, I. S. & Philip, R. (2002). Use of immunostimulants in aquaculture management. In: *Recent Advances in Diagnosis and Management of Diseases in Mariculture -Course Manuel*; 1-5; 7-27 November 2002, Cochin, India

Brown, G.D. (2006). Dectin-1: a signaling non-TLR pattern-recognition receptor. *Nature Reviews Immunology*, Vol.6, pp. 33-43

Brown-Treves, K.M. (2000). Immuno-stimulants, Chapter 19, In: *Applied Fish Pharmacology, Aquaculture Series 3*, K.M. Brown-Treves (Ed.), 251-259, Kluwer Academic Publishers, ISBN 0-412-62180-0, Dordrecht, Netherlands

Burrells, C.; Williams, P.D. & Forno, P.F. (2001a). Dietary nucleotides: a novel supplement in fish feeds: 1. Effects on resistance to disease in salmonids. *Aquaculture*, Vol. 199, pp. 159–169

Burrells, C.; Williams, P.D.; Southage, P.J. & Wadsworth, S.L. (2001b). Dietary nucleotides: a novel supplement in fish feeds: 1. Effects on vaccination, salt water transfer, growth rate and physiology of Atlantic salmon. *Aquaculture*, Vol. 199, pp. 171-184

Cabezas L. R. (2006). Functional genomics in fish: towards understanding stress and immune responses at a molecular level. *PhD Thesis*, 1-223, Departament de Biologia Cellular, Fisiologia i Immunologia, Facultat de Ciències, Universitat Autòmona de Barcelona, Barcolona, Spain

Campos-Perez, J.J.; Ward, M.; Grabowski, P.S.; Ellis, A.E. & Secombes, C.J. (2000). The gills are an important site of iNOS expression in rainbow trout *Oncorhynchus mykiss* after challenge with the Gram-positive pathogen *Renibacterium salmoninarum*. *Immunology*, Vol. 99, pp: 153-161

Cerezuela, R.; Cuesta, A.; Messeguer, J.; Ángeles Esteban, M. (2009). Effects of dietary vitamin D3 administration on innate immune parameters of seabream (*Sparus aurata* L.). *Fish and Shellfish Immunology*, Vol. 26, pp. 243-248

Chaves-Pozo, E.; Mulero, V.; Meseguer, J. & Ayala, A.G. (2005). Professional phagocytic granulocytes of the bony fish gilthead seabream display functional adaptation to testicular microenvironment. *Journal of Leukocyte Biology*, Vol. 75, pp. 345–351

Clem, L.W.; Sizemore R.C.; Ellsaesser, C.F. & Miller. N.W. (1985). Monocytes as accessory cells in fish immune responses. *Developmental and Comparative Immunology*, Vol. 9, pp. 803-809

Cole, A.M.; Weis, P. & Diamond, G. (1997). Isolation and characterization of pleurocidin; an antimicrobial peptide in the skin secretions of winter flounder. *Journal of Biological Chemistry*, Vol. 272, pp. 12008-12013

Corripio-Miyar, Y.; Bird, S.; Tsamopoulus, K. & Secombes, C. J. (2006). Cloning and expression analysis of two pro-inflamatory cytokines; IL-1beta and IL-8, in haddock (*Melanogrammus aeglefinus*). *Molecular Immunology*, Vol. 44, pp. 1361-1373

Cuesta, A.; Vargas-Chacoff, L.; García-López, A.; Arjona, F.J.; Martínez-Rodríguez G.; Meseguer, J.; Mancera, J.M. & Esteban, M.A. (2007). Effect of sex-steroid hormones, testosterone and estradiol, on humoral immune parameters of gilthead seabream. *Fish and Shellfish Immunology*, Vol. 23, pp. 693-700

Cunningham, C.O. (2004). Use of molecular diagnostic tests in disease control: Making the leap from laboratory to field application, Chapter 11, In: *Molecular Aspects of Fish and Marine Biology Vol. 3: Current trends in the study of bacterial and viral fish and shrimp diseases*, K.Y. Leung (Ed.), 292-312, World Scientific Publishing Co. Pte. Ltd., ISBN 981-238-749-8, London, UK

Dalmo R. A. (2002). Immunostimulation of fish. ICES CM 2002/R:09, Cited 05.04.2011. Available from http://www.ices.dk/products/CMdocs/2002/R/R0902.PDF

Dannevig, B. H.; Lauve, A.; Press, C.McL. & Landsverk, T. (1994). Receptor-mediated endocytosis and phagocytosis by rainbow trout head kidney sinusoidal cells. *Fish and Shellfish Immunology,* Vol. 4, pp. 3-18

Dickerson, H.W. & Clark, T.G. (1996). Immune response of fishes to ciliates. *Annual Review of Fish Diseases,* Vol. 6, pp. 107-120

Dorin, D.; Sire, M.F. & Vernier, J.M. (1994). Demonstration of an antibody response of the anterior kidney following intestinal administration of a soluble protein antigen in trout. *Comparative Biochemistry and Physiology*, Vol. 109, pp. 499-509

Duffy, J.E.; Carlson, E.; Li, Y.; Prophete, C. & Zelikoff, J.T. (2002). Impact of polychlorinated biphenyls (PCBs) on the immune function of fish: age as a variable in determining adverse outcome. *Marine Environmental Research*, Vol. 54, pp. 559-563

Dugenci, S.K.; Arda, N. & Candan, A. (2003). Some medicinal plants as immunostimulant for fish. *Journal of Ethnopharmacology,* Vol. 88, No. 1, pp. 99-106

Duncan, P.L. & Klesius, P.H. (1996). Effects of feeding spirulina on specific and nonspecific Immune responses of channel catfish. *Journal of Aquatic Animal Health*, Vol. 8, pp. 308-313

Dupont, N.C.; Wang, K; Wadhwa, P.D.; Culhane, J.F. & Nelson, E.L. (2005). Validation and comparison of luminex multiplex cytokine analysis kits with ELISA: Determinations of a panel of nine cytokines in clinical sample culture supernatants. *Journal of Reproductive Immunology*, Vol. 66, pp. 175-191

Dykova I. (2006). Phylum microspora. In: *Fish Diseases and Disorders, Vol. 1, Protozoon and Metazoan Infections, 2nd Edition*, P.T.K. Woo (Ed.), 205-229, CABI Publishing, ISBN 0851990150, Oxford, UK

Ebran, N.; Julien, S.; Orange, N.; Saglio, P.; Lemaître, C. & Molle, G. (1999). Pore-forming properties and antibacterial activity of proteins extracted from epidermal mucus of fish. *Comparative Biochemistry and Physiology. Part A, Molecular & Integrative Physiology*, Vol. 122, No. 2, pp. 181-189

El-Gohary, M.S.; Safinaz, G.M.; Khalil, R.H.; El-Banna, S. & Soliman, M.K. (2005). Immunosuppressive effects of metrifonate on *Oreochromis Niloticus. Egyptian Journal of Aquatic Research,* Vol. 31, pp. 448-458

Ellis, A.E. (1999). Immunity to bacteria in fish. *Fish and Shellfish Immunology,* Vol. 9, pp. 291- 308

Ellis, A.E. (2001). Innate host defense mechanisms of fish against viruses and bacteria. *Developmental and Comparative Immunology*. Vol. 25, pp. 827-839

Ellsaesser, C.F.; Bly, J.E. & Clem, L.W. (1988). Phylogeny of lymphocyte heterogeneity: The thymus in channel catfish. *Developmental and Comparative Immunology*, Vol. 12, pp. 787-799

Enis-Yonar, M.; Mise-Yonar, S. & Silici, S. (2011). Protective effect of propolis against oxidative stress and immunosuppression induced by oxytetracycline in rainbow trout (*Oncorhynchus mykiss*; W.). *Fish and Shellfish Immunology*, Vol. 31, pp. 318-325

Espenes, A.; Press, C.; Danneving, B.H. & Landsverk, T. (1995). Immune-complex trapping in the splenic ellipsoids of rainbow trout (*Oncorhynchus mykiss*). *Cell and Tissue Research*, Vol. 282, pp. 41-48

EU (European Union) (2007). Council Regulation (EC) No. 834/07 of 28 June 2007 on organic production and labelling of organic products and repealing Regulation (EEC) No. 2092/91. Official Journal of the European Union L 189, 20/07/2007, pp. 1-23

EU (European Union) (2008). Council Regulation (EC) No. 889/08 of 5 September 2008 laying down detailed rules for the implementation of Council Regulation (EC) No. 834/2007 on organic production and labelling of organic products with regard to organic production; labelling and control. Official Journal of the European Union L 250, 18/09/2008, pp. 1-84

Evans, D.L. & Gratzek, J.B. (1989). Immune defense mechanisms in fish to protozoan and helmint infections. *American Zoologist* (new name; *Integrative and Comparative Biology* (*ICB*)), Vol., 29, No. 2, pp. 409-418

Evans, D.L. & Jaso-Friedmann, L. (1992). Nonspecific cytotoxic cells as effectors of immunity in fish. *Annual Review of Fish Diseases*, Vol. 2, No. 1, pp. 109-121

Fast, M.D.; Sims, D.E.; Burka, J.F.; Mustafa, A. & Ross, N.W. (2002). Skin morphology and humoral non-specific defence parameters of mucus and plasma in rainbow trout; coho and Atlantic salmon. *Comparative Biochemistry and Physiology*, Vol. 132, No. 3, pp. 645-57

Fearon, D.T. & Locksley, R. M. (1996). The instructive role of innate immunity in the acquired immune response. *Science*, Vol. 272, pp. 50-53

Fletcher, T.C. (1981). Non-antibody molecules and the defense mechanisms of fish. In: *Stress and Fish*, A.D. Pickering (Ed.), 171-183, Academic Press, ISBN 0125545509, New York, USA

Galina, J; Yin, G.; Ardó, L. & Jeney, Z. (2009). The use of immunostimulating herbs in fish. An overview of research. *Fish Physiology and Biochemistry*, Vol. 35, No. 4, pp. 669- 676

Galindo-Villegas, J. & Hosokawa H. (2004). Immunostimulants: Towards temporary prevention of diseases in marine fish. In: *Avances en Nutrición Acuícola VII. Memorias del VII Simposium Internacional de Nutrición Acuícola*, L. E. Cruz Suárez, D. Ricque Marie, M. G. Nieto López, D. Villarreal, U. Scholz & M. Gonzalez (Eds.), 279-319, 16-19 Noviembre 2004, Hermosillo, Sonara, México

Galindo-Villegas, J.; Fukada, H; Masumoto, T. & Hosokawa, H. (2006). Effect of dietary immunostimulants on some innate immune responses and disease resistance against *Edwardsiella tarda* infection in Japanese flounder (*Paralichthys olivaceus*). *Aquaculture Science*, Vol. 54, No. 2, pp. 153-162

Gannam, A.L. & Schrock, M.R. (2001). Immunostimulants in fish diets, Chapter 10, In: *Nutrition and Fish Health,* L. Chhorn & C.D. Webster (Eds.), 235-266, The Haworth Press, Inc., ISBN 1-56022-887-3, Binghamton, New York, USA

García-Ayala, A. & Chaves-Pozo, E. (2009). Leukocytes and cytokines present in fish testis: A review, Chapter 2, In: *Fish Defenses, Volume 1: Immunology*, G. Zaccone, J. Meseguer, A. García-Ayala, B.G. Kapoor (Eds.), 37-74, Science Publishers, ISBN 978-1-57808-327-5, Enfield, New Hampshire, USA

Giavedoni, L.D. (2005). Simultaneous detection of multiple cytokines and chemokines from nonhuman primates using luminex technology. *Journal of Immunological Methods*, Vol. 301, pp. 89-101

Gil, A., 2002. Modulation of the immune response mediated by dietary nucleotides. *Europen Journal of Clinical Nutrition*, Vol. 56, No. Suppl. 3, pp. S1–S4.

Gildberg, A.; Bogwald, J.; Johansen, A. & Stenberg, E. (1996). Isolation of acid peptide fraction from a fish protein hydrolysate with strong stimulary effect on atlantic salmon (*Salmon salar*) head kidney leucocytes. *Comparative Biochemistry and Physiology Part B: Biochemistry and Molecular Biology*, Vol., 114, No. 1, 97-101

Grimble, G.K. & Westwood, O.M.R. (2000). Nucleotides. In: *Nutrition and Immunology: Principles and Practice*, German; J.B. & Keen; C.L. (Eds.), 135- 144, Humana Press Inc., Totowa, New Jersey, USA

Gudding, R.; Lillehaug, A. & Evensen, Ø. (1999). Recent developments in fish vaccinology. Veterinary *Immunology and Immunopathology*, Vol. 72, No. 1-2), pp. 203-212

Guttvik, A.; Paulsen, B.; Dalmo, R.A.; Espelid, S.; Lund, V. & Bøgwald, J. (2002). Oral administration of lipopolysaccharide to Atlantic salmon (*Salmo salar* L.) fry. Uptake, distribution, influence on growth and immune stimulation. *Aquaculture,* Vol., 212, pp. 35-53

Hamerman, J. A.; Ogasawara, K. & Lanier, L.L. (2005). NK cells in innate immunity. *Current Opinion in Immunology*, Vol. 17, pp. 29-35

Hoffmann, K. (2009). Stimulating immunity in fish and crustaceans: some light but more shadows. *Aqua Culture Asia Pacific Magazine*, Vol. 5, No. 5, pp. 22-25

Hogan, R.J.; Stuge, T.B.; Clem, L.W.; Miller, N.W. & Chinchar, V.G. (1996). Anti-viral cytotoxic cells in the channel catfish /Ictalurus punctatus). *Developmental and Comparative Immunology*, Vol. 20, pp. 115-127

Holland, J.W.; Pottinger, T.G. & Secombes, C.J. (2002). Recombinant interleukin-1 beta activates the hypothalamic-pituitary-interrenal axis in rainbow trout; *Oncorhynchus mykiss. Journal of Endocrinology,* Vol. 175, pp. 261-267

Holland, M.C.H. & Lambris, J.D. (2002). The complement system of teleosts. *Fish and Shellfish Immunology,* Vol. 12, pp. 399-420

Hølvold, L.B. (2007). Immunostimulants connecting innate and adaptive immunity in Atlantic salmon (*Salmo salar*). *Master in Biology-Field of study Marine Biotechnology*, 1- 69, Department of Marine Biotechnology, Norwegian College of Fishery Science, Univetsity of Tromso, Tromsø, Norway

Horsberg, T.E. (2003). Aquatic animal medicine. *Journal of Veterinary Pharmacology and Therapeutics*, Vol. 26, No. 1-2, pp. 39-42

Igawa, D.; Sakai, M. & Savan, R. (2006). An unexpected discovery of two interferon gamma– like genes along with interleukin (IL)-22 and -26 from teleost: IL-22 and -26 genes have been described for the first time outside mammals. *Moleculer Immunology*, Vol. 43, pp. 999-1009

Inoue, Y.; Kamuta, S.; Ito, K.; Yoshiura, Y.; Ototake, M.; Moritomo, T. & Nakanishi, T. (2005). Molculer cloning and expression analysis of rainbow trout (*Oncorhynchus mykiss*) interleukin-10 cDNAs. *Fish and Shellfish Immunology*, Vol. 18, pp. 335-344

Ispir, U. & Dorucu, M. (2005). A study on the effects of levamisole on the immune system of rainbow trout (*Oncorhynchus mykiss*; Walbaum). *Turkish Journal of Veterinary and Animal Sciences*, Vol. 29, pp. 1169-1176

Jansson E. (2002). Bacterial Kidney Disease in salmonid fish: Development of methods to assess immune functions in salmonid fish during infection by *Renibacterium salmoninarum. PhD Thesis*, 1-52, Department of Pathology and Department of Fish, National Veterinary Institute, Swedish University of Agricultural Sciences, Uppsala, Sweden

Jeney, G. & Anderson, D.P. (1993). Enhanced immune response and protection in rainbow trout to *Aeromonas salmonicida* bacterin following prior immersion in immunostimulants. *Fish and Shellfish Immunology,* Vol. 3, pp. 51-58

Jimeno, C.D. (2008). A transcriptomic approach toward understanding PAMP-driven macrophage activation and dietary immunostimulant in fish. *PhD*

Thesis, 1-222, Departament de Biologia Cellular, Fisiologia i Immunologia, Facultat de Ciències, Universitat Autòmona de Barcelona, Barcelona, Spain

Jiye, L.; XiuQin, S.; Fengrong, Z. & LinHua, H. (2009). Screen and effect analysis of immunostimulants for sea cucumber, *Apostichopus japonicus. Chinese Journal of Oceanology and Limnology,* Vol. 27, No. 1, pp. 80-84

Jones, S.R.M. (2001). The occurrence and mechanisms of innate immunity against parasites in fish. *Developmental and Comparative Immunology,* Vol. 25, pp. 841-852

Joosten, P.H.M.; Kruijer, W.J. & Rombout, J.H.W.M. (1996). Anal immunisation of carp and rainbow trout with different fractions of a *Vibrio anguillarum* bacterin. *Fish and Shellfish Immunology,* Vol. 6, pp. 541-551.

King, P.D.; Aldridge, M.B.; Kennedy-Stoskopf, S. & Stott, J.L. (2001). Immunology, Chapter 12, In: *CRC Handbook of Marine Mammal Medicine, 2nd Edition*, L.A. Dierauf & F.M.D. Gulland (Eds.), 237-252, CRC Press LLC, ISBN 0-8493-0839-9, Boca Raton, Florida, USA

Klesius, P.H.; Shoemaker, C.A.; Evans, J.J. & Lim, C. (2001). Vaccines: Prevention of Diseases in aquatic animals, Chapter 17, In: *Nutrition and Fish Health,* L. Chhorn & C.D. Webster (Eds.), 317-335, The Haworth Press, Inc., ISBN 1-56022-887-3 , Binghamton, New York, USA

Klesius, P.H.; Pridgeon, J.W. & Aksoy, M. (2010). Chemotactic factors of *Flavobacterium columnare* to skin mucus of healthy channel catfish (*Ictalurus punctatus*). *FEMS Microbiology Letters*, Vol. 310, pp. 145–151

Kunttu, H.M.T.; Valtonen, E.T.; Suomalainen, L.R.; Vielma, J. & Jokinen, I.E. (2009). The efficacy of two immunostimulants against *Flavobacterium columnare* infection in juvenile rainbow trout (*Oncorhynchus mykiss*). *Fish and Shellfish Immunology,* Vol. 26, pp. 850-857

Kusher, D.I. & Crim, W.C. (1991). Immunosuppression in bluegill (*Lepomis macrochirus*) induced by environmental exposure to cadmium. *Fish and Shellfish Immunology,* Vol. 1, pp. 157-161

Lauridsen, J.H. & Buchmann, K. (2010). Effects of short- and long-term glucan feeding of rainbow trout (Salmonidae) on the susceptibility to *Ichthyophthirius multifiliis* infections. *Acta Ichthyologica et Piscatoria*, Vol. 10, No. 1, pp. 61-66

Lemaître, C.; Orange, N.; Saglio, P.; Saint, N.; Gagnon, I. & Molle, G. (1996). Characterisation and ion channel activities of novel antibacterial proteins from the skin mucosa of carp (*Cyprinus carpio*). *European Journal of Biochemistry*, Vol. 240, No. 1, pp. 143-149

Li, J.; Bardera, D.R.; Zhang, Y.A.; Boshra, H.; Gelman, A.E.; LaPatra, S.; Tort, L. & Sunyer, J.O. (2006). B lymphocyte from early vertebrates have potent phagocytic and microbicidal abilities. *Nature Immunology*, Vol. 7, pp. 1116-1124

Li, J.H.; Shao, J.Z.; Xiang, L.X. & Wen, Y. (2007). Cloning; characterization and expression analysis of puffer fish interleukin-4 cDNA: the first evidence of Th2-type cytokine in fish. *Molecular Immunology*, Vol. 44, pp. 2088-2096

Li, P. & Gatlin, D.M. (2006). Nucleotide nutrition in fish: Current knowledge and future application. *Aquaculture*, Vol., 251, pp. 141-152

Lim, C.; Klesius, P.H. & Shoemaker, A.C. (2001a). Dietary iron and fish health, Chapter 9, In: *Nutrition and Fish Health,* L. Chhorn & C.D. Webster (Eds.), 189-199, The Haworth Press, Inc., ISBN 1-56022-887-3, Binghamton, New York, USA

Lim, C.; Klesius, P.H. & Webster, A.C. (2001b). The role of dietary phosphorus, zinc, and selenium in fish health, Chapter 10, In: *Nutrition and Fish Health,* L. Chhorn & C.D. Webster (Eds.), 201-212, The Haworth Press, Inc., ISBN 1-56022-887-3 , Binghamton, New York, USA

Lorenzen, K. (1993). Acquired immunity to infectious diseases in fish: implications for the interpretation of fish disease surveys. In: *Fish: Ecotoxicology and Ecophysiology,* T. Braunbeck, W. Hanke, H. Segner, pp. 183-196; ISBN 3527300104, Verlag Chemie, Weinheim, New York, USA

Lumlertdacha, S. & Lovell, R.T. (1995). Fumonisin-contaminated dietary corn reduced survival and antibody production by channel catfish challenged with *Edwardsiella ictaluri*. *Journal of Aquatic Animal Health*, Vol. 7, pp. 1-8

Lundén, T.; Miettinen, S.; Lonnstrom, L.G.; Lilius, E. M. & Bylund, G. (1998). Influence of oxytetracycline and oxolinic acid on the immune response of rainbow trout (*Oncorhynchus mykiss*). *Fish and Shellfish Immunolog,;* Vol. 8, pp. 217-230

Lundén, T.; Miettinen, S.; Lonnstrom, L.G.; Lilius, E. M. & Bylund, G. (1999). Effect of florfenicol on the immune response of rainbow trout (*Oncorhynchus mykiss*). *Veterinary Immunology and Immunopathology,* Vol. 67, pp. 317-325

Lundén, T. & Bylund, G. (2002). Effect of sulphadiazine and trimethoprim on the immune response of rainbow trout (*Oncorhynchus mykiss*). *Veterinary Immunology and Immunopathology* Vol. 85 pp. 99-108

Lutfalla, G.; Crollius, H.R.; Stange–Thomann, N.; Jaillon, O.; Mogensen, K. & Monneron, D. (2003). Comparative genomic analysis reveals independent expansion of lineage specific gene family in vertebrates: the class II cytokine receptors and their ligands in mammals and fish. *BMC Genomics*, Vol. 4, pp. 29

Magnadóttir, B. (2006). Innate immunity of fish (overview). *Fish and Shellfish Immunology,* Vol. 20, pp. 137-151

Magnadóttir, B. (2010). Immunological control of fish diseases. *Marine Biotecnology*, Vol. 12, pp. 361-379

Manning, B.B. (2001). Mycotoxins in fish feeds, Chapter 13, In: *Nutrition and Fish Health*, L. Chhorn & C.D. Webster (Eds.), 267-287, The Haworth Press, Inc., ISBN 1-56022-887- 3, Binghamton, New York, USA

Manning, B.B. (2010). Mycotoxins in aquaculture feed. In: *Nutrition and Fish Health*, L. Chhorn & C.D. Webster (Eds.), 267-287, The Haworth Press, Inc., ISBN 1-56022-887- 3 , Binghamton, New York, USA

Manning, M.J. 1994. Fishes. In: *Immunology: A Comparative Approach*, R.J. Turner (Ed.), 69- 100, John Wiley & Sons Ltd., ISBN 0471944009, Chichester, UK

Manning, M.J. & Nakanishi, T. (1996). The specific immune system: Cellular defenses In: *The Fish Immune System: Organism, Pathogen, and Environment*, G. Iwama & T. Nakanishi (Eds.), 159-205, Academic Press, ISBN 0-12-350439-2, San Diego, California, USA

Maqsood, S.; Singh, P.; Samoon, M.H. & Wani, G.B. (2011). Use of immunostimulants in aquaculture systems. Cited 11.07.2011. Available from http://aquafind.com/ articles/Immunostimulants-in-aquaculture.php

Medzhitov, R. & Janeway, C.A. Jr. (2002). Decoding the patterns of self and nonself by the innate immune system. *Science*, Vol. 296, pp. 298-300

Medzhitov, R. (2007). Recognition of microorganisms and activation of the immune response. *Nature*, Vol. 449, pp. 819-826

Meseguer, J.; López-Ruiz, A. & García-Ayala, A. (1995). Reticulo-endothelial stroma of the head-kidney from the seawater teleost gilthead seabream (*Sparus aurata* L): an ultrastructural and cytochemical study. *Anatomical Record*, Vol., 241, pp: 303-309

Miller, N.; Wilson, M.; Bengtén, E.; Stuge, T.; Warr, G. & Clem, W. (1998). Functional and molecular characterization of teleost leukocytes. *Immunological Reviews*, Vol. 166, pp. 187–197

Moore, J.D.; Ototake, M. & Nakanishi, T. (1998). Particulate antigen uptake during immersion immunisation of fish: The effectiveness of prolonged exposure and the roles of skin and gill. *Fish and Shellfish Immunology*, Vol. 8, pp. 393-407

Mulero, I.; Sepulcre, M.P.; Meseguer, J.; Garcia-Ayala, A. & Mulero, V. (2007). Histamine is stored in mast cells of most evolutionarily advanced fish and regulates the fish inflammatory response. *The Proceeding of the National Academy of Science USA* (*PNAS*), Vol. 104, No. 49, pp. 19434–19439

Nakano, M.; Mutsuro, J.; Nakahara, M.; Kato, Y. & Yano, T. (2003). Expansion of genes encoding complement components in bony fish: biological implications of the complement diversity. *Developmental and Comparative Immunology*, Vol. 27, pp. 764- 762

Nayak, S.K. (2010). Probiotics and immunity: A fish perspective. *Fish and Shellfish Immunology*, Vol. 29, pp. 2-14

Nelson, J.S. (2006). *Fishes of the world, 4th Edition*, John Wiley & Sons Inc. Publication ISBN 0- 471-25031-7, New York, USA

Noga, E. J. (2010). *Fish Disease Diagnose and Treatment, 2nd Edition*, Wiley-Blackwell: John Wiley & Sons Inc. Publication, ISBN 978-0-8138-0697-6, Iowa, USA

Nygaard, R.; Husgard, S.; Sommer, A.I.; Leong, J.A. & Robertsen, B. (2000). Induction of Mx protein by interferon and double-stranded RNA in salmonid cells. *Fish and Shellfish Immunology*, Vol. 10, pp. 435-450

Ortuño, J.; Esteban, M.A. & Messeguer, J. (2001). Effects of short-term crowding stress on gilthead seabream (*Sparus aurata* L.) innate immune response. *Fish and Shellfish Immunology*, Vol. 11, pp. 187-197

Ortuño, J.; Cuesta, A.; Rodríguez, A.; Esteban, M.A. & Meseguer, J. (2002). Oral administration of yeast; *Saccharomyces cerevisiae*; enhances the cellular innate immune response of gilthead seabream (*Sparus aurata* L.). *Veterinary Immunology and Immunopathology*, Vol. 85, pp. 41-50

Palaksha, K.J.; Shin, G.W.; Kim, Y.R. & Jung, T.S. (2001). Evolution of non-specific immune components from the skin mucus of olive flounder (*Paralichthys olivaceus*). *Fish and Shellfish Immunology*, Vol. 24, pp. 479-488

Pancer, Z. & Cooper, M.D. (2006). The evolution of adaptative immunity. *Annual Review of Immunology*, Vol. 24, pp. 497-518

Park, I.Y.; Park, G.B.; Kim, M.S. & Kim, S.C. (1998). Parasin I; an antimicrobial peptide derived from histone H2A in the catfish, *Parasilurus asotus*. *FEBS Letters*, Vol. 437, pp. 258-268

Passer, B.J.; Chen, C.H.; Miller, N.W. & Cooper, M.D. (1996). Identification of a T lineage antigen in the catfish. *Developmental and Comparative Immunology*, Vol. 20, pp. 441- 450

Paulsen, S.M.; Lunde, H.; Engstad, R.E. & Robertsen, B. (2003). In vivo effects of β-glucan and LPS on regulation of lysozyme activity and mRNA expression in Atlantic salmon (*Salmo salar* L.) *Fish and Shellfish Immunology*, Vol. 14, No. 1, pp. 39-54

Peatmen, E. & Liu, Z. (2006). CC chemokines in zebrafish: evidence for extensive intrachoromosomal gene duplications. *Genomics*, Vol. 88, pp. 381-385

Peatman, E. & Liu, Z. (2007). Evolution of CC chemokines in teleost fish: a case study in gene duplication and implications for immune diversity. *Immunogenetics,* Vol. 59, pp. 613-623

Peddie, S.; Zou, J. & Secombes, C.J. (2002). Immunostimulation in the rainbow trout (*Oncorhynchus mykiss*) following intraperitoneal administration of Ergosan. *Veterinary Immunology and Immunopathology*, Vol. 86, pp. 101-113

Pedersen, G.M.; Gildberg, A. & Olsen, R.L. (2004). Effects of including cationic proteins from cod milt in the feed to Atlantic cod (Gadus morhua) fry during a challenge trial with *Vibrio anguillarum*. *Aquaculture*, Vol. 233, pp. 31-43

Pellitero, P.A. (2008). Fish immunity and parasite infections: from innate immunity to immunoprophylactic prospects. *Veterinary Immunology and Immunopathology,* Vol. 126, pp. 171-198

Perdikaris, C. & Paschos, I. (2010). Organic aquaculture in Greece: a brief review. *Reviews in Aquaculture*, Vol. 2, No. 2, pp. 102–105

Perera, H.A.C.C & Pathiratne A. (2008). Enhancement of immune responses in Indian carp, Catla carta, following administration of levamisole by immersion. In: *Disease in Asian Aquaculture VI, Fish Health Section*, M.G. Bondad-Reantosa, C.V. Crumlish &

R.P. Subasingle (Eds.), 129-142, Asian Fisheries Society, Manilá, Philippines Piferrer, F.; Beaumont, A.; Falguière, J.C.; Flajšhans, M.; Haffray, P. & Colombo, L. (2009).

Polyploid fish and shellfish: production, biology and applications to aquaculture for performance improvement and genetic containment. *Aquaculture,* Vol. 293, pp. 125-156

Plouffe, D.A.; Hanington, P.C.; Walsh, J.G.; Wilson, E.C. & Belosevie, M. (2006). Comprasion of select innate immune mechanisms of fish and mammals. *Xenotransplantation,* Vol. 12, pp. 226-277

Plumb, J.A. & Hanson, L.A. (2011). *Health Maintenance and Principal Microbial Diseases of Cultured Fishes, 3rd Edition*, Wiley-Blackwell: John Wiley & Sons Inc. Publication, ISBN 978-0-8138-1693-7, Iowa, USA

Powell, J.L. & Loutit, M.W. (2004). Development of a DNA probe using differential hybridization to detect the fish pathogen *Vibrio anguillarum.* Microbial Ecology, Vol. 28, pp. 365-373

Press, C. McL,; Evensen, Ø.; Reitan, L.J. & Landsverk, T. (1996). Retention of furunculosis vaccine components in Atlantic salmon *Salmon solar* L., following different routes of administration. *Journal of Fish Disease*, Vol. 19, 215-224

Press, C.McL. & Evensen, Ø. (1999). The morphology of the immune system in teleost fishes. *Fish and Shellfish Immunology,* Vol. 9, pp. 309-318

Raa, J. (2000). The use of immune-stimulants in fish and shellfish feeds. In: *Avances en Nutrición Acuícola V. Memorias del V Simposium Internacional de Nutrición Acuícola.*

L.E. Cruz-Suárez; D. Ricque-Marie, M. Tapia-Salazar, M.A. Olvera-Novoa & R. Civera-Cerecedo (Eds.). 19-22 Noviembre 2000, Mérita, Yucatán, Mexico

Randelli, E.; Buonocore, F. & Scapigliati, G. (2008). Cell markers and determinants in fish immunology. *Fish and Shellfish Immunology,* Vol. 25, pp. 326-340

Retie, O. (1998). Mast cells/eosinophilic granular cells of teleostean fish: A review focusing on standing properties and functional responses. *Fish and Shellfish Immunology,* Vol. 8, pp. 489-513

Rodriguez-Tovar, L.E.; Speare, D.J. & Markham, R.J. (2011). Fish microsporidia: immune response, immunomodulation and vaccination. *Fish and Shellfish Immunology,* Vol. 30, pp. 999-1009

Rombout, J.H.M.W.; Huttenhuis, H.B.T.; Picchietti, S. & Scapigliati, G. (2005). Phylogeny and ontogeny of fish leucocytes. *Fish and Shellfish Immunology,* Vol. 19, pp. 441-445

Roque, A.; Soto-Rodríguez, S.A. & Gomez-Gil, B. (2009). Bacterial fish diseases and molecular tools for bacterial fish pathogens detection. In: *Aquaculture Microbiology and Biotechnology: Volume 1,* D. Montet & R.C. Ray (Eds), 73-99, Science Publishers, ISBN 978-1-57808-574-3, Enfield, New Hampshire, USA

Sakai, M. (1999). Current research status of fish immunostimulants. *Aquaculture,* Vol. 172, pp. 63-92

Sakai, M.; Yoshida, T. & Kobayashi, M. (1995). Influence of the immunostimulant, EF203, on the immune responses of rainbow trout, *Oncorhynchus mykiss*, to *Renibacterium salmoninarum. Aquaculture*, Vol. 138, no. 1-4, pp. 61-67

Savan, R.; Kono, T.; Igawa, D. & Sakai, M. A. (2005). A novel tumor necrosis factor (TNF) gene present in tandem with the TNF-alpha gene on the same chromosome in teleosts. *Immunogenetics*, Vol. 57, pp. 140-150

Schluter, S.F.; Bernstein, R.M. & Marchalonis, J.J. (1999). Big Bang - emergence of the combinatorial immune system. *Developmental and Comparative Immunology*, Vol. 23, pp. 107-111.

Secombes, C.J. (1996). The nonspecific immune system: Cellular defenses. In: In: *The Fish Immune System: Organism, Pathogen, and Environment*, G. Iwama & T. Nakanishi (Eds.), 63-105, Academic Press, ISBN 0-12-350439-2, San Diego, California, USA

Secombes, C.J.; Manning, M.J. & Ellis, A.E. (1982). The effect of primary and secondary immunization on the lymphoid tissue of the carp, *Cyprinus carpio* L. *Journal of Experimental Zoology*, Vol. 220, pp. 277-287

Secombes, C.J.; Hardie, L.J. & Daniels. G. (1996). Cytokines in fish: An update. *Fish and Shellfish Immunology*, Vol. 6, pp. 291-304

Secombes, C.J.; Wang, T.; Hong, S.; Peddie, S.; Crampe, M.; Laing, K.J.; Cunningham, C. & Zou, J. (2001). Cytokines and innate immunity of fish. *Developmental and Comparative Immunology*, Vol. 25, No. 8-9, pp. 713-723.

Seker, E.; Ispir, U. & Dorucu, M. (2011). Immunostimulating effect of levamisole on spleen and head-kidney leucocytes of rainbow trout (*Oncorhynchus mykiss*; Walbaum 1792). *Kafkas Universitesi Veteriner Fakultesi Dergisi*, Vol. 17, no. 2, pp. 239-242

Shoemaker, C.A.; Klesius, H.P. & Lim C. (2001). Immunity and disease resistance in fish, Chapter 7, In: *Nutrition and Fish Health,* L. Chhorn & C.D. Webster (Eds.), 149-162, The Haworth Press, Inc., ISBN 1-56022-887-3 , Binghamton, New York, USA

Smith, J.V.; Fernandes, J.M.O.; Jones, S.J.; Kemp, G.D. & Tatner, M.F. (2000). Antibacterial proteins in rainbow trout, *Oncorhynchus mykiss. Fish and Shellfish Immunology,* Vol. 10, pp. 243-260

Smith, J.V. & Fernandes, J.M.O. (2009). Antimicrobial peptides of the innate immune system, Chapter 8, In: *Fish Defenses, Volume 1: Immunology*, G. Zaccone, J. Meseguer, A. García-Ayala, B.G. Kapoor (Eds.), 241-275, Science Publishers, ISBN 978-1-57808- 327-5, Enfield, New Hampshire, USA

Soltani, M.; Sheikhzadeh, N.; Ebrahimzadeh-Mousavi, H.A. & Zargar, A. (2010). Effects of *Zataria multiflora* essential oil on innate immune responses of common carp (*Cyprinus carpio). Journal of Fisheries and Aquatic Science,* Vol. 5, pp. 191-199

Subasinghe, R. (2009). Disease control in aquaculture and the responsible use of veterinary drugs and vaccines: The issues; prospects and challenges. In: *Options Méditerranéennes, Series A, No. 86: The Use of Veterinary Drugs and Vaccines in Mediterranean Aquaculture*, C. Rodgers & B. Basurco (Eds.), 5-11, CIHEAM/ FAO, ISBN 2-85352-422-1, Zaragoza, Spain

Tafalla, C.; Aranguren, R.; Secombes, C.J.; Castrillo, J.L.; Novoa, B. & Figueras, A. (2003). Molecular characterisation of sea bream (*Sparus aurata)* transforming growth factor beta1. *Fish and Shellfish Immunology*, Vol. 14, pp. 405-421

Tatner, M. F. & Findlay, C. (1991). Lymphocyte migration and localization patterns in rainbow trout, *Onchorhynchus mykiss,* studies using the tracer sample method. *Fish and Shellfish Immunology*, Vol. 1, pp. 107-117

Torroba, M. & Zapata, A.G. (2003). Aging of the vertebrate immune system. *Microscopy Research and Technique,* Vol. 62, pp. 477– 481

Tort, L.; Balasch, J.C. & Mackenzi, S. (2003). Fish immune system. A crossroads between innate and adaptive responses. *Inmunología,* Vol. 22, No. 3, pp. 277-286

Vallejo, A. N. & Ellis, A. E. (1989). Ultrastructural study of the response of eosinophil granule cells to *Aeromonas salmonicida* extracellular products and histamine liberators in rainbow trout *Salmo gairdneri* Richardson. *Developmental and Comparative Immunology,* Vol. 13, pp. 133-148

Vallejo, A.N.; Miller, N.W. & Clem. L.W. (1992). Antigen processing and presentation in teleost immune responses. *Annual Review of Fish Diseases,* Vol. 2, pp. 73-89

Vatsos, I.; Thompson, K.D & Adams, A. (2003). Starvation of *Flavobacterium psychrophilum* in broth, stream water and distilled water. *Disease of Aquatic Organisms*, Vol. 56, pp. 115-126

Wang, T.; Holland, J.W.; Bols, N. & Secombes, C.J. (2005). Cloning and expression of the first non-mammalian interleukin-11 gene in rainbow trout *Oncorhynchus mykiss. FEBS Journal*, Vol. 272, pp. 1136-1147

Whyte, S.K. (2007). The innate immune response of finfish: A review of current knowledge. *Fish and Shellfish Immunology,* Vol. 23, No. 6, pp. 1127-1151

Wilson, M.; Bengten, E.; Miller, N.W.; Clem, L.W.; Du Pasquer, L. & Warr, G.W. (1997). A novel chimeric Ig heavy chain from a teleost fish shares similarities to IgD. *Proceedings of the National Academy of Sciences of the USA 94, April 1997 Immunology*, Vol. 94, pp. 4593- 4597

Wilson, T. & Carson, J. (2003). Development of sensitive, high-throughput one-tube RT- PCR-enzyme hybridisation assay to detect selected bacterial fish pathogens. *Disease of Aquatic Organisms*, Vol. 54, pp. 127-134

Yano, T. (1996). The nonspecific immune system: Humoral defense. In: *The Fish Immune System: Organism, Pathogen, and Environment*, G. Iwama & T. Nakanishi (Eds.), 105- 157, Academic Press, ISBN 0-12-350439-2, San Diego, California, USA

Yin, G.; Ardó, L.; Thompson, K.D.; Adams, A.; Jeney, Z. & Jeney, G. (2009). Chinese herbs (*Astragalus radix* and *Ganoderma lucidum*) enhance immune response of carp, *Cyprinus carpio*, and protection against *Aeromonas hydrophila. Fish and Shellfish Immunology,* Vol. 26, pp. 140-145

Yoshida, T.; Kruger, R. & Inglis, V. (1995). Augmentation of non-specific protection in African catfish, *Clarias gariepinus* (Burchell), by the long-term oral administration of immunostimulants. *Journal of Fish Disease*, Vol. 18, pp.195–198

Yoshiura, Y.; Kiryu, I.; Fujiwara, A.; Suetake, H.; Suzuki, Y.; Nakanishi, T. & Ototake, M. (2003). Identification and characterisation of Fugu orthologues of mammalian interleukin-12 subunits. *Immunogenetics*, Vol. 55, pp. 296-306

Yousif, A.N.; Albright, L.J. & Evelyn, T.P.T. (1995). Interaction of coho salmon *Oncorhynchus kisutch* egg lectin with the fish pathogen *Aeromonas salmonicida. Diseases of Aquatic Organisms*, Vol. 21, pp. 193-199

Zapata, A.; Diez, B.; Cejalvo, T.; Gutierrez de Frias, C. & Cortes, A. (2006). Ontogeny of the immune system of fish. *Fish and Shellfish Immunology*, Vol. 20, pp. 126-136

Zapata, A.G.; Chibá, A. & Varas, A. (1996). Cells and tissue of the immune system of fish. In: *The Fish Immune System: Organism, Pathogen, and Environment*, G. Iwama & T. Nakanishi (Eds.), 1-62, Academic Press, ISBN 0-12-350439-2, San Diego, California, USA

Zhang, J.Y.; Wu, Y.S. & Wang, J.G. (2004). Advance of phage display antibody library and its' implication prospect in aquaculture. *Journal of Fisheries of China*, Vol. 28, pp. 329–333

Zhang, J.Y.; Wang, J.G.; Wu, Y.S.; Li, M.; Li, A.H. & Gong, X.L. (2006). A combined phage display ScFv library against *Myxobolus rotundus* infecting crucian carp, *Carassius auratus auratus* (L.), in China. *Journal of Fish Diseases*, Vol. 29, pp. 1-7

Zhao, W.; Liang, M. & Zhang P. (2010). Effect of yeast polysaccharide on the immune function of juvenile sea cucumber, *Apostichopus japonicus* Selenka under pH stres. *Aquaculture International*, Vol. 18, pp. 777-786

Zou, J.; Secombes, C.J.; Long, S.; Miller, N.; Clem, L.W. & Chinchar, V.G. (2003). Molecular identification and expression analysis of tumor necrosis factor in channel catfish (*Ictalurus punctatus*). *Developmental and Comparative Immunology*, Vol. 27, pp. 845-858

Zou, J.; Yoshiura, Y.; Dijkstra, J.M.; Sakai, M.; Ototake, M. & Secombes, C. (2004). Identification of an interferon gamma homologue in fugu; *Takifigu rubripes. Fish and Shellfish Immunology*, Vol. 17, pp. 403-409

12 Chapter

ANTIBACTERIAL AGENTS IN AQUACULTURE: UTILIZATION AND IMPACT

1. INTRODUCTION: AN OVERVIEW

Antibacterial chemotherapy has been utilized in aquaculture for more than 60 years. The advent of antibacterials transformed the management of infectious diseases, resulting in a substantial decrease in morbidity and mortality, and facilitating notable improvements in public health. Antibacterials are utilized both prophylactically during periods of increased disease risk and therapeutically in response to disease outbreaks within the system. The elimination of antibacterials from aquaculture pharmaceuticals would result in significant welfare issues. Numerous antibacterial agents exist for veterinary health. Pharmacological research on aquaculture medications has predominantly concentrated on a limited number of antibacterials extensively utilized in aquaculture. The concerns with antimicrobial usage in animal feed are globally acknowledged. At present, there is a widespread belief that veterinary drugs, especially antibacterials, have not consistently been utilized responsibly. In certain instances, chemotherapy may exacerbate health management by inducing toxicity, resistance, residues, and occasionally resulting in public health and environmental repercussions. Consequently, authorities have implemented national laws on the utilization of antibacterials.

2. ANTIBACTERIAL USE IN AQUACULTURE

Aquaculture continues to be the fastest growing animal food producing sector, and aquaculture accounted for 46% of total food fish supply (FAO, 2011). The intensification of aquaculture has led to the promotion of conditions that favour the development of a number of diseases and problems related to biofouling. It is worth remembering the age-old adage that "prevention is better than cure?" and certainly it is possible to devote more attention to preventing the occurrence of disease in fish. Fish may be reared under ideal conditions, in which case the stock are inevitably in excellent condition and without signs of disease (Austin & Austin, 2007). However, disease is a component of the overall welfare of fish (Bergh, 2007). Consequently, a wide range of chemicals are used in aquaculture, including antibacterials, pesticides, hormones, anaesthetics, various pigments, minerals and vitamins, although not all of them are antibacterial agents. As is the case in terrestrial animal production, antibacterials are also used in aquaculture in attempts to control bacterial disease (Burka et al., 1997; Horsberg, 1994; Defoirdt et al., 2011). Usage patterns also vary between countries and between individual aquaculture operations within the same country.

Antibacterials are among the most-used drugs in veterinary medicine (Sanders, 2005). The principal reasons behind the control of infectious diseases in hatcheries are to prevent losses in production; to prevent the introduction of pathogens to new facilities when eggs, fry, or broodstock are moved; to prevent the spread of disease to wild fish via the hatchery effluent or when hatchery fish are released or stocked out; and to prevent the amplification of pathogens already endemic in a watershed (Phillips et al., 2004; Winton, 2001; Lupin, 2009). Antibacterial usage requires veterinary prescription in aquaculture as with usage in terrestrial animals (Sanders, 2005; Prescott, 2008; Rodgers & Furones, 2009). We have limited data about antibacterial use in world aquaculture. For most of the species farmed, we also lack adequate knowledge of the pharmacokinetics (PK) and pharmacodynamics (PD) of administration (Smith et al., 2008). Along with widespread use comes a growing concern about irresponsible use, such as the covert use of banned products, misuse because of incorrect diagnose and abuse owing to a lack of professional advice. That said, there are still not enough approved products for a range of species and diseases in aquaculture (FAO, 2011).

Antibacterials are drugs of natural or synthetic origin that have the capacity to kill or to inhibit the growth of micro-organisms. Antibacterials that are sufficiently non-toxic to the host are used as chemotherapeutic agents in the treatment of infectious diseases amongst humans, animals and plants (Table 1). Drug choices for the treatment of common infectious diseases are becoming increasingly limited and expensive and, in some cases, unavailable due to the emergence of drug resistance in bacteria that is threatening to reverse much of the medical progress of the past 60 years (FAO, 2005). In aquaculture, antibacterials have been used mainly for therapeutic purposes and as prophylactic agents (Shao, 2001; Sapkota et al., 2008; FAO, 2005). The voluntary use of antibacterials as growth promoters in any aspect of aquaculture is generally rare. Prophylactic treatments, when they are employed, are mostly confined to the hatchery, the juvenile or larval stages of aquatic animal production. Prophylactic treatments are also thought to be more common in small-scale production units that cannot afford, or cannot gain access to, the advice of health care professionals. There are no antibacterial agents that have been specifically developed for aquacultural use and simple economic considerations suggest that this will always be the case (Smith et al., 2008; Rodgers & Furones, 2009). Despite the widespread use of antibacterials in aquaculture facilities, limited data is available on the speci c types and amounts of antibacterials used (Sapkota et al., 2008; Heuer et al., 2009). General considerations in the selection and use of antibacterial drugs are given by Figure 1. (Walker & Giguére, 2008). Treatment options will be different for animals that are held in net pens at sea as opposed to those held in an indoor facility or an aquarium. A treatment must also be feasible: an appropriate treatment route for aquarium fish or selected broodstock individuals may be cost- or labour-prohibitive in commercial aquaculture ventures. The stress associated with treatments must be balanced with the need for and the expected benefits of treatment (Smith et al., 2008). Also, before making a decision to treat a group of fish, the following questions should be asked (Winton, 2001):

1. Does the loss-rate, severity or nature of the disease warrant treatment?
2. Is the disease treatable, and what is the prognosis for successful treatment?
3. Is it feasible to treat the fish where they are, given the cost, handling, and prognosis?
4. Is it worthwhile to treat the fish or will the cost of treatment exceed their value?
5. Are the fish in a good enough condition to withstand the treatment?
6. Will the treated fish be released or moved soon, and is adequate withdrawal or recovery time available?

Table 1. Properties of the major classes of antibacterial agents (Modified from (Yan & Gilbert, 2004) and (Defoirdt et al., 2011)).

Antibacterial class	Mode / mechanism of action	Mechanisms of resistance	Multiple resistance[a]	PK-PD relationship[b]
ß-Lactams (penicillins, cephalosporins, and carbapenems)	Bactericidal. Inhibition of the penicillin binding proteins (PBPs) located on the cytoplasmic membrane	ß-lactamase production, PBPs modifications, reduced permeability, and efflux	Yes	Time-dependent
Aminoglycosides (streptomycin and neomycin)	Bactericidal. Protein synthesis inhibition through binding to the 30s subunit of the ribosome	Decreased permeability, efflux, modification of enzymes, and target (ribosome) modification	Yes	Concentration-dependent

Macrolides (erythromycin, tylosin and spiramycin)	Bacteriostatic. Protein synthesis inhibition through binding to the 50s subunit of the ribosome	Target (ribosome) modification of enzymes, decreased permeability and efflux	Yes	Time-dependent
Fluoroquinolones (enrofloxacin and ciprofloxacin)	Bactericidal. Inhibition of DNA gyrase and topoisomerase	Target point mutations decreased permeability, efflux and plasmid mediated mechanism	Yes	Concentration-/time-dependent
Tetracyclines (oxytetracycline and chlortetracycline)	Bacteriostatic. Protein synthesis inhibition at the ribosomal level (interference with peptide elongation)	Efflux, drug detoxification, and ribosome modification	Yes	Time-dependent
Folate synthesis inhibitors (sulphonamides and ormetoprim)	Single bacteriostatic, combination bactericidal. Inhibition of dihydro-pteroate synthase and dihydrofolate reductase	Decreased permeability, formation of enzymes with reduced sensitivity to the drugs	Yes	Concentration-dependent
Phenicols (florfenicol and chloramphenicol)	Bacteriostatic. Inhibit the peptidyltransferase reaction at the 50s subunit of the ribosome	Decreased target binding, reduced permeability, efflux and modifying enzymes	Yes	Time-dependent

[a]Resistance to antibacterials belonging to different classes in at least one of the isolates. [b]Represents a generalisation only, the actual relationship can be variable when an individual drug is involved.

2.1 The pathogen and the host

Organisms responsible for disease in aquatic species include fungi, bacteria, nematodes, cestodes and trematodes as well as parasitic protozoans, copepods and isopods. Some can cause death, while others may stress the affected animal to the point that it becomes more susceptible to additional diseases (Stickney, 2005). Disease forms a part of the lives of wild fish and farmed fish. Often, it is not cultured fish that are most susceptible, due to efficient prophylactic strategies and good culture practices. Unprotected wild fish, as exemplified in the case of salmon lice, will be more susceptible to infections and mortality (Bergh, 2007). The difference between health and disease typically depends on the balance between the pathogen and the host, and that balance is greatly influenced by environmental factors, such as temperature and water chemistry (Winton, 2001). The diagnostic techniques for pathogens that are used range from gross observation to highly technical bimolecular-based tools. Pathogen screening is another health management technique, which focuses on the detection of pathogens in subclinical or apparently healthy hosts (Subasinghe, 2009).

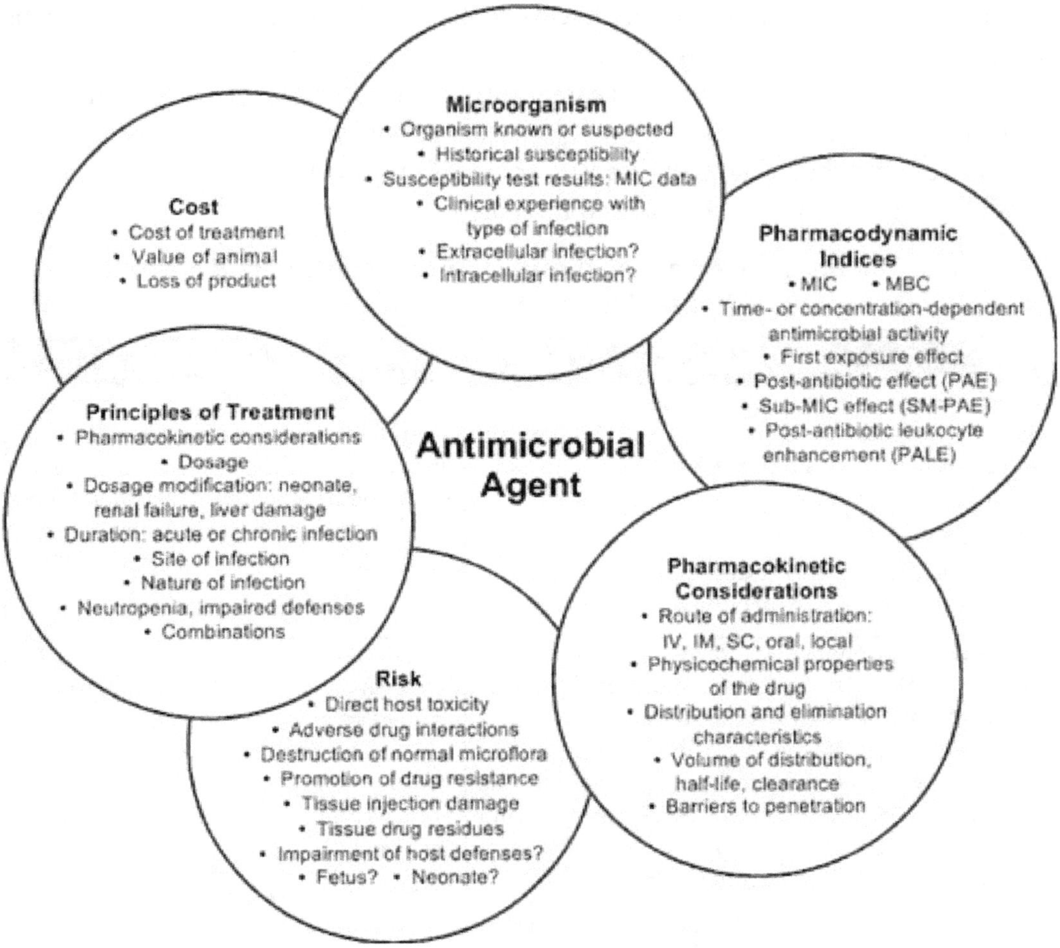

Fig. 1. Some considerations in selecting and using antibacterial drugs (Walker & Giguére, 2008).

The primary pathogens in aquaculture are bacteria and viruses (Shao, 2001). More than 100 bacterial pathogens of fish and shellfish have been reported (Alderman & Hastings, 1998; Winton, 2001). The artificially high host-densities associated with aquaculture are evolutionarily beneficial for pathogens (Bergh, 2007). Bacterial pathogens probably cause more disease problems overall than all other causes combined. In virtually every type of aquaculture, bacterial diseases rank number one amongst aetiological agents. In each type of culture, and for virtually every species, specific bacterial pathogens are responsible for serious disease problems. Gram-negative bacilli are the most frequent cause of bacterial diseases in finfish. Although only a few Gram-positive forms affect finfish, such bacteria cause serious diseases among crustaceans (Meyer, 1991; Rodger, 2010; Roberts, 2004). Whereas similar types of pathogens affect freshwater and marine fish, relatively few pathogens are transmissible from freshwater fish to marine fish, and vice versa (i.e. most pathogens affect either marine or freshwater fish, but not both). This is the rationale for why many freshwater pathogens can be treated with salt, and many marine pathogens can be treated with freshwater (Noga, 2010). Choosing the right drug depends in part on such fac- tors as age, size and the housing of the animal. Common bacterial fish diseases, their definition, aetiology and treatment, as well as control issues, are resumed in Table 2.

Table 2. Common bacterial fish diseases, their aetiology and treatment – control issues (Modified from (Rodger, 2010)).

Disease / Aetiology	Treatment and control
Mycobacteriosis. Mycobacteriosis in fish is a chronic progressive disease caused by certain bacterial species within the genus Mycobacterium. *Mycobacterium marinum, M. fortuitum, M. salmoniphilum and M. chelonae* are all considered pathogenic for fish. All are aerobic, acid-fast, Gram-positive and non-spore forming.	There is no fully effective treatment. Therefore, the best course is to cull and disinfect the premises. Rifampicin in combination with tetracycline (Boos et al., 1995) and clarithromycin may reduce infection (Collina et al., 2002).
Coldwater Diseases & Rainbow Trout Fry Syndrome (RTFS). Bacterial coldwater disease is a serious septicaemic infection of hatchery-reared salmonids, especially young coho salmon, which has also been referred to as peduncle disease. *Flexibacter psychrophilus, Cytophaga psychrophila* and *Flavobacterium psychrophilum* are all terms that have been used for the causal agents of these diseases. Most of the recent classificatory work indicates that *Flavobacterium psychrophilum* is the correct name for these bacteria. These bacteria are Gram-negative.	Broad-spectrum antibacterials have been partially ineffective in controlling an outbreak, but the improvement of the environment and using 3-4 times the recommended doses of antibacterials have shown benefits (Bebak et al., 2007). Florfenicol also appears to be effective for recommended dose regimes.
Bacterial Kidney Disease (BKD). A serious disease of freshwater and seawater fish, farmed and wild salmonids, that results in an acute to chronic systemic granulomatous disease. *Renibacterium salmoninarum* is a Gram-positive diplococcus that grows best at 15- 18°C, is causative agent of BKD and a significant threat to the healthy and sustainable production of salmonid fish worldwide (Wiens et al., 2008).	Chemotherapy (erythromycin) provides limited and only temporary relief. The bacteria can survive and multiply within phagocytic cells.

Enteric Redmouth Disease (ERM). A bacterial septicaemic condition of farmed salmonids, and in particular rainbow trout. There are recent reports amongst channel catfish. *Yersinia ruckeri* is the causal agent and the Gram-negative, motile rod-shaped bacterium is catalase positive and oxidase negative. Several serotypes have been identified.	Broad-spectrum antibacterials are effective in controlling an outbreak, although increasing antibacterial resistance is observed. The bacteria can survive and multiply within phagocytic cells (Tobback et al., 2009; Rykaert et al., 2010).
Furunculosis. Furunculosis is a fatal epizootic disease, primarily of salmonids. It also causes clinical diseases in other fish species, where it is named ulcer disease or carp erythrodermatitis. *Aeromonas salmonicida* is a Gram-negative bacteria. Atypical furunculosis is caused by a slower growing non- pigmenting isolate, *A. salmonicida achromogenes*.	Broad-spectrum antibacterials are effective in controlling an outbreak, but increasing antibacterial resistance is observed.
Piscirickettsiosis. A disease of salmonids caused by *Piscirickettsia salmonis* and a significant disease problem in farmed marine salmonids. *Piscirickettsia salmonis* is a Gram-negative, acid-fast, non-motile, spherical to coccoid, non-capsulated (although often pleomorphic) organism.	Broad-spectrum antibacterial therapy is used, although some resistance is developing. Outbreaks are usually associated with stressful events, such as algal blooms, sudden changes in the environment or grading.
Bacterial Gill Disease (BGD). BGD is an important disease in farmed freshwater salmonids. The bacterium *Flavobacterium branchiophila* causes a chronic, proliferative response in gill tissue. *Flavobacterium branchiophila* is a Gram-negative, long, thin, filamentous rod.	BGD usually responds well to antiseptic and surfactant baths, such as chloramine T and benzalkonium chloride. Providing adequate oxygen is a useful supportive therapy.
Vibriosis. Vibriosis is the term most commonly used to describe infections associated with *Vibrio spp.* In recent years, vibriosis has become one of the most important bacterial diseases in marine-cultured organisms (Stabili et al., 2010). *Vibrio anguillarum, V. ordalii* and other *Vibrio sp.* may cause similar clinical signs in wild and farmed fish. It is Gram-negative, and has straight or slightly curved rods which are motile.	Broad-spectrum antibacterials are effective in controlling an outbreak, but increasing antibacterial resistance is observed. Vaccines are widely used. Caprylic acid may be helpful as an alternative or as an adjunct to antibacterial treatment (Immanuel et al., 2011).
Epitheliocystis. Epitheliocystis is a chronic and unique infection caused by the *Chlamydia spp.* organism and which results in hypertrophied epithelial cells - typically of the gills but sometimes also of the skin - or certain freshwater and marine fishes.	Broad-spectrum antibacterials have been used with some degree of success, though the avoidance of infected fish should be adhered to at all costs.

Tenacibaculosis. An infection of marine fish by *Tenacibaculum maritimum* is common in farmed fish and many species. The bacteria appear to be opportunistic, commonly infecting fish after minor epidermal or epithelial trauma or irritation, and they can rapidly colonise such tissue. It is Gram negative, with slender bacilli which multiply in mats on the damaged tissue.	Oral treatment with broad-spectrum antibacterials is generally successful if the fish are maintained in a low-stress environment.
Francisellosis. Francisellosis is the term used to describe infection associated with *Francisella philomiragia subspecies noatunensis*, which has emerged as a major pathogen of farmed cod. *Francisella spp.* is also a major pathogen of farmed tilapia. It is Gram-negative, with intracellular coccobacilli.	There is no effective treatment due to the intracellular nature of the infection. Removal of the affected fish and the disinfection of the premises and equipment, and fallowing.
Rainbow Trout Gastro-Enteritis (RTGE). RTGE is an enteric syndrome of freshwater farmed rainbow trout, reported in several European countries, and which results in significant economic loss and daily mortalities. It is not fully established and the role of the segmented filamentous bacteria remains unclear, as they are also found in apparently healthy fish. However, *Candidatus arthromitis* may have some role to play in the disease. This bacterium have not yet been cultured in vitro.	Changing the diet-type or the addition of salt to the diet as well as broad-spectrum antibacterials all appear to be effective, once the disease is present. However, none appear to be preventative. Biosecurity is important in preventing the disease from entering a farm.
Red Mark Syndrome (RMS) or Cold Water Strawberry Disease. RMS is an infectious dermatitis of rainbow trout which does not cause mortality but which presents as dramatic haemorrhagic marks on the skin. It is not fully established, although *Flavobacterium psychrophilum* and rickettsia-like organisms have been associated with it.	The lesions will resolve eventually without treatment; however, broad-spectrum antibacterials do induce the rapid healing of the condition. The avoidance of any livestock from infected farms reduces the chance of the introduction of RMS onto a site.

2.2 Antibacterial susceptibility testing of aquatic bacteria

Resistance is a description of the relative insusceptibility of a microorganism to a particular treatment under a particular set of conditions. Therefore, it should be noted that resistance or at least the resistance level depends strongly upon the test type and test conditions, as well as the type of compound and its mode of action (Kümmerer, 2008). The empirical use of antibacterials should be avoided. The use of antibacterials should always be based upon an examination of the clinical case, the diagnosis of a bacterial infection and the selection of a clinically efficacious antibacterial agent. However, in certain situations (such as when the animal is seriously ill or where there is an outbreak with a high mortality or a rapid spread) therapy may be initiated on the basis of clinical signs (Guardabassi & Kruse, 2008; Smith et al., 2008). The target organisms must be known or shown to be susceptible, and adequate concentrations must be shown to reach the target (Phillips et al., 2004). A definitive diagnosis requires the isolation and identification of the causative organism, preferably from three to five infected fish (Smith et al., 2008). Samples for a bacteriologic culture should be collected from the actual site of infection before administering an antibacterial drug (Walker & Giguére, 2008). Currently, a wide range of standardised methods are available (Smith et al., 2008; Guardabassi & Kruse, 2008; CLSI, 2006a, 2006b). It should be expected that there will be differences between the bacteria isolated and their antibacterial sensitivities, between freshwater and saltwater fish, between different taxa of fish, and possibly even between different species of fish (Mulcahy, 2011). Furthermore, due to the varying activity spectrum of the different compounds in some tests, microbial population dynamics may overrule their effects in some populations. They may thereby mask effects (Kümmerer, 2008). The discrepancies between testing methods may also require further studies (Kum et al., 2008). Differences in the measurement of zone sizes by individual scientists also represent a possible source of inter-laboratory variation (Nic Gabhainn et al., 2004).

2.3 The treatment route

In intensive fish farming, the antibacterials used to treat bacterial infections are administrated generally by either water-borne or oral means, or else through injection (Shao, 2001; Treves-Brown, 2001; Zounkova et al., 2011). Agents that are intended to treat diseases must reach therapeutic levels in target tissues. It is always advisable to perform a bioassay of a small number of individuals before treating any fish species without a known history of response to the treatment. A bioassay can be performed by placing five or six fish in an aquarium that has the treated pond water. The fish should be observed for 1-2 days before treatment so as to be sure that none have died from the stress of collection. Fish should never be left unattended during treatment and, if an adverse response occurs, the drug should be immediately removed by transferring the fish to clean water or diluting the treatment water. It is necessary to take the presence of these additives into account when calculating the active drug quantity required for any treatment (Noga, 2010; Rodgers & Furones, 2009). Adequate plans for detoxification and the removal and disposal of used drugs must be in place before treatment is begun (e.g., ammonia and nitrite levels must be monitored closely during therapy). When hospitalisation is completed, the aquarium, the filter, and all other materials in contact with the hospitalisation aquarium should be disinfected before re-use. Used drugs must be disposed of responsibly. Deteriorated or otherwise uncontrolled water quality poses particular challenges to farmed fish and their surroundings. Outputs from these systems can further harm local wildlife and the ecosystem (Cottee, 2009). There is no one specific drug application method which is better than other; rather, the method of treatment should be based on the specific situation encountered. Here, experience is exceptionally valuable. A fish health professional or other knowledgeable source should be consulted if one is unfamiliar with the disease or treatment proposed (Winton, 2001). Methods for the application of antibacterials to fish are resumed in Table 3.

Table 3. Methods for application of antibacterials to fish (Haya et al., 2005).

Method of application	Comments
Oral route (on food)	Needs palatable components; minimal risk of environmental pollution
Bioencapsulation	Needs palatable compounds; minimal risk of environmental pollution
Bath	Need for a fairly lengthy exposure to the compound, which must be soluble or capable of being adequately dispersed; problem of the disposal of spent drug
Dip	Brief immersion in a compound, which must be soluble or capable of being adequately dispersed; problem of disposal of the dilute compound
Flush	Compound added to a fish holding facility for brief exposure to fish; must be soluble or capable of being adequately dispersed; poses a problem of environmental pollution
Injection	Feasible for only large and/or valuable fish; usually requires prior anaesthesia; slow; negligible risk of environmental pollution
Topical application	Feasible for the treatment of ulcers on valuable/pet fish

2.3.1 Water medication

The water-borne route is the most common method for administering treatments to fish and it has distinct advantages, such as being relatively non-stressful and easy to administer. Drugs are added to water for two distinct purposes. The first and most obvious one is so that the drug will be absorbed by, and so medicate, the fish; the second is to kill the free- living and, hence, transmissible stages of parasites (Treves-Brown, 2001). Seawater fish drink significant amounts of water and may absorb large amounts of a drug via the gastrointestinal tract (Noga, 2010). Application by the water-borne route becomes necessary if the fish refuse to eat, and, therefore, would be unlikely to consume any medicated food. With these methods, the fish are exposed to solutions/suspensions of the drug for a predetermined period. This may be only briefly, i.e., a few seconds duration (a "dip") or for many minutes to several hours (a "bath") (Haya et al., 2005). Waterborne antibacterial treatments will vary depending upon the animal and its holding conditions. Treating fish by applying the drug to the water avoids stressing the fish by handling (Reimschuessel & Miller, 2006). However, there are disadvantages. Relative to other treatment routes, dosing is less precise (too little or too much). Baths and dips are not as effective as some of the other treatment methods – particularly for systemic infections – because of generally poor internal absorption of the antibacterial being used. Water-borne treatments are mainly used for surface-dwelling (skin and gill) pathogens, including parasites, bacteria and water moulds. Certain species, such as scaleless fish are often especially sensitive to water-borne treatments (Rodgers & Furones, 2009). Antibacterials which are absorbed from the water include chloramines, dihydrostreptomycin, enrofloxacin, erythromycin, flumequine, furpyrinol, kanamycin, oxolinic acid, oxytetracycline, nifurpirinol, sulphadimethoxine, sulphadimidine, sulpha-monomethoxine, sulphanilamide, sulphapyridine, sulphisomidine and trimethoprim. Antibacterials that are absorbed poorly or not at all include chloramphenicol and gentamicin (Reimschuessel & Miller, 2006). With bath-type treatments, more antibacterials are required when compared with oral (feed) treatments or injections. Bath treatments are also not recommended for recirculation systems or aquarium systems using biological filters. The accurate calculation of the volume of water in the tank, pond or cage is also required (Rodgers & Furones, 2009). If both short- and long-term exposures are probably equally feasible and effective, it is preferable to use a short-duration drug exposure. The advantages of this type of treatment lie in reduced waste (and thus reduced expense) and less environmental contamination (Reimschuessel & Miller, 2006). Even where absorption is known to occur, the technique does have some important disadvantages. In particular, in most cases less than 5% of the administered dose will be absorbed by the fish. In this case, the technique is wasteful, expensive (at least twenty times the dose required by the fish must be provided) and environmentally undesirable (Treves-Brown, 2001).

2.3.2 Oral medications

In food fish or ornamental aquaculture, many of the bacterial diseases of fish can be successfully treated with medicated feeds, and it is usually the preferred method of treatment. However, care must be taken because some of the causes of disease – such as stress – can lead to treatment failures or the recrudescence of disease after the completion of treatment (Rodgers & Furones, 2009). Fish in ponds are best treated using oral medications. However, sick fish may not eat, and withholding food for 12-24 hours may increase the acceptance of a medicated feed (Reimschuessel & Miller, 2006). The incorporation of an antibacterial in the feed is usually via a powdered premix in conjunction with a binder, such as gelatine (up to 5%), fish or vegetable oil (Shao, 2001). The dosage required for treatment with a medicated feed depends upon the original level of active ingredient/kg fish body weight. The dosage rates used in medicated feeds will vary according to the specific antibacterial used, but usually the rate is based on a number of grams per 100 kg of fish per day. The exact dosage will also require the number and average weight of the fish to be treated, as well as a daily feeding rate and consideration of whether the fish are marine or fresh water species. It is also important that treated fish must not be harvested for food use until a specified withdrawal period has elapsed (Rodgers & Furones, 2009). One problem for the treatment of marine species is that antibacterials have been shown to be less effective in seawater, which is related to their reduced bioavailability, e.g., tetracycline has a low bioavailability in fish (< 10%) due to binding with sea-water-borne divalent cations such as Mg^{2+} and Ca^{2+}. It is noteworthy that non-bioavailable tetracyclines contaminate the environment (Toutain et al., 2010). The bioavailability of some aquaculture drugs in salmon held in seawater is shown in Table 4. (Rodgers & Furones, 2009). The dosage can vary within certain limits and depending upon the feeding rate. It is usually best to use a feed that has enough medication so that feeding at a rate of 1% of body weight per day will give the needed dosage (Noga, 2010). Absorption from the intestinal tract may vary between species. Saltwater fish will drink and, therefore, drugs may bind cations in the water in their intestinal tracts, affecting bioavailability (Reimschuessel & Miller, 2006). For particular applications, like the treatment of young larvae and fry, some success has been obtained with the bio-encapsulation of drugs in live feeds, especially with artemias. Other innovative methods of oral delivery – like microspheres or coated beads – offer the possibility of protecting fragile molecules from deterioration in the gastric juices, carrying them up to their target sites in the intestine. Though still quite recent, large developments are expected from these innovative technologies in the near future (Daniel, 2009).

Table 4. Examples of reduced bioavailability for some aquaculture drugs in seawater (Rodgers & Furones, 2009).

Antibacterial	Bioavailability (%)
Oxytetracycline	1
Amoxicillin	2
Sarafloxacin	2
Oxolinic acid	30
Flumequine	45
Sulfadiazine	50
Trimethoprim	96
Florfenicol	97

2.3.3 Injection

The injection of antibacterials can be a more effective treatment for bacterial infections than the use of a medicated feed, particularly for advanced infections and as the best way of being sure of the given dose (Douet et al., 2009). However, it is usually only practical for valuable individual fish, such as brood stock or ornamental fish, rather than for fish in large-scale production facilities. Injection quickly leads to high blood and tissue levels of antibacterials (Yan & Gilbert, 2004; Haya et al., 2005). Normally, an individual fish will also need to be anaesthetised before treatment. Typical injection sites include the intraperitoneal (IP) cavity and the intramuscular route (IM) (Rodgers & Furones, 2009; Treves-Brown, 2001). Disadvantages include the stress imposed by capturing the fish and, for aquarium fish, the need to bring the fish to the clinic for every injection, since the owner is usually unable to perform the treatment (Noga, 2010). The IP route is the widely used route for injection. Fish should be fasted for 24 hours prior to injection. The landmarks for an IP injection are the pelvic fins and the anus. All fish should be at least 35g. Improper injection can lead to peritoneal adhesion, ovulation problems, mortality from injection, reduced efficacy, side effects (local reactions), reduced carcass quality and therapy failure (Treves-Brown, 2001; Noga, 2010). IM injection is best used only on fish more than 13 cm long or else more than 15g. The best site is the dorsal musculature just lateral to the dorsal fin. Only relatively small amounts can be injected (0.05 ml/50 grams of fish). Injections should be done slowly. The IM route has the disadvantage of causing damage to carcass quality and has the potential of forming sterile abscesses (Noga, 2010). The volume required for the injection of antibacterials is based on the weight of fish to be treated, the recommended dosage for the antibacterial being used and its supplied concentration (Rodgers & Furones, 2009). This is usually expressed as:

Volume of antibacterial required = recommended dosage (mg/kg) × weight of fish (kg)/supplied solution concentration (mg/ml)

2.3.4 Topical application

The topical application of drugs to fish is rare. Anaesthesia is an essential preliminary procedure. Topical treatments are usually only necessary for more valuable individual fish, such as ornamental varieties or brood stock. Ointments containing antibacterials have sometimes been used in fish surgeries, applied to the sutures and incision site. Commercial antibacterial ointments are most commonly used (Mulcahy, 2011). Open sores or ulcers that are secondarily infected by bacteria or water moulds can be treated. A cotton swab should be dipped in a drug solution and then used to gently touch the lesion, allowing the solution to soak the lesion via capillary action. Nevertheless, it is possible that ulcers may heal themselves with improved water quality and the elimination of parasites (Treves-Brown, 2001; Haya et al., 2005; Noga, 2010).

2.3.5 Water treatment

Disinfection can reduce the risk of disease transmission within aquaculture facilities, and from facilities to the environment, by deactivating or destroying pathogens with disinfecting agents. Disinfection can be done routinely, but also in response to the outbreak of specific diseases (Winton, 2001). In this procedure the drug is applied to all the water in the aquarium. It is, therefore, not applicable to antibacterial drugs, as these would inactivate the filter (Treves-Brown, 2001).

2.4 Dosage

PK and PD data has allowed the design of therapeutic regimens, with the PK/PD variable providing the most appropriate surrogate for drug effectiveness being dependent upon several factors (Rigos & Troisi, 2005; Martinez & Silley, 2010; Toutain et al., 2010). Within some species there may be considerable differences both within and between breeds in PK and PD profiles; veterinary pharmacogenetics aims to identify genetic variations (polymorphisms) as the origin of differences in the drug response of individuals within a given species. These between- and within-species differences in drug response are largely explained by variations in drug PK and PD, the magnitude of which varies from drug to drug (Toutain et al., 2010). We will not be able to apply the full power of the PK/PD approach to either the design of treatment regimens that minimise the development of resistance or the setting of clinical breakpoints that provide an empirical definition of resistance (Smith et al., 2008). There is a considerable amount of information available on the PK of various antibacterials delivered by different routes to different species of fish, but little information about the plasma levels of antibacterials that are required to be of benefit for the implanted fish or the calculation of the dosage of antibacterial required to obtain a positive benefit (Mulcahy, 2011). It is important to remember that it is the host immune system that is ultimately responsible for success in combating bacterial disease (Martinez & Silley, 2010). If one is unsure about the dose to use, it is usually best to start with the lower recommended dose. If the disease does not respond adequately, repeat the treatment with a higher dose. For oral medications, dosage varies with feed intake. Fish that are eating less need a higher percentage of the drug in their diet, but there are limits on the legally allowable amount as well as practical considerations, since some drugs are unpalatable at high doses (e.g., many antibacterials) (Noga, 2010; Winton, 2001). Drug dosage regimens also are host-dependent. Fish species reared in warm water may absorb, metabolise and excrete drugs at a different rate (often faster) than those in cold water. The salinity of the holding water also affects drug kinetics. Fish kept in saltwater drink the water while freshwater fish do not. Thus, antibacterials in the gastrointestinal tract of fish species held in saltwater may bind cations, which can reduce their uptake (Smith et al., 2008; Toutain et al., 2010). This is especially true for antibacterials – such as the tetracyclines – that have low bioavailability even in freshwater. The half-lives of drugs in fish are highly dependent upon the dosage regimen, the route and the temperature. Therefore, these parameters are included in the Phish-Pharm Database and should be considered when administering antibacterials to fish. Table 5 shows some drug dosages that have been reported for fish (Reimschuessel et al., 2005). It is important to realise that the dosages listed in Table 5 may not have been shown to be safe or effective in all fish species. No generalisations are possible. Successful therapy often depends on maintaining adequate blood levels over a course of seven to ten days. Temperature is a very important factor in deciding on the dose and treatment intervals (Toutain et al., 2010).

Table 5. Half-lives and dosages of antibacterials in fish (Reimschuessel et al., 2005).

Drug	Species	$t_{1/2}$ (hr)	Dosage	Route[b]	°C
Amoxicillin	Atlantic salmon	120	12.5 mg/kg sd[a]	IM	13
	Atlantic salmon, sea bream	14-72	40-80 mg/kg sd	IV/PO	16-22
Chloramphenicol	Carp	48-72	40 mg/kg sd	IP	9
Ciprofloxacin	Carp, rainbow trout, African catfish	11-15	15 mg/kg sd	IM/IV	12-25
Difloxacin	Atlantic salmon	16	10 mg/kg sd	PO	11
Enrofloxacin	Atlantic salmon, red pacu, rainbow trout, sea bass, sea bream	24-105	5-10 mg/kg sd	IM/ IV/P O	10-26
Erythromycin	Chinook salmon	120	0.1 g/kg 21 d	PO	10
Florfenicol	Atlantic salmon	12-30	10 mg/kg sd	IV/PO	10-11
	Cod	39-43	10 mg/kg sd	IV/PO	8
Flumequine	Eel	255	9 mg/kg sd	IM	23
	Atlantic halibut, brown trout, corkwing wrasse, Atlantic halibut,	21-96	5-25 mg/kg sd	IP/IV/ PO	5-25
	Atlantic salmon, cod, goldsinny wrasse, sea bass, sea bream, turbot				
	Eel	208-314	10 mg/kg sd	IV/PO	23
	Rainbow trout	285-736	5 mg/kg sd	IV/PO	13 vs 3
Furazolidone	Channel catfish	1-24	1 mg/kg sd	IV/PO	24
Gentamicin	Channel catfish, brown shark, goldfish	12-54	1-3.5 mg/kg sd	IC/IM	20-25
	Toadfish	602	3.5 mg/kg sd	IM	19

Drug	Species	t1/2 (hr)	Dosage	Route[b]	°C
Miloxacin	Eel	35	30-60 mg/kg sd	IV/PO	27
Nalidixic acid	Rainbow trout, amago salmon	21-46	5-40 mg/kg sd	IV/PO	14-15
Nifurstyrenate	Yellowtail	2	100 mg/kg sd	PO	23
Ormetoprim	Atlantic salmon, channel catfish, rainbow trout, hybrid striped bass	4-25	4-50 mg/kg sd	IV/PO	10-28
Oxolinic acid	Atlantic salmon, corkwing wrasse, channel catfish, cod, rainbow trout, red sea bream, sea bass	15-87	4-20 mg/kg sd	IP/IV	8-24
	Atlantic salmon, cod, rainbow trout	82-146	25-75 mg/kg sd	PO	5-8
	Atlantic salmon, gilthead sea bream, rainbow trout, sharpsnout, sea bream, turbot	13-48	10-40 mg/kg up to 10d	PO	9-19
Oxytetracycline	African catfish, carp, rainbow trout, red pacu, sockeye salmon	63-95	5-60 mg/kg sd	IM	12-25
	African catfish, Atlantic salmon, ayu, carp, Chinook salmon, eel, rainbow trout, red pacu, sea bass, sea bream, sharpsnout, sea bream	6-167	5-60 mg/kg sd	IV	8-25
	Arctic charr	266-327	10-20 mg/kg sd	IV	6
	Atlantic salmon, ayu, black sea bream, carp, channel catfish, eel, perch, rainbow trout, sea bass, sea bream, hybrid striped bass, summer, flounder, walleye	43-268	10-100 mg/kg up to 10d	PO	7-27
	Arctic charr, sockeye salmon, Chinook salmon	428-578	10-100 mg/kg sd	PO	6-11
Piromidic acid	Eel, goldfish	24	5 mg/kg sd	PO	26
Sarafloxacin	Atlantic salmon, cod	12-45	10-15 mg/kg sd	IV/PO	8-24

Streptozotocin	Toadfish	24	50 uCi	IV	
Sulphachlor-pyridazine	Channel catfish	4-5	60 mg/kg sd	IC/PO	22
Sulfadiazine	Atlantic salmon, carp, rainbow trout	26-96	25-200 mg/kg sd	IV/PO	8-24
Sulfadimethoxine	Atlantic salmon, channel catfish, rainbow trout, hybrid striped bass	1-48	25-200 mg/kg sd	IV/PO	10-20
Sulphadimidine	Carp, rainbow trout	18-57	100-200 mg/kg sd	IV/PO	10-20
Sulphametho-xypyridazine	Rainbow trout	72	200 mg/kg sd	PO	13
Sulphamono-methoxine	Rainbow trout, yellowtail	5-33	100-400 mg/kg sd	IV/PO	15-22
Sulphanilamide	Rainbow trout	36	200 mg/kg sd	PO	13
Sulfathiazole	Rainbow trout	60	200 mg/kg sd	PO	13
Thiamphenicol	Sea bass	21	30 mg/kg 5d	PO	19
Tobramycin	Brown shark	48	1-2.5 sd	IM	25
Trimethoprim	Atlantic salmon, carp, rainbow trout	21-50	1-100 mg/kg sd	IV/PO	8-24
Vetoquinol	Cod	79	25 mg/kg sd	PO	8
	Atlantic salmon	16	40 mg/kg sd	PO	10

[a]sd: single dose. [b]Abbreviations, IM: intra muscular, IV: intra venous, PO: per os (oral), IP: intra peritoneal, IC: intra coelom, uCi: a unit of radioactivity (Curie= 3,7x10^{10} disintegration per second).

2.5 Drug metabolism in fish

Liver is the primary organ for the detoxification of drugs in fish. Similarities exist in the metabolism of drugs by fish and mammals. The metabolism of aquaculture antibacterials by the cytochrome P450 system could affect their activation, tissue distribution and elimination rates, and determine the persistence of residues as well as the length of the withdrawal period before the fish can be used for human consumption (Moutou, 1998). The elimination rate of antibacterials from fish tissues varies greatly with the temperature. The temperature dependency of drug PK is an important consideration for drug residues. The elimination half-life of antibacterial drugs increases significantly as the temperature decreases. Ideally, the drug dose should be adjusted according to the water temperature, but in clinical practise the dose is normally fixed (Toutain et al., 2010). However, unmetabolised oxytetracycline can be passed unabsorbed through the body of treated sparids and then excreted via the faeces into the local marine environment (Rigos & Troisi, 2005; Rigos et al., 2004).

2.6 Duration of antibacterial treatment

It is universally recognised that a drug must be present in a sufficient concentration for an adequate length of time at the site of the infection, although the variables affecting the length of treatment have not yet been fully defined (Walker & Giguére, 2008). The responses of different types of infections to antibacterial drugs vary, and clinical experience with many infections is important in assessing the response to the treatment. For serious acute infections, treatment should last at least 7 to 10 days. If no response is seen by that time, both the diagnosis and treatment should be reconsidered (Walker & Giguére, 2008). It is important to remember that it is the host immune system that is ultimately responsible for any success in combating bacterial diseases (Martinez & Silley, 2010).

2.7 Failure of antibacterial therapy

Treatment failure has many causes. The selected antibacterial may be inappropriate because of misdiagnosis, poor drug diffusion at the site of the infection, inactivity of a given drug at the site of infection, failure to identify the aetiological agent including inaccurate results of laboratory tests, resistance of pathogens, intra-cellular location of bacteria, metabolic state of the pathogen, or errors in sampling. Other factors that may contribute are inadequate dosage or the use of drugs with low bioavailability. When failure occurs, diagnose must be reassessed and proper samples collected for laboratory analysis. Patient factors such as the persistence of foreign bodies, neoplasia, and impairment of host defences are important to consider. It is important also to ensure that persons medicating their own animals comply with dosing instructions (Walker & Giguére, 2008; Winton, 2001; Treves - Brown, 2001; Noga, 2010).

3. TREATMENT OPTIONS IN VARIOUS AQUACULTURE SYSTEMS

Another important factor influencing treatment is the type of culture system. The four major types of culture system are aquaria, ponds, cages and flow-through systems (Noga, 2010). The main factors that may influence a treatment's success are given in Table 6.

Table 6. Major types of culture systems influencing the diseases' treatment (Table established from (Noga, 2010)).

Aquaria	Ponds	Cages	Raceways
The most highly controllable culture systems for maintaining temperature, biological filtration and oxygen. Amenable to various waterborne treatments. Ease of manipulability.	Influenced by natural factors such as light, temperature and rainfall. Natural biological cycles are less controllable. Interventional strategies are more limited compared with aquaria.	Susceptible to the vagaries of natural environmental changes. Water-borne treatments are possible in such systems, but are much more difficult. The fish that need to be treated in such systems must have their cage enclosed. Alternatively, the fish must be treated in a closed system (e.g., a bath treatment) or the medications must be delivered orally.	Raceways and other flow-through systems are the least manipulable systems by virtue of the constant and rapid water turnover. Similar adverse environmental consequences can follow such treatments. Flow-through systems are even more limited than cages in the ability to use water-borne treatments.

4. LEGAL USE OF ANTIBACTERIALS

A number of international and regional codes of practice, agreements and technical guidelines exist for aquatic animals (Subasinghe, 2009). The drugs available for use and their treatment protocols are tightly regulated. The consumers of fish – and particularly in the world's richer economies – are increasingly demanding that retailers guarantee that the fish which they offer are not only of a high quality and safe to eat, but also that they derives from fisheries that are sustainable (FAO, 2011). As health threats have appeared, management practices have evolved and fish husbandry has greatly improved over the past 20 years, resulting in a reduction in the use of some chemicals, and particularly the use of antibacterials in most jurisdictions (Burridge et al., 2010). The banning of any antibacterial usage in animals based upon the "precautionary principle" in the absence of a full quantitative risk assessment is likely to be wasted at best, and even harmful at worst, both to animal and human health (Phillips et al., 2004). The antibacterials used in veterinary medicine are only prescribed by veterinarians in the European Union (EU). The prescription scheme could be discussed and improved, and non-approved and even banned antibacterials are purchased "over-the-counter" (without the need for a prescription) or their use is undeclared in fish feed formulations. The use of specifically banned antibacterials in aquaculture is a violation of regulations (Lupin, 2009). The user safety data included on labelling and packaging inserts should provide sufficient information for such occupational safety assessments to be made (Alderman & Hastings, 2009). Before approval, drugs are assessed for the definition of their maximum residue limits (MRLs) (Table 7), and their environmental impact and efficacy (Sanders, 2005). MRLs are generated by a number of bodies, such as the EU, and more globally within the framework of the FAO/WHO Codex Alimentarius Commission, which is advised scientifically by the JECFA (Joint FAO/WHO Expert Committee on Food Additives). The use of antibacterial agents in food animal species, including fish, is controlled by regulations, particularly in Europe and the USA.

Table 7. Main antibacterial compounds having fixed MRLs (Modified from (Daniel, 2009)).

Antibacterial	Species*	Tissue**	MRL	Comments
Amoxicillin	All FPS	Muscle	50 pg/kg	
Ampicillin	All FPS	Muscle	50 pg/kg	
Benzylpenicillin	All FPS	Muscle	50 pg/kg	
Chlortetracycline	All FPS	Muscle	100 pg/kg	
Cloxacillin	All FPS	Muscle	300 pg/kg	
Colistine	All FPS	Muscle	150 pg/kg	
Danofloxacin	All FPS	Muscle	100 pg/kg	
Dicloxacillin	All FPS	Muscle	300 pg/kg	
Difloxacin	All FPS	Muscle	300 pg/kg	
Enrofloxacin	All FPS	Muscle	100 pg/kg	Enro.+ciprofloxacin
Erythromycin	All FPS	Muscle	200 pg/kg	Erythromycin A
Florfenicol (Fish)	Fish	Muscle+skin	1000 pg/kg	
Flumequine	Fish	Muscle+skin	600 pg/kg	
Lincomycin	All FPS	Muscle	100 pg/kg	
Neomycin (Incl. Framycetin)	All FPS	Muscle	500 pg/kg	Neomycin B
Oxacillin	All FPS	Muscle	300 pg/kg	
Oxolinic Acid	Fish	Muscle+skin	100 pg/kg	
Oxytetracycline	All FPS	Muscle	100 pg/kg	
Paromomycin	All FPS	Muscle	500 pg/kg	
Sarafloxacin (Fish & Poultry)	Salmonids	Muscle+skin	30 pg/kg	
Spectinomycin	All FPS	Muscle	300 pg/kg	
Sulphonamides (All)	All FPS	Muscle	100 pg/kg	
Tetracycline	All FPS	Muscle	100 pg/kg	
Thiamphenicol	All FPS	Muscle	50 pg/kg	
Tilmicosine	All FPS	Muscle	50 pg/kg	
Trimethoprim	All FPS	Muscle	50 pg/kg	
Tylosin	All FPS	Muscle	100 pg/kg	

*All FPS: all food producing species (with some exclusions and depending on each compound).
**For all fish MRLs, the target tissues "muscle" or "muscle and skin" shall be understood as "muscle and skin in natural proportions."

The approval process is very costly and time consuming, and the sales potential for the aquaculture market in global terms is limited, which in some cases has meant a certain lack of interest on the part of pharmaceutical companies for developing new antibacterials and registering them (Alderman & Hastings, 2009; Rodgers & Furones, 2009). In the USA, the regulatory authority for the approval of Veterinary Medicinal Products (VMPs) is the FDA (FDA, 2009, 2011). The body that is responsible for the authorization procedure in the EU is the European Medicines Agency (EMEA) and the European Commission, or else the national competent authorities in the EU Member States (depending on the procedure chosen for the marketing authorisation application). The EMEA's Committee for Medicinal Products for Veterinary Use (CVMP) carries out the scientific evaluation (Sanders, 2005; Prescott, 2008; Valois et al., 2008; Alderman, 2009). Compared to agricultural use and medicinal use, the market for aquaculture antibacterials is fairly small and the approval process can be expensive. The availability of antibacterial agents for aquacultural use is affected by the setting of MRLs. However, these withdrawal times are based on studies that are mainly performed on fish held in temperate freshwater. The excretion of a drug by a fish can vary greatly with its environmental conditions, and especially the temperature (Daniel, 2009; Noga, 2010). Because of the variability of drug excretion, especially with temperature, a rule of thumb called "degree days" has been advocated for estimating the required withdrawal time. If the data does not indicate a temperature effect on depletion, then a day-based withdrawal can be accepted (Alderman & Hastings, 2009).

5. PROBLEMS ASSOCIATED WITH ANTIBACTERIAL USE IN AQUACULTURE

Consumers demand guarantees that their food has been produced, handled and sold in a way that is not dangerous to their health, and which respects the environment and addresses various other ethical and social concerns (FAO, 2011). Even if the occurrence, effects and fate of antibacterials have been considered from the perspective of scientific interest, little is still known about the actual risk to both humans and the environment (Kemper, 2008). However, medicines legislation requires that user-safety be assessed in the safety package and that the product label must include advice and warnings to the user, giving guidance for safe use. Any hazards associated with feed medication – whether in feed mills or on farms – must be considered, as must any hazards to the final user (the fish farm staff) (Alderman & Hastings, 2009). The majority of fishers and aquaculturists are in developing countries, and mainly in Asia which has experienced the largest increases over recent decades, reflecting the rapid expansion of aquacultural activities (Sapkota et al., 2008; FAO, 2011; Smith et al., 2008). Fish diseases are generally coupled with cultured fish and viewed as a result of aquaculture (Bergh, 2007). As a consequence, it is probable that the majority of antibacterial use in world aquaculture is not associated with any classification of the target bacterium or of its susceptibility to the range of available antibacterials (Smith et al., 2008). There is also a need for assurance that the usage will not harm animals or humans (Phillips et al., 2004). With an increase in consumers' recognition of the health benefits associated with seafood consumption, the volume of fisheries and aquaculture products consumed is expected to rise (Storey, 2005). There is little doubt that aquaculture production will continue to grow (Asche et al., 2008). The world food supply will probably have to double in quantity and increase in quality over the next 30–50 years as populations and incomes rise. The demand for fish as food will probably double and could even more than double (Pullin et al., 2007). Consequently, an increase in the number of problems associated with aquaculture production may be expected.

5.1 Toxicity to the host

Antibacterials that are sufficiently non-toxic to the host are used as chemotherapeutic agents in the treatment of the infectious diseases of humans, animals and plants. Direct host toxicity is the most important factor limiting drug dosage. Tolerance studies must be carried out to determine the safety of the product to the target fish species (Alderman & Hastings, 2009). Also, it is important for the clinician to report adverse drug events to legal authorities. Antibacterial agents can have a wide variety of damaging effects on the host, including: (1) direct host toxicity; (2) adverse interactions with other drugs; (3) interference with the protective effect of normal host microflora or the disturbance of the metabolic function of microbial flora in the digestive tract of herbivores; (4) the selection or promotion of antibacterial resistance; (5) tissue necrosis at injection sites; (6) drug residues in animal products that are intended for human consumption; (7) impairment of the host's immune or defence mechanisms; and (8) damage to foetal or neonatal tissues (Guardabassi & Kruse, 2008; Mulcahy, 2011). Nonetheless, the most used aquaculture antibacterial agent oxytetracycline may have genotoxic and ecotoxic effects in aquatic ecosystems (Zounkova et al., 2011). The selective toxicity of antibacterials is variable. Some agents, such as beta- lactams, are generally considered to be safe, whereas others, such as the aminoglycosides, are potentially toxic (Guardabassi & Kruse, 2008).

5.2 Resistance of aquatic bacteria

The capacity of bacteria to adapt to changes in their environment and thus survive is called resistance. Drug choices for the treatment of common infectious diseases are becoming increasingly limited and expensive and, in some cases, unavailable due to the emergence of drug resistance in bacteria (FAO, 2005). In general, aquatic bacteria are not different from other bacteria in their responses to exposure to antibacterial agents, and they are capable of transferring antibacterial resistance genes to other bacteria (Heuer et al., 2009). The WHO has long recognised that antibacterial use in food animals – which seems to outweigh antibacterial use for human therapy in many countries – contributes importantly to the public health problem of antibacterial resistance (WHO, 2011). The resistance of pathogenic bacteria to antibacterials is a growing problem in human and veterinary medicine, and antibacterial use in fish – especially in aquaculture – is an area of increasing concern over health risks (Kemper, 2008; Mulcahy, 2011). The fact that some of the bacteria that cause infections in fish belong to the same genera as the bacteria causing infections in humans is likely to increase the probability of the spread of antibacterial resistance from aquaculture to humans (Heuer et al., 2009). The continued use of subtherapeutic levels of antibacterials to prevent disease increases the likelihood of establishing populations of multiply resistant strains of pathogenic bacteria. These may ultimately result in outbreaks of disease which cannot be controlled by antibacterial therapy (Mulcahy, 2011; Roberts, 2004). Also, the selection and use of inappropriate antibacterials, and the use of insufficient dosages, incorrect routes of application, incorrect dosing frequencies and administering antibacterials for an insufficient time period, are ways to select for antibacterial-resistant bacteria (Mulcahy, 2011). The excessive use of antibacterials in fish aquaculture is increasing the resistance in bacteria that can infect both humans and animals (Burridge et al., 2010; Kümmerer, 2010; Defoirdt et al., 2011). It is not only the direct therapeutic use of antibacterials, but also their indirect contact with them which might enhance the resistance of bacteria: not taking into account the bacteria's origins, resistance genes have been isolated from human pathogens, bacteria of animal origin and even environmental bacteria (Kemper, 2008; Martinez & Silley, 2010; Martinez, 2009). The consequences of increasing resistance in bacteria and the diminishing impact of therapeutic drugs reach far beyond the geographic origins of antibacterial compounds and are, therefore, of global concern (Kemper, 2008).

Antibacterials exhibit different activity spectra and mechanisms of action. It has been recognised for some time that susceptibility to antibacterials varies markedly both between different groups of organisms and within these groups (Kümmerer, 2008). A large variety of antibacterial resistance mechanisms have been identified in bacteria and several different mechanisms may be responsible for resistance to a single antibacterial agent in a given bacterial species. Antibacterial resistance mechanisms can be classified into four major categories (shown in Table 3.1 by the asterisk): (1) the antibacterial agent can be prevented from reaching its target by reducing its penetration into the bacterial cell; (2) general or specific efflux pumps may expel antibacterial agents from the cell; (3) the antibacterial agent can be deactivated by modification or degradation, either before or after penetrating the cell; or (4) the antibacterial target may be modified so that the antibacterial cannot act on it anymore, or else the microorganism's acquisition or activation of an alternate pathway may render the target dispensable (see Table 8) (Boerlin & White, 2008; Nikaido, 2009).

Drug resistance may be natural or acquired (Roberts, 2004; Douet et al., 2009). Some organisms have always been resistant to a particular agent by the nature of their physiology or biochemistry (i.e., inherent or intrinsic resistance); others have acquired resistance as a result of the application of antibacterials by humans (i.e., acquired resistance) (Kümmerer, 2008). Resistance to antibacterials may be acquired by the mutation of a chromosomal gene which modifies the structure of the ribosomal target or by the infection of the cell with a resistant R-factor plasmid. Plasmids are extrachromosomal circular DNA molecules capable of autonomous replication (Alderman & Hastings, 1998; Boerlin & White, 2008; Defoirdt et al., 2011). Once they are integrated in successful gene-transmission elements, antibacterial resistance genes can persist and spread even in the absence of antibacterials (Martinez, 2009). Multidrug resistance in bacteria occurs with the accumulation – on resistant R plasmids or transposons – of genes, with each coding for resistance to a specific agent, and/or by the action of multidrug efflux pumps, each of which can pump out more than one drug-type (Nikaido, 2009). The demonstration of R-factor transfer to fish pathogens was first shown with certain strains of *Aeromonas salmonidae.* Also, transferable R-factor plasmids in drug-resistant strains were shown with *Aeromonas hydrophila, Vibrio anguillarum,* marine *Vibrio sp., Edwardsiella tarda* and *Patteurella piscicida* (Alderman & Hastings, 1998). Tetracycline-resistance genes are found even in small farms which rarely use antibacterials. The copy numbers of tetA, tetC, tetH, and tetM genes (tetR reported by Seyfried et al., (2010)) remain elevated at farms over the surveillance period of four years in the absence of any selection pressure from tetracycline or even other antibacterials (Schmitt & Römbke, 2008; Tamminen et al., 2010). The continued introduction of tetracycline-resistant organisms from the hatchery to the stream, even after a significant time period had elapsed since the use of antibacterials, indicates the presence of reservoirs of organisms or unknown sources of resistance (Stachowiak et al., 2010) as well as other aquatic bacteria, and also illustrates that these bacteria can act as reservoirs of resistance genes that can be further disseminated. Ultimately, resistance genes in the aquatic environment may reach human pathogens and thereby add to the burden of antibacterial resistance in human medicine (Heuer et al., 2009).

Table 8. Examples of resistance mechanisms (Boerlin & White, 2008).

Antibacterial agent	Resistance mechanism	*	Examples of genetic determinant
Tetracycline	Inducible efflux of tetracycline in *E. coli* and other *Enterobacteriaceae*	2	tetA, tetB, tetC
	Ribosomal protection in Gram-positive bacteria	4	tetO, tetM
Chloramphenicol	Efflux in *Enterobacteriaceae*	2	cmlA, floR
	Acetylation in *Enterobacteriaceae*	3	catA
ß-lactams	ß-lactamases in *Enterobacteriaceae, Staphylococcus aureus*	3	bla_{TEM}, bla_{SHV},bla_{CMY-2},bla_{Z}
Oxacillin, methicillin	Alternate penicillin-binding proteins in Staphylococcus aureus	4	mecA
Imipenem	Decreased porin formation in *Enterobacter aerogenes* and *Klebsiella spp.*	1	Mutations
Aminoglycosides	Phosphorylation, adenylation, and acetylation of aminoglycosides in Gram-negative and -positive bacteria	3	Numerous genes with a broad variety of specificities
Streptomycin	Modification of ribosomal proteins or of 16s rRNA in *Mycobacterium spp.*	4	Mutations
Macrolides, lincosamides, streptogramins	Methylation of ribosomal RNA in Gram-positive organisms	4	ermA, ermB, ermC
Macrolides, streptogramins	Staphylococcus spp.	2	vga(A), msr(A)
Fluoroquinolones	DNA topoisomerases with low affinity to quinolones	4	Mutations in gyrA, gyrB, parC, parE
Sulphonamides	Bypass of blocked pathways through additional resistant dihydropteroate synthase in Gram-negative bacteria	4	Sul1, sul2, sul3
Trimethoprim	Bypass of blocked pathways through additional resistant dihydrofolate reductase	4	Diverse dfr genes

Note:Thisisbynomeansacomprehensivelistofalltheresistancemechanismsforeachcategoryof antibacterials listed. *: Numbers 1, 2, 3 and 4 refer to mechanisms listed in the text.

Aquaculture is thought to stimulate the spread and stability of antibacterial resistance in the environment (Sapkota et al., 2008). Commercial fish production facilities could be a source of antibacterial-resistant microorganisms to receiving waters at times when there is no active use of antibacterials as a result of cross-resistance induced by biocides (Stachowiak et al., 2010). It has been shown that antibacterial-resistant bacteria are more likely to occur in the water and sediment associated with aquaculture. Already, in several areas of the world, this is beginning to take place. The comparison of predicted antibacterial concentrations to published minimum inhibitory concentrations suggests that antibacterials in wastewater – but probably not antifungals – may select for low-level antibacterial resistance (Kostich & Lazorchak, 2008). Also, the presence of R-factor-infected populations of bacteria in aquaculture systems may lead to the transfer of antibacterial resistance to other micro-organisms, including potential human pathogens (Roberts, 2004; Cabello, 2006). Both the percentage and level of bacterial resistance to drugs was higher when drugs were administered as medicated feed. In addition, the duration of the resistance was longer when medicated feed was the mode of administration. The presence of feed residue in the aquatic system would have an important effect for the generation and maintenance of the drug resistance of bacteria in sediment (Yu et al., 2009).

These results call for the development of better management strategies for fish farming so as to prevent the emergence of resistant gene pools in the sediments of aquaculture facilities, and to promote the disappearance of established resistant gene pools (Tamminen et al., 2010). Principles for the prudent use of antibacterials should be developed and awareness of the problem of antibacterial resistance should be raised by informing the public (FAO, 2005). The most effective and direct approach is thought to be the reasonable use of antibacterials in health protection and agriculture production (Zhang et al., 2009). Without a doubt, a promising approach for proper risk-assessment and management would be the reduction of the emission of antibacterials into the environment, whether of human or veterinary medical origin. In either case, it may not be appropriate to assume that terminating the use of antibacterials will lead to a rapid decrease in resistant organisms (Stachowiak et al., 2010). The appropriate use of antibacterials in livestock production will preserve the long-term efficacy of existing antibacterials, support animal health and welfare, and limit the risk factors of transferring antibacterial resistance to animals and humans (Kemper, 2008). Whatever is done, the competent surveillance of disease and antibacterial resistance, as well as the repeated refinement of risk analyses, are a necessity if we are to concentrate our efforts to limit the effects of antibacterial resistance on what is shown to work in practice (Phillips et al., 2004; Sanders, 2005). In general, the emergence of resistance to antibacterials is a highly complex process, which is not yet fully understood with respect to the significance of the interaction of bacterial populations and antibacterials, even in a medicinal environment (Kümmerer, 2010). In the EU, the EMA works for the development of a harmonised approach to the surveillance of antibacterial usage in animals and the collection of data from EU Member States (WHO, 2011). Also, research projects should be encouraged which aim at the better understanding of the mechanisms of the emergence and spread of resistance within a species, and from animal to human and the environment (FAO, 2005).

5.3 Aquatic food residues

The case of the residues of antibacterial substances in fish and fish products represents, in practice, a complex problem for society and regulators, and particularly in developing countries where regulations and the possibilities for enforcing them are scarce (Cabello, 2006; Lupin, 2009). In addition to selecting for antibacterial resistance, the heavy prophylactic and therapeutic use of antibacterials in aquaculture environments can lead to elevated antibacterial residues in ponds, marine sediments, aquaculture products, wild fish and other natural aquatic environments that are impacted by aquaculture facilities (Sapkota et al., 2008). Also, the use of large amounts of antibacterials that have to be mixed with fish food creates problems for industrial health and increases the opportunities for the presence of residual antibacterials in fish meat and fish products (Cabello, 2006). Withdrawal times are recommended, and in many countries they are legally enforced for some drugs, and especially antibacterials. The Food Animal Drug Avoidance Databank (FARAD) assists veterinarians in estimating residue-depletion times for antibacterial agents that are administered at doses in excess of label recommendations (Walker & Giguére, 2008). A good rule of thumb for the withdrawal time is 500 degree days. Thus, if the mean daily water temperature after treatment is 10°C, the withdrawal period should be at least 50 days (10 x 50 = 500), while at 25° C, the withdrawal period would be 20 days (Noga, 2010). Obviously, this can only be a rough estimate of the elimination-rate because temperatures fluctuate diurnally and from day-to-day and other factors besides temperature affect elimination- rates. Note also that 500 degree days might not be sufficient in some cases (Treves - Brown, 2001). Therefore, the accurate and sensitive determination of antibacterial residues is now a necessity. In order to protect human health, the EU and other regulatory authorities worldwide have established MRLs for antibacterial residues in animal products entering the human food chain (Cañada-Cañada & Pena, 2009; Lupin, 2009). Research projects should be promoted on pharmacology and the PK of antibacterials in aquatic species in order to provide a more exact approach to establishing MRLs' values (Table 7) (FAO, 2005).

5.4 Environmental impact of antibacterial use in aquaculture

Aquaculture is so integrally linked to the surrounding environment that if sustainable practices are not employed, the degradation of the surrounding environment will ultimately lead to the degradation of the industry itself (Bergh, 2007). The wellbeing of the environment – in cases of disease and treatment – is related to two aspects of biota conservation; the transmission of microbial pathogens to wild populations and the pollution from chemotherapeutics (Grigorakis, 2010). The extensive use of veterinary pharmaceuticals is supposed to represent a daunting public health risk, resulting not only in the emergence and spread of resistant bacteria, but also in other human, animal and environmental impairments (Kemper, 2008). The input of resistant bacteria into the environment from different sources seems to be the most important basis of resistance in the environment. The possible impact of resistant bacteria on the environment is not yet known and the health risks of active pharmaceutical ingredients remain poorly understood (Kümmerer, 2010). The physicochemical fate and environmental concentrations of antibacterials in soil has been the subject of a number of recent studies. During recent years, significant attention has been paid to the occurrence of drugs in the environment. Several classes of antibacterials have been detected in field soils, and their sorption behaviour and degradation have been studied to a large extent (Schmitt & Römbke, 2008; Zounkova et al., 2011). In general, farmed fish is as safe and nutritious as wild-caught species, but there are public health hazards associated with ignorance, abuse and the neglect of aquaculture technology. Numerous small fish ponds increase the shoreline of ponds, causing higher densities of mosquito larvae and cercaria, which can increase the incidence and prevalence of lymphatic filariasis and schistosomiasis (Lessenger, 2006). Fish production can generate considerable amounts of dissolved effluents, which potentially affect water quality in the vicinity of the farms and, due to rapid dilution, also at larger scales (km-scale) (Costanzo et al., 2005; Holmer et al., 2008). High antibacterial load in sediments and in concentrations potent enough to inhibit the growth of bacteria have been reported for aquaculture (Kümmerer, 2008).

Tetracycline has a low bioavailability in fish (< 10%), due to binding with sea-water-borne divalent cations such as Mg^{+2} and Ca^{+2}. It is noteworthy that non-bioavailable tetracyclines contaminate the environment (Rigos & Troisi, 2005). However, it has been shown that residues of oxytetracycline in marine sediments were very stable over a period of months (Toutain et al., 2010). Often, the existing data used to assess the environmental effects of antibacterials is not adequate for the establishment of how long bacteria maintain antibacterial resistance in the absence of continued selective pressure for that resistance (Kümmerer, 2008). Also, in order to minimise the possible risks of antibacterials in dust, the use of antibacterials in livestock farming should be strictly reduced to therapeutic use (Hamscher & Hartung, 2008). In one study (Wei et al., 2011) conducted in China (Jiangsu Province) – the biggest aquaculture producer (FAO, 2011) – contamination with antibacterials indicated that ten veterinary antibacterials around farms were found in animal wastewaters, eight antibacterials were detected in pond waters, and animal farm- effluents and river water samples were contaminated by nine antibacterials. The most frequently detected antibacterials were sulphamethazine (75%), oxytetracycline (64%), tetracycline (60%), sulfadiazine (55%) and sulphamethoxazole (51%). This research has demonstrated that animal wastewater is a major source of pollution of veterinary antibacterials. By applying the animal wastewater to agricultural soils, the antibacterials might contaminate the soils and surrounding water systems, thus posing a serious threat to humans and wildlife (Figure 2) (Boxall et al., 2004; Wei et al., 2011). Antibacterials may be detected in effluent entering receiving waters and be detectable 500m from the source (Costanzo et al., 2005). There is very little information about the chronic toxicity or the bioaccumulation potential of pharmaceuticals in biota and food chains (Christen et al., 2010). Not much is known about the occurrence, fate and activity of metabolites (Kümmerer, 2010). Another study showed that more than 30 antibacterial substances have been found in sewage influent and effluent samples, in surface waters and even in ground and drinking water (Kemper, 2008). At the same time, with antibacterials, disinfectants, and heavy metals being released into water, they might exert selective activities as well as ecological damage in water communities, resulting in antibacterial resistance (Baquero et al., 2008). For example, the exposure of eels to pollution during their development is inducing changes on the biomarkers involved in physiological functions that are determinants for the survival and performance of the eels, namely biotransformation enzymes and antioxidative stress defences, and these alterations may have negative effects on sexual development. In addition, the mechanisms used to face chemical stress need energy which is probably allocated from other functions, such as tissue repair, growth and weight increase, and which are determinants for a successful migration into the reproduction area (Gravato et al., 2008).

Ideally, aquaculture operations would be planned with background knowledge of the ecosystems in which the facilities will operate as well as knowledge of the potential environmental, social, and economic effects (both positive and negative) that could be incurred, and the cost:benefit ratio associated with operating, given knowledge of that background (Bergh, 2007). Environmental observations and models can then be combined with effective aquaculture husbandry practices so as to manage environmental risks from all sources (Hargrave et al., 2005). Hopefully, research on aquaculture-environment interactions has progressed remarkably during recent years, particularly in the framework of EU-funded projects, which have provided useful information for the understanding of various ecosystem processes affected by the presence and operation of fish farms (Holmer et al., 2008).

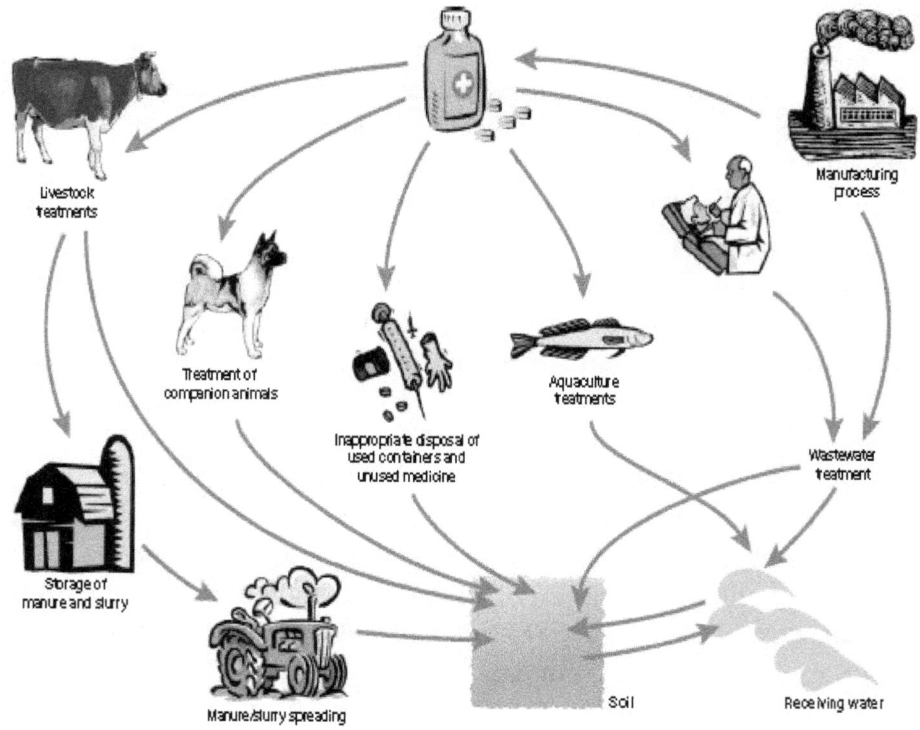

Fig. 2. Routes of pharmaceuticals entering the environment (Boxall et al., 2004).

6. ANTIBACTERIAL USAGE SUGGESTIONS IN AQUACULTURE

When it is apparent that a treatment is necessary, the following check-lists may be useful (Winton, 2001):

Before treating:

1. Accurately determine the water-volume, flow-rate, and temperature.
2. Accurately determine the number and total weight of fish in the rearing unit.
3. Confirm the identity, expiration date, and active ingredient concentration of the regulated product to be applied.
4. Double-check treatment calculations. Beware of confusion from mixing metric and standard units.
5. Have aeration devices ready for use if needed.
6. If treated water is to be discharged, make sure all appropriate permits are in place and regulatory authorities have been notified.
7. If possible, conduct a bioassay on a small group of fish before treating the entire population in the rearing unit.

When treating:

1. Dilute the regulated product with rearing water before applying it (or follow product directions).
2. Ensure that the regulated product is well-mixed and evenly applied in the rearing units.
3. Observe the fish closely and frequently during treatment for signs of distress.
4. Monitor the temperature and dissolved-oxygen levels in the rearing unit during treatment.
5. Except for oral treatments, discontinue feeding during treatment. Fish are unlikely to feed during treatment, and uneaten feed will foul the system and may reduce the efficacy of some treatments.
6. Discontinue treatment and restore normal culture conditions if the fish become distressed.

After treating:

1. Observe the fish frequently for at least 24 hours following the treatment.
2. Do not stress the treated fish for at least 48 hours.
3. Recheck the fish to determine the efficacy of the treatment

Judicious antibacterial use principles for veterinarians are discussed and concluded in Table 9 (FDA, 2009).

Table 9. Judicious antibacterial use principles for veterinarians (resumed from (FDA, 2009)).

The food fish veterinarian should:
Accept responsibility for helping clients design management, immunization, production unit and nutritional programmes that will reduce the incidence of disease and the need for antibacterial treatment.
Use antibacterial drugs only within the confines of a valid veterinarian-client-patient relationship, including both the dispensing and issuing of prescriptions and veterinary feed directives. Extra-label usage should be consistent with regulatory agency laws, regulations and policies.
Properly select and use antibacterial drugs. Veterinarians should participate in continuing education programmes that include therapeutics and the emergence and/or development of antibacterial resistance.
Have strong clinical evidence of the identity of the disease's aetiology, based upon history, clinical signs, necropsy, laboratory data, and/or past experience before recommending an antibacterial drug treatment.
Treat food fish with antibacterial drugs according to the product label recommendations (including indication, dosage, duration, fish species and environmental conditions).
Choose an antibacterial drug and treatment regimen based on the available laboratory and label (including package insert) information, additional data in the literature, and consideration of the pharmacokinetics, spectrum of activity and pharmacodynamics of the drug.

Use antibacterial drugs with a specific clinical outcome(s) in mind, including a specific target for population morbidity and/or mortality-rate reduction.
Specific outcome criteria will prevent an unnecessarily long therapy and indicate when the current therapy is no longer effective.
Determine the production population pathogen susceptibility at the first indication of increasing morbidity or mortality, and monitor the therapeutic response so as to detect changes in microbial susceptibility and in order to evaluate antibacterial selections.
Routine necropsy examination of fish populations should be periodically performed, including antibacterial susceptibility testing and update historical information for developing treatment and control protocols.
Use products that have the narrowest spectrum of activity and known effectiveness *in vivo* against the pathogen causing the disease problem.
Choose antibacterial drugs of lesser importance in human medicine, if these receive future food fish use approval, and do not choose an antibacterial for which the emergence of resistance is expected in an advanced stage.
Use, whenever possible, an antibacterial drug labelled to treat the condition diagnosed.
Do not use combination antibacterial drug therapy unless there is information to show that this decreases or suppresses the target organism resistance development.
Do not compound antibacterial drug formulations.
Do not use antibacterial drugs to treat cases with a poor chance of recovery.
Do not use antibacterial drugs prophylactically.
Ensure proper on-farm drug use and protect antibacterial drug integrity through proper handling, storage and observation of the expiration date.
Prescribe, dispense or write a Veterinary Feed Directive for drug quantities appropriate to the production-unit size and expected need using the approved formulation.
Work with producers and/or facility fish health management personnel so as to ensure that farm personnel receive adequate training on the use of antibacterial drugs, including indications, diagnoses, dosages, withdrawal times, the route of administration, storage, handling and accurate record-keeping.
Work closely with all other fish health experts involved in fish population health management at the fish production facility.

7. CONCISE SUMMARY

The occurrence of disease in aquaculture fish populations significantly impacts the wellbeing of the afflicted fish and endangers the welfare of those uninfected. Significant amounts of antibacterials are utilized in aquaculture in several countries, frequently without expert consultation or oversight. As a result, numerous issues are linked to the application of antibacterials in aquaculture. Further research is required to ascertain the effects of utilizing substantial amounts of antibacterials. Given the swift expansion and significance of the aquaculture sector in several global locations, along with the prevalent, intensive, and frequently unregulated application of antibacterial agents in animal production, more measures are necessary to avert the emergence and dissemination of antimicrobial resistance in aquaculture. Moreover, the development of safer and more efficacious pharmaceuticals is essential, alongside enhancements in husbandry and management practices that will diminish the reliance on such medications. Nonetheless, the absence of veterinary pharmaceuticals would hinder aquaculture food production. Moreover, aquaculturists and their veterinarians must ensure that fish are maintained in optimal health and welfare conditions. Governments, farmers, and veterinarians collectively bear the responsibility to guarantee the prudent use of medications.

8. REFERENCES

Alderman, D. J. (2009). Control of the Use of Veterinary Drugs and Vaccines in Aquaculture in the European Union. In *The Use of Veterinary Drugs and Vaccines in Mediterranean Aquaculture*, edited by Basurco Rogers C.J., B. Zaragoza: CIHEAM - IAMZ.

Alderman, D. J. & Hastings, T. S. (1998). Antibiotic Use in Aquaculture: Development of Antibiotic Resistance - Potential for Consumer Health Risks. *International Journal of Food Science and Technology* no. 33:139 -155.

Asche, F., Roll, K. H. & Tveterås, S. (2008). Future Trends in Aquaculture: Productivity Growth and Increased Production. In *Aquaculture in the Ecosystem*, edited by Marianne Holmer, Kenny Black, Carlos M. Duarte, Nuria Marbà and Ioannis Karakassis, 271-292. Springer Netherlands.

Austin, B. & Austin, D. A. (2007). Control. In *Bacterial Fish Pathogens*, 337-404. Springer Netherlands.

Baquero, F., Martínez, J. L. & Cantón, R. (2008). Antibiotics and antibiotic resistance in water environments. *Curr Opin Biotechnol* no. 19 (3):260-265.

Bebak, J. A., Welch, T. J., Starliper, C. E. Baya, A. M. & Garner, M.M. (2007). Improved husbandry to control an outbreak of rainbow trout fry syndrome caused by infection with Flavobacterium psychrophilum. *J Am Vet Med Assoc* no. 231 (1):114-6.

Bergh, O. (2007). The dual myths of the healthy wild fish and the unhealthy farmed fish. *Dis Aquat Organ* no. 75 (2):159-64.

Boerlin, P. & White, D. G. (2008). Antimicrobial Resistance and Its Epidemiology. In *Guide to Antimicrobial "Use": in Animals*, edited by Lars Bogø Jensen Luca Guardabassi, Hilde Kruse, 27 - 43. Oxford, UK: Blackwell Publishing Ltd.

Boos, S., Schmidt, H., Ritter, G. & Manz, D. (1995). Effectiveness of oral rifampicin against mycobacteriosis in tropical fish. *Berl Munch Tierarztl Wochenschr* no. 108 (7):253-5.

Boxall, A. B., Fogg, L. A., Blackwell, P. A., Kay, P., Pemberton, E. J. & Croxford, A. (2004). Veterinary medicines in the environment. *Rev Environ Contam Toxicol* no. 180:1- 91.

Burka, J. F., Hammell, K. L., Horsberg, T. E., Johnson, G. R., Rainnie, D. J. & Speare, D. J. (1997). Drugs in salmonid aquaculture – A review. *Journal of Veterinary Pharmacology and Therapeutics* no. 20 (5):333-349.

Burridge, L., Weis, J. S., Cabello, F., Pizarro, J. & Bostick, K. (2010). Chemical use in salmon aquaculture: A review of current practices and possible environmental effects. *Aquaculture* no. 306 (1-4):7-23.

Cabello, F. C. (2006). Heavy use of prophylactic antibiotics in aquaculture: a growing problem for human and animal health and for the environment. *Environmental Microbiology* no. 8 (7):1137-1144.

Cañada-Cañada, F., Muñoz de la Peña, A. & Espinosa-Mansilla, A. (2009). Analysis of antibiotics in fish samples. *Analytical and Bioanalytical Chemistry* no. 395 (4):987- 1008.

Christen, V., Hickmann, S., Rechenberg, B. & Fent, K. (2010). Highly active human pharmaceuticals in aquatic systems: A concept for their identification based on their mode of action. *Aquat Toxicol* no. 96 (3):167-81.

(CLSI), Clinical and Laboratory Standards Institute. (2006a). Methods for Antimicrobial Disk Susceptibility Testing of Bacteria Isolated From Aquatic Animals. In *Approved Guideline.* Wayne, Pennsylvania, USA.

(CLSI). 2006b. Methods for Broth Dilution Susceptibility Testing of Bacteria Isolated From Aquatic Animals. In *Approved Guideline.* Wayne, Pennsylvania, USA.

Collina, G., Morandi, L., Lanzoni, A. & Reggiani, M. (2002). Atypical cutaneous mycobacteriosis diagnosed by polymerase chain reaction. *Br J Dermatol* no. 147 (4):781-4.

Costanzo, S. D., Murby, J. & Bates, J. (2005). Ecosystem response to antibiotics entering the aquatic environment. *Mar Pollut Bull* no. 51 (1-4):218-23.

Cottee, S. & Petersan, P. (2009). Animal Welfare and Organic Aquaculture in Open Systems. *Journal of Agricultural and Environmental Ethics* no. 22 (5):437-461.

Daniel, P. (2009). Drugs and Chemicals in Aquafeeds: The problems and solutions. In *The Use of Veterinary Drugs and Vaccines in Mediterranean Aquaculture*, edited by Basurco Rogers C.J., B. Zaragoza: CIHEAM - IAMZ.

Defoirdt, T., Sorgeloos, P. & Bossier, P. (2011). Alternatives to antibiotics for the control of bacterial disease in aquaculture. *Current Opinion in Microbiology* no. 14 (3):251- 258.

Douet, D.G., Le Bris, H. & Giraud, E. (2009). Environmental Aspects of Drug and Chemical Use in Aquaculture: An Overview. In *The Use of Veterinary Drugs and Vaccines in Mediterranean Aquaculture*, edited by Basurco Rogers C.J., B. Zaragoza: CIHEAM - IAMZ.

FAO. (2005). Responsible Use of Antibiotics in Aquaculture. edited by FAO Fisheries Technical Paper. No.469. Rome, Italy.

FAO. (2011). The State of World Fisheries and Aquaculture 2010. edited by FAO Fisheries and Aquaculture Department. Rome, Italy.

FDA. *Judicious Use of Antimicrobials for Aquatic Veterinarians*. U.S. Food and Drug Administration 2009 [cited 11.07.2011]. Available from http://www.fda.gov/AnimalVeterinary/SafetyHealth/ AntimicrobialResistance/JudiciousUseofAntimicrobials/ucm095473.htm

FDA. (2011). Aquaculture Drugs. In *Fish and Fishery Products Hazards and Controls Guidance*, 183 -208. Rockville, USA: Center for Food Safety and Applied Nutrition.

Gravato, C, Melissa, F., Anabela, A., Joana, S. & Lúcia, G. (2008). Biomonitoring Studies Performed with European Eel Populations from the Estuaries of Minho, Lima and Douro Rivers (NW Portugal). In *Advanced Environmental Monitoring*, edited by Young J. Kim and Ulrich Platt, 390-401. Springer Netherlands.

Grigorakis, K. (2010). Ethical Issues in Aquaculture Production. *Journal of Agricultural and Environmental Ethics* no. 23 (4):345-370.

Guardabassi, L. & Kruse, H. (2008). Principles of Prudent and Rational Use of Antimicrobials in Animals. In *Guide to Antimicrobial Use in Animals*, edited by Lars Bogø Jensen Luca Guardabassi, Hilde Kruse, 1-12. Oxford, UK: Blackwell Publishing Ltd.

Hamscher, G. & Hartung, J. (2008). Veterinary Antibiotics in Dust: Sources, Environmental Concentrations, and Possible Health Hazards. In *Pharmaceuticals in the Environment*, edited by Klaus Kümmerer, 95-102. Springer Berlin / Heidelberg.

Hargrave, B., William, S. & Paul, K. (2005). Assessing and Managing Environmental Risks Associated with Marine Finfish Aquaculture. In *Environmental Effects of Marine Finfish Aquaculture*, edited by Barry Hargrave, 433-461. Springer Berlin / Heidelberg.

Haya, K., Burridge, L., Davies, I. & Ervik, A. (2005). A Review and Assessment of Environmental Risk of Chemicals Used for the Treatment of Sea Lice Infestations of Cultured Salmon. In *Environmental Effects of Marine Finfish Aquaculture*, edited by Barry Hargrave, 305-340. Springer Berlin / Heidelberg.

Heuer, O. E., Kruse, H., Grave, K., Collignon, P., Karunasagar, I. & Angulo, F. J. (2009). Human Health Consequences of Use of Antimicrobial Agents in Aquaculture. *Clinical Infectious Diseases* no. 49 (8):1248-1253.

Holmer, M., Hansen, P.K., Karakassis, I., Borg, J.A. & Schembri, P.J. (2008). Monitoring of Environmental Impacts of Marine Aquaculture. In *Aquaculture in the Ecosystem*, edited by Marianne Holmer, Kenny Black, Carlos M. Duarte, Nuria Marbà and Ioannis Karakassis, 47-85. Springer Netherlands.

Horsberg, T. E. (1994). Experimental methods for pharmacokinetic studies in salmonids. *Annual Review of Fish Diseases* no. 4:345-358.

Immanuel, G., Sivagnanavelmurugan, M. & Palavesam, A. (2011). Antibacterial Effect of Medium-chain fatty Acid: Caprylic Acid on Gnotobiotic *Artemia franciscana nauplii* Against Shrimp Pathogens *Vibrio harveyi* and *V. parahaemolyticus*. *Aquaculture International* no. 19 (1):91-101.

Kemper, N. (2008). Veterinary antibiotics in the aquatic and terrestrial environment. *Ecological Indicators* no. 8 (1):1-13.

Kostich, M. S. & Lazorchak, J.M. (2008). Risks to aquatic organisms posed by human pharmaceutical use. *Science of The Total Environment* no. 389 (2-3):329-339.

Kum, C., Kirkan, S., Sekkin, S., Akar, F. & Boyacioglu, M. (2008). Comparison of in vitro antimicrobial susceptibility in Flavobacterium psychrophilum isolated from rainbow trout fry. *J Aquat Anim Health* no. 20 (4):245-51.

Kümmerer, K. (2008). Effects of Antibiotics and Virustatics in the Environment. In *Pharmaceuticals in the Environment*, edited by Klaus Kümmerer, 223-244. Springer Berlin /Heidelberg.

Kümmerer, K. (2010). Pharmaceuticals in the Environment. *Annual Review of Environment and Resources* no. 35 (1):57-75.

Lessenger, J. E. (2006). Diseases from Animals, Poultry, and Fish. In *Agricultural Medicine*, edited by James E. Lessenger, 367-382. Springer New York.

Lupin, H. M. (2009). Human Health Aspects of Drug and Chemical Use in Aquaculture. In *The Use of Veterinary Drugs and Vaccines in Mediterranean Aquaculture*, edited by Basurco Rogers C.J., B. Zaragoza: CIHEAM - IAMZ.

Martinez, J. L. (2009). Environmental pollution by antibiotics and by antibiotic resistance determinants. *Environmental Pollution* no. 157 (11):2893-2902.

Martinez, M. & Silley, P. (2010). Antimicrobial Drug Resistance. *Handb Exp Pharmacol* (199):231-268.

Meyer, F. P. (1991). Aquaculture disease and health management. *J Anim Sci* no. 69 (10):4201-8.

Moutou, K. A., Burke, M. D. & Houlihan, D. F. (1998). Hepatic P450 monooxygenase response in rainbow trout (*Oncorhynchus mykiss* [Walbaum]) administered aquaculture antibiotics. *Fish Physiol Biochem* no. 18 (1):97-106.

Mulcahy, D. (2011). Antibiotic use during the intracoelomic implantation of electronic tags into fish. *Reviews in Fish Biology and Fisheries* no. 21 (1):83-96.

Nic Gabhainn, S., Bergh, O., Dixon, B., Donachie, L., Carson, J., Coyne, R., Curtin, J., Dalsgaard, I., Manfrin, A., Maxwell, G. & Smith, P. (2004). The precision and robustness of published protocols for disc diffusion assays of antimicrobial agent susceptibility: an inter-laboratory study. *Aquaculture* no. 240 (1-4):1-18.

Nikaido, H. (2009). Multidrug Resistance in Bacteria. *Annual Review of Biochemistry* no. 78 (1):119-146.

Noga, E. J. (2010). *Fish Disease Diagnose and Treatment.* Second ed. Iowa, USA: Wiley - Blackwell.

Phillips, I., Casewell, M., Cox, T., De Groot, B., Friis, C., Jones, R., Nightingale, C., Preston, R. & Waddell, J. (2004). Does the use of antibiotics in food animals pose a risk to human health? A critical review of published data. *J Antimicrob Chemother* no. 53 (1):28-52.

Prescott, J.F. (2008). Antimicrobial use in food and companion animals. *Animal Health Research Reviews* no. 9 (Special Issue 02):127-133.

Pullin, R., Froese, R. & Pauly, D. 2007. Indicators for the Sustainability of Aquaculture. In *Ecological and Genetic Implications of Aquaculture Activities*, edited by Theresa M. Bert, 53-72. Springer Netherlands.

Reimschuessel, R. & Miller, R. (2006). Antimicrobial Drug Use in Aquaculture. In *Antimicrobial Therapy in Veterinary Medicine*, edited by J.F. Prescott, Baggot, J.D., Walker, R.D., Dowling, P.M., 593-606. Iowa, USA: Blackwell Publishing Professional.

Reimschuessel, R.L., Stewart, E., Squibb, K., Hirokawa, T., Brady, D., Brooks, B., Shaikh & Hodsdon, C. (2005). Fish drug analysis--Phish-Pharm: a searchable database of pharmacokinetics data in fish. *AAPS J* no. 7 (2):E288-327.

Rigos, G., Nengas, I., Alexis, M. & Athanassopoulou, F. (2004). Bioavailability of oxytetracycline in sea bass, Dicentrarchus labrax (L.). *J Fish Dis* no. 27 (2):119-22.

Rigos, G. & Troisi, G. (2005). Antibacterial Agents in Mediterranean Finfish Farming: A Synopsis of Drug Pharmacokinetics in Important Euryhaline Fish Species and Possible Environmental Implications. *Reviews in Fish Biology and Fisheries* no. 15 (1):53-73.

Roberts, R.J. (2004). The Bacteriology of Teleosts. In *Fish Pathology*, edited by R.J. Roberts, 297 - 331. Philadelphia, USA: W.B. Saunders.

Rodger, H. D. (2010). Fish Disease Manual. Marine Institute and the Marine Research Sub- Programme of the National Development Plan. Original edition, PBA/AF/08/003.

Rodgers, C.J. & Furones, M.D. (2009). Antimicrobial Agents in Aquaculture: Practice, needs and issues. In *The Use of Veterinary Drugs and Vaccines in Mediterranean Aquaculture*, edited by Basurco Rogers C.J., B. Zaragoza: CIHEAM - IAMZ.

Ryckaert, J., Bossier, P., D'Herde, K., Diez-Fraile, A., Sorgeloos, P., Haesebrouck, F. & Pasmans, F. (2010). "Persistence of Yersinia ruckeri in trout macrophages." *Fish Shellfish Immunol* no. 29 (4):648-55.

Sanders, P. (2005). Antibiotic Use in Animals—Policies and Control Measures Around Europe. In *Antibiotic Policies*, edited by Ian Gould and Jos Meer, 649-672. Springer US.

Sapkota, A., Sapkota, A. R., Kucharski, M., Burke, J., McKenzie, S., Walker, P. & Lawrence, R. (2008). Aquaculture practices and potential human health risks: current knowledge and future priorities. *Environ Int* no. 34 (8):1215-26.

Schmitt, H. & Römbke, J. (2008). The Ecotoxicological Effects of Pharmaceuticals (Antibiotics and Antiparasiticides) in the Terrestrial Environment – a Review. In *Pharmaceuticals in the Environment*, edited by Klaus Kümmerer, 285-303. Springer Berlin/Heidelberg.

Seyfried, E., Newton, R., Rubert, K., Pedersen, J. & McMahon, K. (2010). Occurrence of Tetracycline Resistance Genes in Aquaculture Facilities with Varying Use of Oxytetracycline. *Microbial Ecology* no. 59 (4):799-807.

Shao, Z.J. (2001). Aquaculture pharmaceuticals and biologicals: current perspectives and future possibilities. *Adv Drug Deliv Rev* no. 50 (3):229-43.

Smith, P.R., Le Breton, A., Horsberg, T.E. & Corsin, F. (2008). Guidelines for Antimicrobial Use in Aquaculture. In *Guide to Antimicrobial Use in Animals*, edited by Lars Bogø Jensen Luca Guardabassi, Hilde Kruse, 207- 218. Oxford, UK: Blackwell Publishing Ltd.

Stabili, L., Gravili, C., Boero, F., Tredici, S. & Alifano, P. (2010). Susceptibility to Antibiotics of *Vibrio sp.* AO1 Growing in Pure Culture or in Association with its Hydroid Host *Aglaophenia octodonta (Cnidaria, Hydrozoa) Microbial Ecology* no. 59 (3):555-562.

Stachowiak, M., Clark, S., Templin, R. & Baker, K. (2010). Tetracycline-Resistant *Escherichia coli* in a Small Stream Receiving Fish Hatchery Effluent *Water, Air, & Soil Pollution* no. 211 (1):251-259.

Stickney, R.R. (2005). Aquaculture. In *Encyclopedia of Coastal Science*, edited by Maurice L. Schwartz, 33-38. Springer Netherlands.

Storey, S. (2005). Challenges with the development and approval of pharmaceuticals for fish. *AAPS J* no. 7 (2):E335-E343.

Subasinghe, R. (2009). Disease Control in Aquaculture and the Responsible use of Veterinary Drugs and Vaccines: The Issues, Prospects and Challenges. In *The Use of Veterinary Drugs and Vaccines in Mediterranean Aquaculture*, edited by Basurco Rogers C.J., B. Zaragoza: CIHEAM - IAMZ.

Tamminen, M., Karkman, A., Lohmus, A., Muziasari, W. I., Takasu, H., Wada, S., Suzuki, S. & Virta, M. (2010). Tetracycline Resistance Genes Persist at Aquaculture Farms in the Absence of Selection Pressure. *Environmental Science & Technology* no. 45 (2):386- 391.

Tobback, E., Decostere, A., Hermans, K., Ryckaert,J., Duchateau, L., Haesebrouck, F. & Chiers, K. (2009). Route of entry and tissue distribution of Yersinia ruckeri in experimentally infected rainbow trout Oncorhynchus mykiss. *Dis Aquat Organ* no. 84 (3):219-28.

Toutain, P. L., Ferran, A. & Bousquet-Melou, A. (2010). Species differences in pharmacokinetics and pharmacodynamics. *Handb Exp Pharmacol* (199):19-48.

Treves - Brown, K.M. (2001). *Applied Fish Pharmacology*. Edited by G. Poxton Michael. 3 vols. Vol. 3, *Aquaculture*. The Netherlands: Kluwer Academic Publishers.

Valois, A.A., Endoh, Y.S., Grein, K. & Tollefson, L. (2008). Geographical Differences in Market Availability, Regulation and Use of Veterinary Antimicrobial Products. In *Guide to Antimicrobial Use in Animals*, edited by Lars Bogø Jensen Luca Guardabassi, Hilde Kruse, 59 - 76. Oxford, UK: Blackwell Publishing Ltd.

Walker, R.D. & Giguére, S. (2008). Principles of Antimicrobial Drug Selection and Use. In *Antimicrobial Therapy in Veterinary Medicine*, edited by J.F. Prescott, Baggot, J.D., Walker, R.D., Dowling, P.M., 107-117. Iowa, USA: Blackwell Publishing Professional.

Wei, R., Ge, F., Huang, S., Chen, M. & Wang, R. (2011). Occurrence of veterinary antibiotics in animal wastewater and surface water around farms in Jiangsu Province, China. *Chemosphere* no. 82 (10):1408-1414.

WHO. (2011). *Tackling Antibiotic Resistance From a Food Safety Perspective in Europe.* Edited by World Health Organization. Copenhagen, Denmark.

Wiens, G. D., Rockey, D. D., Wu, Z., Chang,J., Levy,R., Crane,S., Chen,D.S., Capri,G.R., Burnett,J.R., Sudheesh, P.S., Schipma,M.J., Burd,H., Bhattacharyya,A., Rhodes, L.D., Kaul,R. & Strom,M.S. (2008). Genome sequence of the fish pathogen Renibacterium salmoninarum suggests reductive evolution away from an environmental Arthrobacter ancestor. *J Bacteriol* no. 190 (21):6970-82.

Winton, J. R. (2001). Fish Health Management. In *Fish Hatchery Management*, edited by G. Wedemeyer, 559 - 640. Bethesda, USA: American Fisheries Society.

Yan, S. S. & Gilbert, J.M. (2004). Antimicrobial drug delivery in food animals and microbial food safety concerns: an overview of in vitro and in vivo factors potentially affecting the animal gut microflora. *Adv Drug Deliv Rev* no. 56 (10):1497-521.

Yu, D., Yi,X., Ma,Y., Yin,B., Zhuo,H., Li,J. & Huang, Y. (2009). Effects of administration mode of antibiotics on antibiotic resistance of Enterococcus faecalis in aquatic ecosystems. *Chemosphere* no. 76 (7):915-920.

Zhang, X. X., Zhang, T. & Fang, H. (2009). Antibiotic resistance genes in water environment. *Applied Microbiology and Biotechnology* no. 82 (3):397-414.

Zounkova, R., Klimesova, Z., Nepejchalova, L.,Hilscherova, K. & Blaha, L. (2011). Complex evaluation of ecotoxicity and genotoxicity of antimicrobials oxytetracycline and flumequine used in aquaculture. *Environ Toxicol Chem* no. 30 (5):1184-9.

Index

www.ingramcontent.com/pod-product-compliance
Lightning Source LLC
LaVergne TN
LVHW081938200926
848484LV00009B/2463
* 9 7 8 1 7 8 7 1 5 6 2 8 9 *